THE FUTURE OF TESTING

Buros-Nebraska Symposium on Measurement & Testing

Volume 2

Series Editor

JAMES V. MITCHELL, JR.

Buros Institute of Mental Measurements
and
Department of Educational Psychology
University of Nebraska-Lincoln

THE FUTURE OF TESTING

Edited by
Barbara S. Plake
University of Nebraska-Lincoln

Joseph C. Witt
Louisiana State University

LAWRENCE ERLBAUM ASSOCIATES, PUBLISHERS
1986 Hillsdale, New Jersey London

Lawrence Erlbaum Associates, Inc., Publishers
365 Broadway
Hillsdale, New Jersey 07642

Library of Congress Cataloging-in-Publication Data
Main entry under title:

The Future of testing.

(Buros-Nebraska symposium on measurement & testing ; v. 2)
Includes bibliographies and indexes.
1. Psychological tests. 2. Psychometrics. I. Plake, Barbara S. II. Witt, Joseph C. III. Series.
BF176.F87 1986 371.2′6 85-20552
ISBN 0-89859-589-4

Printed in the United States of America
10 9 8 7 6 5 4 3 2 1

Contents

Preface

Tests are constructed and used to facilitate assessment and understanding of human beings in all their multifaceted complexity. Hence, testing by its very nature is both a scientific and a social endeavor.

The interplay between testing and society has resulted in both praise and criticism from concerned citizens, psychologists, educators, and numerous other professional and consumer groups. For over 40 years, Oscar K. Buros, as Director of The Institute of Mental Measurements and Editor of the *Mental Measurements Yearbooks,* contributed immensely to this interplay between testing practices and societal issues. On March 19, 1978, Oscar Buros died. Luella Buros, his wife and lifelong helpmate, completed the work on *The Eighth Mental Measurements Yearbook* with the support of the Institute's devoted staff. She also took steps to relocate the Institute to ensure the continuation of the Institute's scholarly work and services for test consumers. The new Buros Institute of Mental Measurements is now at the University of Nebraska—Lincoln.

An important objective of the new Buros Institute is to conduct an extended outreach effort that will help communicate more effectively with test users about contemporary issues in testing. Thus, it was the combination of recent social issues focusing on testing and our desire to fulfill more vigorously the mission of the Buros Institute that motivated the development of an annual scholarly symposium and this series on measurement and testing.

We intend each symposium and volume in this series to present state-of-the-art knowledge that will contribute to the *improvement of test construction and test usage*. Such a schema will incorporate topics across a broad spectrum such as theoretical models of human behavior, test standardization procedures, social and legal factors in testing, administration of testing programs, and test-based decision making. Thus, the series will be focused thematically and yet be flexible enough to integrate current and future measurement and testing issues into its schema.

The success of the Buros–Nebraska symposium and this volume is the result of the efforts of many individuals. We thank Luella Buros for having faith in us to carry on and extend a tradition that has become so important to the measurement field and to test users. Barbara Plake, as editor of the first volume in the series, made conceptual and editorial contributions that were of critical importance to the success of the series. Finally, we want to thank Larry Erlbaum for his support, encouragement, and commitment to the project and to its timely completion.

Series Editor
James V. Mitchell, Jr.
Lincoln, Nebraska

VOLUME ACKNOWLEDGMENTS

We are hopelessly in debt to a number of individuals who, individually and collectively, have dramatically influenced the quality of this volume. Although it is difficult to know where to start and where to end in acknowledging the many who have contributed in innumerable ways to this volume, this listing here should make it abundantly clear that this book was developed as a result of the gifts of other people.

Our most immediate gratitude must go to the individuals who contributed chapters here. Their dedication to scholarship and academic craftsmanship made our task a pleasure. We extend a warm thank you to the volume authors for their cooperation, persistence, and patience during the editing and revision process.

A number of individuals connected with the Buros Institute of Mental Measurement and Testing are richly deserving of recognition. James V. Mitchell, Director of the Buros Institute, has been intergrally involved in this project since its inception. Also, we are grateful to members of the Buros Advisory Committee within the Department of Educational Psychology:

Robert D. Brown
Roger H. Bruning
David N. Dixon
Terry B. Gutkin
James V. Mitchell, Jr.
Kenneth D. Orton
Wayne C. Piersel
Royce R. Ronning
Toni E. Santmire
Steven L. Wise

A number of dedicated professionals gave their time and talents to review and provide comments on the chapters of this book. William E. Coffman, Frank Farley, Charles Golden, Ronald Hambleton, Jo Ida Hansen, Robert Linn, Joseph Mattarazzo, Gerald J. Melican, Kevin Murphy, Anthony J. Nitko, Herbert C. Rudman, Dr. Frank Schmidt, Lyle Schoenfelt, Barbara Showers are gratefully acknowledged. In addition Lester Andrews provided essential editorial assistance.

Finally, but most especially, we would like to thank our families. Barbara Plake thanks her family, especially her husband and daughters, for their encouragement, sacrifice and understanding. In addition she wishes to express gratitude to her mother (Mrs. John K. Sterrett), brother (Wayne E. Sterrett), and sister (Margaret Sterrett-Bartels), for their love, support, and enthusiasm for her creative activities.

Joe Witt wishes to thank his wife, Larie, and his daughter, Sarah, for their understanding and patience during the completion of this volume.

Series Dedication

At the combined annual meeting of the American Educational Research Association and the National Council on Measurement in Education in April of 1980, Luella Buros, widow of Oscar Buros, was presented with a plaque honoring the achievements of her husband. The inscription read as follows:

"TRIBUTE"

"Whereas Oscar K. Buros established the series of Mental Measurements Yearbooks, and continued publishing these brilliantly over the last 40 years of his life; and

"Whereas these yearbooks have achieved recognition as classic contributions to the theory and practice of educational and psychological measurement, and of great benefit to our various professions—

"Therefore, we the undersigned officers of the American Educational Research Association and the National Council on Measurement in Education do hereby pay public tribute to the memory of Professor Buros, and to the high principles of quality and integrity which he represented in his work and in his life."

It was most fitting that Luella Buros received this tribute on behalf of her husband, because she had helped him in many ways from the very beginning of the series, particularly with matters pertaining to business and design. Oscar Buros had dedicated the *Third* and *Seventh Mental Measurements Yearbooks* to his wife, and when she completed *The Eighth Mental Measurements Yearbook* after his death, she dedicated it "To the memory of my beloved husband: Oscar Krisen Buros." The new Buros Institute of Mental Measurements will be dedicating both *Tests in Print III* and *The Ninth Mental Measurements Yearbook* to Oscar Buros. For the books in this symposium series, however, we decided to take a different path and dedicate the volumes to a remarkable couple who had a loving, happy, and productive relationship over so many years. We therefore dedicate this volume to:

Oscar and Luella Buros

Contributors

BARBARA S. PLAKE
University of Nebraska—Lincoln

JOSEPH C. WITT
Louisiana State University
Baton Rouge, LA

GENE GLASS
University of Colorado
Boulder, CO

GALE ROID
Western Psychological Services
Los Angeles, CA

NANCY S. COLE
University of Pittsburgh
Pittsburgh, PA

RONALD R. BERK
Johns Hopkins University
Baltimore, MD

MICHAEL T. KANE
ACT
Iowa City, IA

JAY ZISKIN
Professor Emeritus
California State University
Los Angeles, CA

RAYMOND DEAN
Ball State University
Muncie, IN

JOHN L. HOLLAND
Johns Hopkins University
Baltimore, MD

MARY L. TENOPYR
American Telephone and Telegraph Company
New York, NY

Barbara S. Plake:

To my father, Dr. John K. Sterrett, who will put this volume in his glass bookcase . . . in heaven.

Joseph C. Witt:

For my mentors, Robert M. Adams, Nancy Kerr, and Lee Meyerson, who provided excellent role models of an academic psychologist.

1 Prediction of the Future from the Present: An Introduction

Barbara S. Plake
University of Nebraska-Lincoln

Joseph C. Witt
Louisiana State University

> *The extent of an agent's capacity for inference, its powers to use a given fact as a sign of something not yet given, measures the extent of its ability systematically to enlarge its control of the future.*
> —Dewey, 1917

Dewey (1917) has encouraged us to nourish and advance our science in two ways. In the most typical way, knowledge advances in an additive sense whereby new information (usually data) is added to what is already known in an existing area of inquiry. A second, less common but nonetheless important method demands qualitative rather than quantitative change in the way we think about problems and the addition of knowledge. In other words, these two alternatives can be summed up by considering the first as looking for new answers to old questions and the second as looking for ways to ask new questions. This book is concerned with both of these approaches of considering the future of testing and measurement.

When a field or endeavor is really understood, it is possible for our future actions to be governed by a knowledge which forces us beyond preconception into a new or renewed awareness of the largeness of possibility (Berry, 1983). Dewey put the notion this way:

> A being which can use given and finished facts as signs of things to come, which can take given things as evidence of absent things, can, in that degree, forecast the future; it can form reasonable expectations. It is capable of achieving ideas; it is possessed of intelligence. For use of the given or finished to anticipate the consequence of processes going on is precisely what is meant by 'ideas,' by 'intelligence.' (Dewey, 1917, p. 21)

Without an overall plan of the desired directions for the future, development and advancement would be relegated to a haphazard status. Such a status is incompatible with the goals of science. Thus, even though advancement may be still possible, an overriding plan is not known or considered by the persons in the positions to be concerned with and potentially responsible for the research and intellectual efforts needed to achieve these advancements and directions. Development and advancement of a field is not so much the desired outcome of research and effort as the concentration of the activities or research and effort to identified goals and objectives. Advancement by goals and objectives requires a focus not only on the current status of the field, but also of the needed directions for the field. Also, by considering the probable directions of the field, undesirable but likely outcomes might be identified. It may then be possible to devise a plan to change the present path and thereby avoid or diminish the impact of this undesirable outcome. Therefore, preparation and preplanning by conscious effort may, in fact, alter the directions of the future.

Authors of the chapters which follow were charged with two tasks. First, to summarize the present state of their respective fields in an additive sense and then to use this information to make informed predictions. In doing this analysis many of the authors also identified new questions which heretofore had not been asked.

The overriding purpose of this volume of the Buros-Nebraska Series on Measurement and Testing is to provide a vision of what may be probable directions for the field of measurement and testing. Through the collection of chapters which address a variety of dimensions of the future of testing, it is possible to identify areas of present concern and to identify potentially important areas for dedication of energies in the future.

OVERVIEW OF THE CHAPTERS

The chapters of this volume are organized into three major sections: (1) Theoretical and methodological directions; (2) Educational and academic/-professional directions; and (3) Clinical, counseling and organizational directions. Therefore Section I contains chapters that address some of the major theoretical issues and methodological concerns and advancements that are expected to have a dramatic impact on the future directions of measurement and testing. Section II applies some of these advancements and areas of needed theoretical development to the applied areas in educational and academic/-professional settings. Then, Section III again looks at the applications of the theoretical and methodological concerns and directions but in this case considering how they apply to the areas of clinical, counseling, and organizational psychology.

Section I: Theoretical and Methodological Directions

The chapters that comprise Section I address two very important theoretical and methodological advancements which promise to have a major impact on the future of measurement and education. The first chapter, "Testing Old, Testing New: Schoolboy Psychology and the Allocation of Intellectual Resources," is authored by Gene Glass. His chapter provides the measurement field with the challenge of bringing psychology closer to the heart of psychological and educational measurement. Echoing the concerns and directions voiced of Glaser (1981), Dr. Glass recommends that the measurement field take a closer look at the *whys* of testing rather than concentrating exclusively on the psychometric *hows* of measurement and testing.

The other chapter in Section I, "Computer Technology in Testing" by Gale Roid, focuses on the technological advancements in measurement related to the integration and utilization of computers, both mainframe and micro. The chapter identifies probable uses of the computer in test development, administration, score reporting, interpretation, and utilization. Viewing computers as probably the most pervasive technological advancement in measurement and testing, Roid identifies both areas of promise and problems related to the incorporation of computer technology in the test development and utilization process.

Together these two chapters make a unique combination. Glass's chapter focuses on the theoretical underpinnings of test development and test utilization while Roid's chapter focuses almost exclusively on the methods of test development and test utilization made potentially possible by computer technology. Therefore, these two chapters address issues of both the *whys* (Glass) and the *how* (Roid) as they relate to the future of measurement and testing.

Section II: Educational and Academic/Professional Directions

The three chapters in Section II address the future of measurement and testing in the areas of educational and academic/professional settings. The first of these chapters is authored by Nancy S. Cole, Chapter 4: "Future Directions for Educational Achievement and Aptitude Testing." In her chapter, Dr. Cole reviews current issues and trends in educational and aptitude testing, focusing on theoretical and technological issues. After establishing a foundation upon which to make reasonable predictions for future directions, Dr. Cole addresses how some of these recent theoretical and technological advances may influence the directions of the future of measurement and testing. In particular, Dr. Cole identifies theoretical developments in the conceptualization of intellectual processing (e.g., Messick, 1984; Sternberg, 1984) and technological advances in computerized testing as potent areas for future impact on the field. Dr. Cole concludes her chapter with a glimpse into a classroom of the future which incorporates these theoretical and technological advancements.

Chapter 5 of the volume is authored by Dr. Ronald A. Berk and continues on the theme of the future of measurement and testing in educational settings, "Minimum Competency Testing: Status and Potential." Dr. Berk's chapter addresses the potential future developments for minimum competency testing in the field of education. Consistent with the structure found in Dr. Cole's chapter on aptitude and achievement testing in education, Dr. Berk begins his chapter by addressing the current status on minimum competency testing, identifying a number of theoretical and technological problems which need to be remedied in the future. Dr. Berk identifies the *Revised Joint Technical Standards for Educational and Psychological Tests* (1985) as one major source of future impact on minimum competency testing in educational settings. In addition, Dr. Berk suggests several directions for future research which will be aimed at solving some of the technological and theoretical concerns for Minimum Competency Testing.

The third and final chapter in Section II addresses the application of testing to the area of licensure and certification examinations: "The Future of Testing for Licensure and Certification Examinations." Michael T. Kane considers the utilization of examination results for licensure and certification for professional fields such as medicine and law. In his chapter he points to several unique theoretical and technical problems and issues which impact on testing for licensure and certification. A major contribution of Dr. Kane's chapter is the integration of legal and procedural issues and concerns into a research agenda for the future.

Together the three chapters in Section II of the volume, then, address the application of many of the theoretical and technological concerns and advancements identified in Section I by Glass and Roid. All three authors of Section II reemphasize the need for theoretical development in measurement. In addition, these three authors provide specific technological applications of computers to their respective areas within educational and academic/professional assessment.

Section III: Clinical, Counseling, and Organizational Directions

The four chapters in Section III of the volume continue to address the future of measurement and testing but in this case these chapters focus on the areas of clinical, counseling and industrial assessment.

Section III begins with Jay Ziskin's chapter on the "Future of Clinical Assessment." Describing the current status of clinical assessment as paradoxical, Ziskin asserts that on the one hand evaluations performed by clinical psychologists are under attack from experts, the courts, and even the general public. On the other hand, the vast majority of employers and university training programs place a heavy emphasis of knowledge of psychological testing. In reviewing the current status of clinical assessment, Ziskin describes a number of formidable problems which are now facing clinical practice including the lack of an empirically validated knowledge base, the lack of an adequate classification system (i.e., DSM-II), and the presence of test bias in numerous instruments. Despite the rather gloomy present state of affairs, Ziskins' projections see hope for the future of clinical assessment. This hope stems from Ziskins' projections for a future where clinicians will see an increased use of computers, will utilize a new diagnostic classification system, and will have a better developed research base so that assessment can be better utilized for planning treatment.

Although technically a subspecialty of clinical assessment, the rapidly evolving area of neuropsychological assessment was allocated an entire chapter in this volume because of its current preminence and future promise. Raymond S. Dean's examination of this area begins with an overview of various systems of neuropsychological assessment classified along a qualitative-quantitative dimension. Research in the neurosciences has helped to elucidate brain-behavior relationships and is increasing the degree to which we can draw valid inferences from neuropsychological test data. Because physical diagnostic techniques are replacing neuropsychological diagnosis in some areas, Dean suggests that a major challenge for the future is the use of assessment data to predict and facilitate adaptation of the neurologically impaired.

In Chapter 9 John Holland provides an overview of interest testing and suggests the field is wrestling with four important issues: (a) How to make inventories available to more clients versus the maintenance of professional standards, (b) How to create inventories with more valid, influential, and satisfying effects, (c) How to insure equity in testing, and (d) How to integrate interest testing with other interventions. It is concluded that none of these problems is resolved easily. Accordingly, Holland recommends a number of areas which require careful scrutiny and research. The proximal goals center around improving the psychometric characteristics of the various scales and studying consumer satisfaction with interest inventories. Holland suggests this research may lead us to the development of inventories which are more practical and which emphasize inventories as interventions.

Finally, Mary Tenopyr describes the many difficulties which influence measurement in work settings. According to Dr. Tenopyr, a primary cause of these difficulties stems from a lack of knowledge on the part of employers who are responsible for implementing and evaluating a testing program. For the future, she suggests an education and research program for employers which would emphasize a reconceptualization of validity, the development of an appropriate system of constructs, clarifications regarding the process of job analysis, development of preformance measurement techniques which are both easy to use and reflective of actual job performance, the development of alternatives to paper—pencil tests (e.g., interviews, work samples), and clarification of differential prediction providing for a melding of theory and data. The majority of the chapter is organized consistent with a reconceptualization of validity.

The chapters in Section III each identify a relatively homogeneous set of current problems. Each emphasizes the need for increased emphasis on test validity. More specifically each seems to be much in line with the theme initiated by Glass in Chapter 2: that testing should be tied more closely to theory than to pure psychometrics and that tests should serve the consumer to a greater extent. Each of the chapters serves to guide future endeavors by specification of the probable courses their respective fields may take. It seems clear that computers, the reduction of test bias, and research directed toward improvement of test validity will be in the measurement future.

CONCLUDING COMMENTS

The field of testing and measurement is a conservative science. New answers to old questions continue to accumulate. Still, many answers have not proved satisfying even to the measurement community, much less to the lay public. At a time when psychometricians are asking increasingly sophisticated questions about the technical properties of tests, some critics are wondering aloud whether tests should be utilized at all. The contributions of the chapters in this volume cause us to imagine a future which is the synthesis of all that is good, and some of what is bad, in the present condition of measurement. Whether this future is reasonable, of course, is problematic; new problems loom on the horizon of the new theories and technologies. However, to the extent that the process of advancing the science of measurement and testing is in fact influenced by what is good and bad about the present, we shall witness a future which is, in Dewey's words, "intelligently constructed."

REFERENCES

Berry, W. (1983). *Standing by words.* San Francisco: North Point Press.

Dewey, J. (1917). *Creative intelligence: Essays in the pragmatic attitude.* New York: Henry Holt and Co.

Glaser, R. (1981). The future of testing: A research agenda for cognitive psychology and psychometrics. *American Psychologist, 36,* 923–936.

Messick, S. (1984). Abilities and knowledge in educational achievement testing: The assessment of dynamic cognitive structures. In B. S. Plake, (Ed.), *Social and technical issues in testing: Implications for test construction and usage (Buros-Nebraska symposium on measurement and testing)* (pp. 155–172). Hillsdale, NJ: Lawrence Erlbaum Associates.

Sternberg, R. J. (1984). What cognitive psychology can (and cannot) do for test development. In B. S. Plake, (Ed.), Social and technical issues in testing: Implications for test construction and usage (Buros-Nebraska symposium on measurement and testing) pp. 39–60. Hillsdale, NJ: Lawrence Erlbaum Associates.

THEORETICAL AND METHODOLOGICAL DIRECTIONS

2 Testing Old, Testing New: Schoolboy Psychology and the Allocation of Intellectual Resources

Gene V Glass
University of Colorado, Boulder

There will be no divining of the future here—no megatrends, no reference to Orwell (except this one). What little we truly know about the future does not bear mentioning. Nor shall I refer to micro-computers, data banks, and other gaudy paraphernalia that the future holds for us. The future is best seen in a rear-view mirror. We can at least hope to see more clearly the recent past we have traversed. Appearances of rapid change are usually superficial. If we see the past and present clearly, we will know as much of our future as it is ever permitted us to know.

Kenneth Boulding (1968) contends that the discovery of knowledge is absolutely unpredictable, since knowledge is the one thing that if we could predict when we would discover it we would have it already. The evolution of a technology, if it were not really radically new, might be predicted with some success; but then, if we can't predict the discovery of new knowledge and we can only predict changes in ordinary technology, then surely we have little idea of the important changes that lie ahead. New knowledge is revolutionary; technology is Establishment. The discovery of knowledge upsets things, changes the way lives are led. Technology serves old entrenched interests and established institutions. The glacial evolution of testing in this century reveals the source of its momentum—new technologies are moving it, not new knowledge. Testing is the conservative wing of the Social Science party.

A Point of View: Abstracted Empiricism

The most revealing perspective to assume for viewing the evolution and the current condition of testing is that which affords the clearest picture of how testing relates to the basic disciplines in the study of human behavior. In the last

100 years, testing has moved gradually from the center to the periphery of the behavioral and social sciences. Once an integral part of the best thinking on human development and behavior, testing has progressively grown more inbred and dissociated from the leading theoretical positions in psychology and the social sciences. In its position at the margin, testing has come to serve more faithfully the goals of its own professional subculture and of a particular political subculture (i.e., its own intellectual Establishment and the professional-managerial Establishment) than to serve the ends of science and the true aims of education. To fulfill its promise, testing must find its way back to the center of psychological thinking.

My message here does little more than echo a theme sounded by Anne Anastasi (1967) in her 1966 presidential address to Division 5 of the American Psychological Association.

> . . . psychological testing is becoming dissociated from the mainstream of contemporary psychology. Those psychologists specializing in psychometrics have been devoting more and more of their efforts to refining the techniques of test construction, while losing sight of the behavior they set out to measure. Psychological testing today places too much emphasis on testing and too little on psychology. As a result, outdated interpretations of test performance may remain insulated from the impact of subsequent behavior research. It is my contention that the isolation of psychometrics from other relevant areas of psychology is one of the conditions that have led to the prevalent public hostility toward testing (p. 297). Although the very essence of psychological testing is the measurement of behavior, testing today is not adequately assimilating relevant developments from the science of behavior. . . .
>
> It is noteworthy that the term "test theory" generally refers to the mechanics of test construction, such as the nature of the score scale and the procedures for assessing reliability and validity. The term does not customarily refer to psychological theory about the behavior under consideration. Psychometricians appear to shed much of their psychological knowledge as they concentrate upon the minutiae of elegant statistical techniques. Moreover, when other types of psychologists use standardized tests in their work, they too show a tendency to slip down several notches in psychological sophistication (p. 300).

Anastasi saw several reasons for this unfortunate dissociation of psychological measurement from psychological theory. Increasing specialization in all disciplines has lessened the chances that one individual will be conversant with both the technical rigmarole that has come to characterize modern psychometrics and the theories of psychology themselves. The expense involved in developing major tests militates against changing them; thus the tests of today reflect the psychology of yesterday. Psychometricians have capitulated to unrealistic public demands for short cuts and magic—psychological theory is often ravaged in the process.

Four years ago, also on the occasion of an American Psychological Association Division 5 presidential address, Robert Glaser (1981) voiced the same concern expressed by Anastasi 15 years earlier: Testing is estranged from its roots in psychological theory. In the written version of his address entitled "The Future of Testing: A Research Agenda for Cognitive Psychology and Psychometrics," Glaser attempted to present ". . . areas of social concern in which education and testing might profit from coordination with potentially helpful areas of psychological research" (p. 935). Glaser applied his understanding of recent advances in cognitive psychology to a critique of existing school testing practices and pointed toward useful applications of recent developments in the psychology of learning and thinking. The work of Brown and Burton (1978) and of Siegler (1976) was suggested as a basis for ways of diagnosing failures in learning and intellectual performance. Herbert Simon's (Simon & Chase, 1973) imaginative research on the nature of expertise was advanced as a beginning in the assessment of differences in knowledge structures and cognitive processes between novices and experts. The work of Hunt, Sternberg, and others suggested to Glaser new views on the assessment of aptitudes with the ultimate goal of altering and building those abilities that early-day psychologists were prone to accept as being immutable.

My proposed point of criticism would seem *ad hoc* and unconvincing if it were said to apply somehow uniquely to the problems of measurement and testing. Fortunately, such is not so. C. Wright Mills (1959) argued forcefully that the schism between method and theory is everywhere evident in the social sciences. Aimless fact finding unguided by worthy conceptual analysis was christened "abstracted empiricism" by Mills. Once severed from worthwhile theoretical thinking, abstracted empiricism follows a bureaucratic course of development.

I doubt that I can advance any more helpful message than to commend once more to your attention the wisdom in the observations of my respected colleagues Anastasi and Glaser. It will give me the greatest satisfaction, in addition, if I can convince you to entertain an even broader scope of relevant psychological theory than they imagined as being a proper mooring place for psychological and educational measurement. But before making that attempt, permit me first to recount briefly how testing came to assume its current condition, which more people agree with each passing year is in need of repair.

Psychology and Psychometrics in 1980: The Golden Days

Testing as we have known it for the past 75 years was originally the tool of psychologists and social scientists living through the denouement of the great Western European empires. The social scientists of the first two decades of this century were, with varying degrees of consciousness, Social Darwinists. The cultural relativist anthropologists, such as Boaz and later Mead, are the exceptions that prove the rule. Regardless of one's contemporary political leanings, it is difficult today to read Galton or Terman without blanching at the coarse ugliness of it, e.g., Galton (1892) analyzing the genetic superiority of one (19th-century English) village over its neighbor. But enough has been written about this embarrassing era in the history of psychometrics (Block & Dworkin, 1976; Fallows, 1980; Gould, 1981), and I do not bring it up here again to heap insult on contumely.[1] I want instead to praise testing of that time; for in spite of its grossness then, it had something that it lost soon after and is missing today. At least the concern with measurement of human behavior in the early stages of its history was allied with the best thinking of the time on psychology and sociology, i.e, with evolution by natural selection, with the attempt to apply ideas of biological evolution to culture and society. This was more true, of course, of the European psychometricians than the Americans; as Sizer (1970, p. 15) observed: ". . . Americans engineered the idea of mental testing and adapted late nineteenth-century European theories to the realities of a more modern America. Terman, Thorndike, and the rest were pioneers, but more as engineers than as theoreticians." It would indeed be an act of insensitive second-guessing to think that Galton and Spearman and Goddard and Terman and the others were wrong *and* should have known better. They may well have been wrong, just as our best theories will seem puerile to future scientists, but they were the leading psychologists and social scientists of their day and testing was their most useful tool. It was, I submit, testing's golden era, and it has not known their like since. Perhaps, as with the triumphs of a precocious child, testing's early successes led to its current difficulties. The technology of testing was quickly wedded to the burgeoning field of statistical methods. The discipline began to grow specialized and esoteric. In the 1920s, testing entered a stage of hyper-rationalization from which it has never re-emerged. Multiple factor analysis plumbed the "vectors of the mind" (Thurstone, 1935) with machinery (centroids, tetrads, reference systems, etc.) beyond the ken of psychologist and social scientist. It is of more than passing significance that the increase in technical sophistication of the testing movement had little effect on tests themselves or the theories on which they were based. I cannot, for example, discern Thurstone or Holzinger's lineaments in any of the contemporary tests of intelligence. Indeed, the modern intelligence scale would seem a familiar artifact if placed in the hands of a suddenly reincarnated psychologist of the Edwardian period. I know of no science—save perhaps, anthropology or history—about which the same could be said.

The development of psychometrics from its early triumphs to the modern era parallels, peculiarly enough, the birth and growth of a bureaucratic agency—

[1]For the most recent chapter in this historical controversy, see Snyderman and Herrnstein (1983), who argue that, in spite of the racist character of much of the psychometrics of the early part of this century, it is doubtful that the research directly influenced the passage of the Immigration Act of 1924. Their apologia reminds one of the man who, when accused of murdering three men and a dog, forthwith produced (before the court) in his defense the dog alive.

there are, indeed, lest anyone doubt it, bureaucracies and bureaucrats of ideas. It is a natural and human failing when one has access to and control over specialized information to exert that control against change (Selznick, 1953). Technological society's emphasis on expertise and specialization produce trained incompetence, the narrowing of the scope permitted for intellectual experimentation. Psychometricians, with control over arcane corners of mathematics, exemplify forces such as these (Andresky, 1972). As Robert Merton (1975) observed, the specialist begins to resist change because of vested interest in the current structure, of a society or culture, whether material or intellectual. "Adherence to the rules, originally conceived as a means, becomes transformed into an end-in-itself; there occurs the familiar process of displacement of goals, whereby an instrumental value becomes a terminal value . . ." (p. 28).

This displacement of goals took place in testing between about 1940 and 1960 as nearly as I can judge. Spurred by what we are told were the great victories for testing in World War II (Chase, 1948), all of which were atheoretical and pragmatic, the discipline of psychometrics began to take shape apart from psychology. It turned its back on psychology and reached instead for an independent, autonomous set of "principles of measurement" that transcended everything in particular. I remember my delight in discovering in 1960 that I could read with nearly complete comprehension the great treatise on psychometrics, Gulliksen's *Theory of Mental Tests* [1950], simply because I had overlearned college math and knew less than nothing at all about psychology.

The development of reliability and validity theory—two of the greatest achievements of the psychometric movement—can be viewed as an over-ambitious attempt to axiomatize a discipline. I view Cronbach's work over the past two decades (and in particular his collaboration with Meehl [Cronbach & Meehl, 1955]) as an attempt to correct the errant ways of methodologists who felt they could safely leave substance behind in the search for the abstract foundations of measurement. It took a while to drive home the point that the validity of testing is a quality of a complex use of information; it is not a property of a random variable. It has taken longer to make the point that reliability is no different, and just as a test has no validity, so it has no realiability either. Measurements have meanings, and they permit or obstruct thinking to various degrees. The process by which measurements are taken, as well as the ideas that gave rise to the measurements, are only judged in accord with how both—ideas and measurements—lead toward greater understanding. The relationship is reciprocal: constructs and observations, meanings and methods. The message is Cronbach and Meehl's.

By the early 1960s, it is fair to say that testing in psychology and education was severed from its roots in the study of human behavior. Indeed, testing flourishes today where the environment is starkly atheoretical (education) and withers in precisely those locales where thinking about human behavior is freshest and most exciting (psychology, psychiatry).

The Modern Testing Establishment

What then has become of testing and measurement? Estranged from the behavioral and social sciences, grown mathematically elaborate and worshipful of the "general principles of measurement," what has become of testing since it reached maturity and autonomy in the academies of post-World War II America and Europe? In short, it sold out to the highest bidder; it went Establishment. In the world of work and in education, testing stepped forward to play the role of gatekeeper and management tool in the processing of human lives for the meritocracy: professional lightning rod attracting and grounding the anger of the excluded and discarded; factotum for society's dirty work.

The testing industry and, more regrettably, the discipline of measurement in the Academy no longer serve science in its search for understanding nor education in its search to educe from individuals that which is best in them. Increasingly they serve a national and international system of processing people by number, managing the flow of bodies from institution to institution, documenting the expected progress of pupils through the vast educational "system." For its efforts, the psychometric industry with its academic support system acquires huge caches of unexpended income.

The lack of articulation between measurement and substantive theory is particularly serious in education, that most pragmatic and atheoretical of all disciplines where testing is applied. Achievement test batteries are designed around what is thought to be the content of the school curriculum as determined by surveys of textbooks, teachers and other tests. Textbooks and curricula are designed, on the other hand, in part around the content of tests. One cannot discern which side leads and which follows; each side influences the other, yet nothing assures us that both are tied to an intelligent conceptualization of what an educated person ought to be.

Considering the prominence of testing in contemporary American schools, it is amazing to realize how useless testing seems to those closest to the core of schooling: teachers and pupils. It is scarcely any secret that teachers regard standardized ability and achievement tests as an irrelevance or worse. They complain that they learn nothing from the results that wasn't obvious before the test was given; the scores give no clue as to what should be done to eradicate ignorance or take advantage of talent and skill. Testing is a transaction between the testing companies and school administrators, state education officials, government agencies, lawyers and other middlemen in the system of schooling.

Tests are not used by educators to decide how children should be educated because they are not designed for such purposes and are virtually worthless toward such ends (Hawkins, 1977). This fact greatly concerned Oscar Buros (1977) who decried the drifting away of achievement tests from what was taught in a course toward the goal of predicting individual differences in attainment at higher grades. After two decades of research on aptitude-treatment interactions, it remains unclear whether one ought to teach so as to utilize the person's strongest aptitude, or teach so as to compensate for the weakest aptitude, or both or neither; and perhaps the meager harvest of so much research should be blamed on the primitive concepts of aptitude on which the tests are based. Glaser (1981) has complained for decades that tests are useless for deciding what it is that a child can and cannot do, hence the need to reference the scale that a test creates to the criterion of skills, knowledges, and understandings that comprise whatever it is that we think of as facility in reading, math, science and the like. The distortion of this eminently sensible call for "criterion referenced testing" (now repeated in hopes of productive hybridization of education and cognitive psychology [Glaser, 1981]) into item banks for behavioral objectives with cut-off scores for mastery is one of the more unfortunate inventions of modern psychometrics. My colleague David Hawkins at the University of Colorado reached back to Dewey in making a similar argument in his brilliant little book, *The Science and Ethics of Equality* (1977, p. 75):

> . . . we need a framework of general ideas adequate to the developmental perspective in which all important abilities and talents should be viewed. This means that we should dig under the surface of those tests which have provided the empirical basis for so much of statistical psychometrics, and specifically the various intelligence tests. . . . I do not want to beat the IQ tests over the head. They are useful in their way, though as John Dewey said more than once, they are of little use to good teachers, who need both a refinement and an immediacy of discrimination in their daily work with children which global test averages do not provide.

Gullickson [1983] surveyed 30 education professors and 400 teachers to record their priorities for the content of educational measurement courses. Of 50 topics rated for desired emphasis in a college course, the greatest discrepancies between professors and teachers emerged on these topics: Teachers wished for "great emphasis" to be placed on ways of "interviewing pupils and parents," "observing pupils' work habits," evaluating "class discussions" and "interpersonal relationships"; the professors rated these topics as deserving only "slight emphasis." (In fact, these items were among the top ten rated items by the teachers and among the professors' bottom ten! The professors' highest rated topics were calculating the mean and variance and calculating correlation coefficients.)

The contemporary problem of "learning disabilities" is a case that can be advanced in behalf of the argument that testing in education follows the wrong lights and serves the wrong masters. Measurement of LD is virtually uniformly pursued across the United States today. IQ and achievement test scores are compared and "significant discrepancies" are tagged as evidence of LD (Shepard, Smith, & Vojir, 1983; Smith, 1982). The use of available published tests to do this work is encouraged by many factors: they are legally defensible, they seem scientific and unimpeachable to parents, they are cheap and quick. But they are being used to measure what was called ''under-achievement'' three decades ago. Learning disability or lack of motivation? Can anyone tell the difference? Well, of course, but not in the naive and mechanical way that these notions are being translated into numbers by clinicians and psychometricians. It doesn't matter in the least whether one regards LD as ''euphemizing'' or not; the periodic purging of dysphemisms like ''retardate'' or ''slow learner'' would only seem unnecessary to one who had never been called such things. The whole idea of ''learning disabilities'' is that obstacles to learning are more variegated and worthy of detailed analysis than being ascribed to dullness or low intelligence. It is manifest that learning involves the hardware of the brain and of the communication channels; and when learning goes wrong, these as well as what a child has experienced in both his distant and recent history are implicated. From this point, the work of psychiatry, medicine, cognitive psychology and the like must begin, and what we know of the technology of measurement may have a contribution to make. But the absurdly premature translation of ill-formed concepts of LD into IQ versus achievement discrepancies was a managerial expedient of a particularly mindless sort, motivated by legal, political and professional interests. Unfortunately, it was all too typical of how these interests have used educational measurement.

Testing has found a new market in minimal competence testing in education and in licensure and certification in the workplace. I regard both of these with disappointment or disdain (Getz & Glass, 1979; Glass, 1978; Glass, 1979; Hogan, 1983; Olson, 1983). I need not go into details of these unseemly businesses. Perhaps it is enough to go on record again as believing that both applications of testing serve crass political ends: One, the extension of centralized political control of education; the other, the protection of economic self-interest in the workplace. Neither (and this goes for academic selection as well) is based on proof of utility that would justify its negative consequences in denial of opportunities or restriction of free trade.

Recently after reviewing and integrating the findings of over 500 psychotherapy outcome experiments, my colleagues and I (Smith, Glass, & Miller, 1980, p. 187) remarked thus on the state of the research art in this area:

> Psychotherapy-outcome research lacks nothing by way of differentiated interventions. The literature on treatment is a veritable pharmacopoeia of prescriptions. The design of controlled experimentation has been refined to a science that is within the grasp of any researcher who owns a table of random digits and recognizes the difference between blind and sighted assessment. However, the measurement of outcomes seems to have been abandoned at a primitive stage in its development. Rating scales are thrown together with little concern expressed for their psychometric properties. Venerable paper-and-pencil tests . . . with roots planted vaguely in no particular theory of pathology or treatment are used to hunt for effects of short-term and highly specialized brands of psychotherapy. A superfluity of instru-

ments exists, and too little is known about them to prefer one to another. Little is known about their structure, and less is known about their sensitivity to treatment.

Thinking back to both of the Coleman et al. (1966; 1982) studies (of equality of educational opportunity and of public and private schools), the Follow-Through evaluation (House, Glass, McLean, & Walker, 1978) and more, I worry that modern testing and measurement—with their overweaning attention to the pragmatic, the conventional, the traditional—have too often not merely failed to reveal the complexity and subtlety of human experience, but have actually denigrated the value of attempts to improve it. The telescoping of the immense variety in the Follow-Through models down into a standardized basic skills test and two dubious ''affective'' measures was a travesty for which the psychometric community might hold certain bureaucrats responsible (House et al., 1978), but the bureaucrats in question are quick to respond to such charges that they chose the measures from among the best that the psychometric arts had to offer.

BRINGING MEASUREMENT BACK TO PSYCHOLOGY: TOWARD A SOLUTION

Case Study Research

The growing significance of the naturalistic or ''case-study'' methodology in the social sciences poses important new problems for measurements; and if these problems are accorded the attention they deserve, the benefits may accrue not just to naturalistic methods but to measurement itself throughout the social sciences. The translation of experience into an observational record—everywhere the fundamental problem in measurement—always requires the imposition of some explanatory, theoretical structure. We tend to forget or ignore this fact in the established areas of testing and measurement, and then we accept the theoretical structure (bequeathed to our generation by faculty psychology or ''self-concept'' psychology) not as supposition but as reality. This bit of self-deception is more difficult to maintain in naturalistic research where experience is more complex and different theoretical systems still compete for favor. The naturalistic scene focuses our attention on several of the problems that need to be addressed by measurement theorists across the entire range of behavioral and social sciences: the necessary tie between theoretical structure and observation; the problems of not oversimplifying, of doing justice to the complexity of human systems (whether they be at the level of the individual, the group or a whole society); the difficulty of ''slicing up the raw behavioral flux'' and translating it into observations or numbers (Meehl, 1978); the unique problems attendant upon the use of human observers of human behavior (i.e., the problems of *Verstehen*

(Dilthey, 1959–68) or countertransference as behavioral science method as Devereux (1967) has written of it). These, I submit, are the key methodological problems that measurement must face if it is to further—rather than retard or play only a superfluous role in—the progress of the social sciences.

Theoretical Possibilities

While I take some comfort in knowing that my observations about what is wrong with testing today were seen earlier by my esteemed colleagues Anastasi and Glaser among other references, I shall derive more satisfaction if I can broaden ever so slightly the range of psychological theory to which testing and measurement would do well to attend. My colleagues have emphasized cognition and its role in learning. It is true that Anastasi mentioned the importance of attending to developments in personality theory, but she singled none out, advising her listener rather to keep abreast of relevant research in clinical and social psychology.

I wish to go further and counsel my fellow educationists and psychometricians that the most exciting and productive thinking outside the area of cognitive psychology is virtually untouched by psychometrics and that we could do less well than to turn our attention there when we seek the path back to the best scientific thinking on human behavior. The ''affective domain,'' as it is so inappropriately named, was not addressed by Anastasi or Glaser, and I take it as a favor that they have left it to me to extend their thesis into this challenging domain of human motives, desires, fears, wishes, antipathies, hopes, and frustrations.

Anastasi (1967) bowed in the direction of ''personality'' assessment in her 1967 address, but it seemed a hesitant gesture. Although she urged psychometricians to keep abreast of clinical psychology, her vagueness on the matter of what in particular in that vast area was worth attending to reflects another dissociation of theory and practice in American psychology, namely, the schism between psychology and psychiatry, or even the parallel split between academic and clinical psychology (splits that have their own political and intellectual roots).

Ever since psychiatry emerged from Bedlam and the scientific dark ages, the dominant theoretical perspective has been psychoanalytic, Freudian. Strangely, in spite of the antipathy with which psychodynamic theory is viewed by American academic psychologists, across the hall their colleagues in clinical psychology honor it as the pre-eminent theoretical system (Garfield & Kurtz, 1974, 1977). The extension of Freud's work made by such investigators as Hartmann, Spitz, Jacobson, Mahler and now the younger generation of ego-psychologists has produced a most impressive and far-reaching psychological theory of human development and behavior (both ''normal'' and pathological). Permit me to declare myself. Coupled with neurophysiology and behavioral genetics, the psychoanalytic perspective represents our best hope for a comprehensive and useful account of the development and psychology of human beings. And yet, the link between psychoanalysis and psychometrics is virtually nonexistent.

Not only is any connection between psychoanalytic psychology and psychometrics impossible to discern, but the latter sometimes rather grandly imagines that it has disproved the former. A methodological critique or a factor analysis of Rorschach responses, or a horse race between statistical and clinical predictions will not deal with the challenge that neo-analytic ego psychology presents to empirical methodology. It is a curious fact unknown to nearly all who cite Meehl (1954) as the refutation of "clinical insight" (and then by implication "psychoanalytic theory") that Paul Meehl's theoretical leanings are self-proclaimed as psychoanalytic.[2] This is not the place to defend psychoanalysis against the charge that it is "unscientific"—a charge made about equally often as the charge that it is false, the two charges being contradictory, at least by Popper's criterion of what constitutes a scientific proposition. The defense of the scientific status of psychoanalysis was presented some time ago in Hook (1959). I only wish to add here that it is scientific in precisely the most significant way and in the way in which too many psychological theories are inadequate, viz., the conclusions of psychoanalysis (e.g., the meaning of dreams, parapraxes, the operation of defense mechanisms, and the like) are "risky" propositions (in the Popperian sense), meaning that they are not independently derivable from common sense.

Psychoanalytic Psychology

It is my opinion that psychoanalytic psychology, particularly in its modern forms, represents the most significant challenge and opportunity that testing and measurement in the social and behavioral sciences could assume. It promises to change fundamentally the way in which psychometrics is pursued outside the narrow area of assessing aptitude and learning. If studied seriously, psychoanalytic psychology could lead to new techniques of observation and at least new concepts of the relationship between manifest behavior and the enduring psychology of the individual.

I see this possible relationship only vaguely myself. Perhaps an example or two will help bring these generalities into better focus.

The measurement of "self-concept" is one of the most active areas of psychometric concern, and yet the construct as embodied in modern tests is a hoary and naive thing scarcely developed any further than William James's (1890) thinking nearly a century ago. In Wiley's (1961) famous treatments of self-concept measurement, though she treats the theoretical foundations of the construct with respect, they are revealed to be little more than vague, commonsensical sketches of the conditioning of Philistine self-satisfaction by a pair of bland

[2] "I am confident that psychoanalytic concepts will be around after rubber band theory, transactional theory, attachment theory, labeling theory, dissonance theory, attribution theory, and so on, have subsided into a state of innocuous desuetude. . . . At the very least, psychoanalysis is an interesting theory, which is more than I can say about some of the 'theories' that are currently fashionable" (Meehl, 1978, p. 817).

parents molding little lumps of clay with praise and kisses. One current "theory," which is attracting psychometric attention (Marsh, Smith, & Barnes, 1983), is a seven-part model of the self-concept: The self-concept is hypothesized to comprise (a) the physical ability self, (b) the appearance self, (c) the peer relations self, (d) the parent relations self, (e) the reading self, (f) the mathematics self, and (g) the school subjects self. In a 10,000-word research report on this theory, one reads of factor analysis, multitrait/multimethod matrices, self-reports versus ratings by others, discriminant validity, divergent validity, halo bias, social desirability response sets, and on and on. About the only reference to psychological theory is, "An implicit assumption of most theorists is that self-concept is multifacted" (p. 334). Further, it is said to be formed out of experience with the environment and interactions with "significant others." Clearly, this is a picture of psychometrics running amuck! Big methodological guns loaded with folk wisdom and truisms. This factor analytic mincing could go on forever; we could, of course, equally well discover "automobile self-concept," "favorite football team self-concept," or even "preferred Baskin-Robbins flavor self-concept."

The problem with this empirical hustle and bustle is that it is thoroughly innocent of any serious theory about the psychological sense of self: how it develops, what it is, how it can become sick, how to make it well. How adequately do the Rogerian theory and the other commonsense accounts of "self-concept" stand up to the "risky" tests of explanatory scope which alone will separate idle psychologizing from respectable theory, such risky tests as accounting for the ephemeral sense of identity of a thoroughly decompensated schizophrenic, the sense of gender identity so confused as to cause a man wilfully to multilate his body surgically and chemically, the dangerous line between positively cathected self-representations and neurotic narcissism, the splintered selves of a multiple personality, or the more ordinary feelings of depression and emptiness in a child whose every act prompts nothing but praise and reward from the "significant others" of his environment? Unless our theories reach this far, they are in jeopardy of being not so much false as uselessly redundant with ordinary common sense.

It will surprise many to learn perhaps that one can scarcely find any reference in all of Freud's voluminous writings to a "self-concept," and that the term does not even appear in the lexicographic bible of psychoanalysis, Laplanche and Pontalis's *The Language of Psychoanalysis* (1973). Furthermore, in the writings of the neo-Freudians, those things that William James once thought of and the person-in-the-street now thinks of as the "self-concept" have been resolved into an extremely complex braid of developmental strands (including, among other things, identity formation *a la* Mahler) through stages of autism, merged self and object representations, differentiated representations, "practicing" and rapprochement subphases to gender identify and separation-individuation; or (*a la* Kohut) the formation of the ego ideal from disillusionment with the grandiose self whose roots reach to the stage of infantile primary narcissism; or (*a la* S. Freud, A. Freud, Mahler, Jacobson and nearly every neo-Freudian) the construction of personal identity through identification with the loved or hated object. Neo-Freudian ego psychology has made exciting advances in those areas referred to colloquially as ''self-concept'' or ''self-esteem''; the best didactic treatment of the past forty years of this research is the impressive two-volume work by Gertrude and Rubin Blanck, *Ego Psychology: Theory and Practice* (1974) and *Ego Psychology II: Psychoanalytic Developmental Psychology* (1979). This corpus of research can be commended to the attention of psychometricians; it has been attended to by at least one such, but more about that later.

In her summation on the state of self-concept measurement, Ruth Wiley (1961, pp. 317 ff.) criticized much of the psychometric work she reviewed on account of its theoretical naiveté. She was more generous than I might have been in the same situation (she being rather charitable toward some trivial conceptualizations), and her own grasp of the role of ''self'' in psychodynamic theory was weak. Nonetheless, she did identify the yawning gap between psychometric practice and psychoanalytic theory: ''. . . certain psychologists have thought that self-concept research yields weak or equivocal results because the theory does not systematically include the unconscious self concept, or other unconscious cognitive and dynamic processes'' (Wiley, 1961, p. 319). Although she thereafter goes on to place an unhealthy emphasis on the criterion of predictive validity for deciding whether new and unusual constructs (like the Freudian unconscious) should be allowed into the test battery, her sense of the seriousness of the omission of the unconscious from consideration of ''self-concept'' seems completely justified. Indeed, the situation is typical of academic psychology's long relationship with Freudian psychology. Everywhere theorists and practitioners wish to cook the Freudian omelet without breaking the Freudian eggs. It cannot be done; the unconscious (whose existence is proved daily in our actions and nightly in our dreams) is the cornerstone of psychoanalysis and it cannot be locked in the closet like some shameful secret if psychology (and psychometrics) are to form any meaningful connection with psychoanalysis.

Consider Jane Loevinger's (Loevinger & Wessler, 1970) work on the measurement of ego development. Loevinger, an early-day quantitative psychologist and psychometrician, has spent the last three decades engaged in an ambitious attempt to develop measures of some of those emotional-cognitive processes talked about by the neo-Freudians, and which she covers with the title ''ego development.'' Her efforts took off from a thorough understanding of traditional psychometric technique and its deficiencies for capturing what the ego psychologists were writing about: (a) there exists no one-to-one correspondence between a particular behavioral act and a level of ego development; (b) many strands of ego development occur simultaneously and one bit of behavior may reflect more than one strand; (c) no error-free way exists of distinguishing signs of developmental levels from signs of non-developmental correlates; (d) each individual displays behavior at more than one level of ego maturity; (e) a behavioral sign may be discriminating in only one direction; and so on. Ironically, the one difficulty that Loevinger clearly identified and sought to overcome in her choice of psychometric format (sentence completion) and the voluminous and demanding scoring guides is precisely the point on which her critics allege that she departed from psychoanalytic theory. Loevinger wrote:

> . . . no behavioral task can be guaranteed to display just what one wants to know about ego level. Neither a structured test nor an unstructured test carries a guarantee. If the test is structured, the investigator is projecting his own frame of reference rather than tapping the frame of reference of his subjects, which is the very thing that reveals their ego level. If the test is unstructured, one cannot control what the subject will choose to reveal (p. 9).

Loevinger must have felt she was steering a safe middle course through this dilemma by choosing incomplete sentences to be completed by the examinee (e.g., "When I get mad . . . ," "When they avoid me . . . ," "When they talked about sex, I . . .") and by struggling heroically with the free-form productions that result; but Gertrude Blanck (1976), whose understanding of Freud and the neo-Freudians is widely honored, blistered Loevinger for her misunderstanding of theory and for her choice of method: "The methodology . . . is simplistic, not alone by comparison (with psychoanalytic observational studies), but in its own right. Sentence completion as a research tool cannot be taken seriously because it relies on conscious responses and overlooks their unconscious determinants" (p. 803).

I can agree enthusiastically at a general level with what Mischel called for in his 1977 paper, "On the Future of Personality Measurement," viz., a broader assessment of persons functioning in their environment. I can endorse wholeheartedly and accept as my own his prediction that,

> In the future, measurement hopefully will be directed increasingly at the analysis of naturally occurring behaviors observed in the interactions among people in real-life settings. . . . The future of personality measurement will be brighter if we can move beyond our favorite pencil-and-paper and laboratory measures to include direct observation as well as unobtrusive nonreactive measures to study lives where they are really lived and not merely where the researcher finds it convenient to look at them (p. 248).

Hear! Hear! And yet. . . . What kind of peeping does Mischel have in mind? And does he realize that if the dignity of the individuals involved is respected, they will ultimately be the source of information about their own lives and what it is like to live them; the assessment of personality might better resemble a psychiatric interview than a bugged room with one-way mirrors. Although Mischel seems to recognize this somewhat and says that personality measurement must rely increasingly on self-reports, he finally (and disappointingly from my perspective) asserts his identity as an American academic psychologist and repudiates the unconscious, even suggesting obliquely that a couple of experiments and his own textbook have disproved its existence. Lord! Don't these psychologists ever sleep? And if they do, do they never dream?

A "Reflection on Schoolboy Psychology"

It is with regard to the role of unconscious processes that the relationship between psychoanalysis and psychometrics will be determined. If psychometrics continues to view the unconscious with arm's length suspicion as something unsavory or pathological or unscientific, then the opportunities for useful collaboration will be few.

It is, I believe, through a largely unconscious process of identification that we become truly educated. We grow to be like those we love. Our teachers give us an identity—not facts, not even a significant amount of whatever golden things lie at the highest level of Bloom's taxonomy. We are, each of us, living out lives that we took from someone else, someone we have loved and whose image we hold close by being what they were. Freud (1914) said as much in his "Some Reflections on Schoolboy Psychology."

> . . . it is hard to decide whether what affected us more and was of greater importance to us was our concern with the sciences that we were taught or with the personalities of our teachers. . . . We courted them or turned our backs on them, we imagined sympathies and antipathies in them which probably had no existence, we studied their characters and on theirs we formed or misformed our own. They called up our fiercest opposition and forced us to complete submission; we peered into their little weaknesses, and took pride in their excellences . . . we can now understand our relation to our schoolmasters. These men, not all of whom were in fact fathers themselves, became our substitute fathers (p. 242).

There is in these instances of unconscious identification more about the true course of education,[3] the way it shapes and molds and occasionally transforms us, than there is in all the behavioral objectives and mastery quizzes and standardized tests that ever were written. Children take more from the adults of their world than knowledge or training or even chromosomes. Through a process of identification, which springs largely from unconscious motives, they take a way of living that reaches to every corner of their lives. The dynamics of identification form some of the more interesting themes in that unstudied genre of liter-

[3]The true aims of education, as Michael Oakeshott [1972, p. 40] characterized them, are ". . . initiation into the mysteries of a human condition; the gift of self-knowledge and of a satisfying intellectual and moral identity."

ature perhaps best called "teacher fiction" (e.g., the relationship of Godfrey St. Peter and Tom Outland in Willa Cather's *The Professor's House,* or the loathesome dynamics in Muriel Spark's *The Prime of Miss Jean Brodie*). We must look in such places as these (the unconscious origins of identification, for one) with such peculiar instruments as interviews and freely flowing association and, yes, even dreams, if we wish to find the secrets of how children come to assume their individual adult forms.

There are two sides to this matter of the enduring impressions that education sometimes leaves. I spoke of one side, that seen by the marble. Now let the sculptor speak. Another man, Loren Eiseley, who after publishing *The Immense Journey* was acclaimed scientist and poet by the likes of Auden, once described[4] the view from the lectern:

> Now, for many years an educator, I often feel the need to seek out a quiet park bench to survey mentally that vast and nameless river of students which has poured under my hands. In pain I have meditated: "This man is dead—a suicide. Was it I, all unknowingly, who directed, in some black hour, his hand upon the gun?" "This man is a liar and a cheat. Where did my stroke go wrong?" Or there comes to memory the man who, after long endeavors, returned happily to the farm from which he had come. Did I serve him, if not in the world's eye, well? Or the richly endowed young poet whom I sheltered from his father's wrath—was I pampering or defending—and at the right or the wrong moment in his life? Contingency, contingency, and each day by word or deed the chisel falling true or blind upon the future of some boy or girl.
>
> Ours is an ill-paid profession and we have our share of fools. We, too, like the generation before us, are the cracked, the battered, the malformed products of remoter chisels shaping the most obstinate substance in the universe: the substance of man. Someone has to do it, but perhaps it might be done more kindly, more precisely, to the extent that we are consciously aware of what we do—even if that thought sometimes congeals our hearts with terror. Or, if we were more conscious of our task, would our hands shake or grow immobilized upon the chisel?
>
> I do not know. I know only that in these late faint-hearted years I sometimes pause with my hand upon the knob before I go forth into the classroom. I am afflicted in this fashion because I have come to follow Dewey in his remarks that "nature is seen to be marked by histories." As an evolutionist I am familiar with that vast sprawling emergent, the universe, and its even more fantastic shadow, life. Stranger still, however, is the record of the artist who creates the symbols by which we live. As Dewey has again anticipated, "No mechanically exact science of an individual is possible. An individual is a history unique in character." "But," he remarks, "constituents of an individual are known when they are regarded not as qualitative, but as statistical constants derived from a series of operations" (1962, p. 25).

[4]In a book dedicated to Letta May Clark, Eiseley's English teacher at University of Nebraska High School: *The Mind As Nature,* No. 5 in the John Dewey Society Lectureship, New York, Harper & Row, 1962.

Eiseley went on to examine creativity: ". . . that enigma to which the modern student of educational psychology is devoting more and more attention" (1962, p. 28). In 1962 when Eiseley wrote these words, educational psychologists did indeed aspire to measure and explore such "statistical constants" as creativity and motivation and trust and perseverance and "social competence." I am old enough to remember when psychometricians spoke unashamedly of their aspirations to capture more of human experience than the IQ. Now they seldom confess such ambitions, content instead, it seems, to refine endlessly the mathematical foundations of measuring nothing in particular. And this concerns me more than anything else about the present condition of testing and its future: that its disciples no longer share any sense of wonder or fascination about the development of thought, the nurture of talent, the mysteries of human personality. Without the aspiration to understand human growth and behavior, testing will drift further from those sciences that keep such aspirations alive.

REFERENCES

Anastasi, A. (1967). Psychology, psychologists and psychological testing. *American Psychologist, 22,* 297–306.

Andresky, S. (1972). *Social sciences as sorcery*. London: Deutsch.

Blanck, G. (1976). An ambitious undertaking. *Contemporary Psychology, 21,* 801–803.

Blanck, G., & Blanck, R. (1974). *Ego psychology: Theory and practice*. New York: Columbia University Press.

Blanck, G., & Blanck, R. (1979). *Ego psychology II: Psychoanalytic developmental psychology*. New York: Columbia University Press.

Block, N. J., & Dworkin, G. (Eds.) (1976). *The IQ controversy*. New York: Pantheon Books.

Boulding, K. E. (1968). Expecting the unexpected: The uncertain future of knowledge and technology (pp. 158–175. In K. E. Boulding, *Beyond economics: Essays on society, religion and ethics*. Ann Arbor: University of Michigan Press.

Brown, J. S., & Burton, R. R. (1978). Diagnostic models for procedural bugs in basic mathematical skills. *Cognitive Science, 2,* 155–192.

Chase, S. (1948). *The proper study of mankind*. New York: Harper and Row.

Coleman, J. S., Campbell, E. A., Hobson, C. J., McPartland, J., Mood, A. M., Weinfeld, F. D., & York, R. L. (1966). *Equality of educational opportunity*. Washington, D.C.: U.S. Government Printing Office.

Coleman, J. S., Hoffer, T., & Kilgore, S. (1982). *High school achievement: Public, Catholic and private schools compared*. New York: Basic Books.

Cronbach, L. J., & Meehl, P. E. (1955). Construct validity in psychological tests. *Psychological Bulletin, 52,* 281–302.

Devereux, G. (1967). *From anxiety to method in the behavioral sciences*. The Hague: Mouton & Co.

Dilthey, W. (1959–68). *Gesammelte Schriften*. 14 volumes. Stuttgart: Teubner.

Eiseley, L. C. (1962). *The mind as nature*. New York: Harper and Row.

Fallows, J. (1980, Feb.). The tests and the "brightest." *Atlantic*, pp. 37–48.

Freud, S. (1914). Some reflections on schoolboy psychology (pp. 241–250, Vol. 13). In *Totem and taboo and other works (1913–14). Standard Edition of the Complete Psychological Works of Sigmund Freud*. James Strachey (Ed.), 24 vols. London: Hogarth Press, 1953–66.

Galton, F. (1892). *Hereditary genius: An inquiry into its laws and consequences.* Bath, England: Pitman Press.

Garfield, S. L., & Kurtz, R. (1974). A survey of clinical psychologists: Characteristics, activities and orientations. *The Clinical Psychologist, 28,* 7–10.

Garfield, S. L., & Kurtz, R. (1977). A study of eclectic views. *Journal of Consulting and Clinical Psychology, 45,* 78–83.

Getz, J., & Glass, G. V. (1979). Lawyers and courts as architects of educational policy. *High School Journal, 62,* 181–186.

Glaser, R. (1981). The future of testing: A research agenda for cognitive psychology and psychometrics. *American Psychologist, 36,* 923–936.

Glass, G. V. (1978). Standards and criteria. *Journal of Educational Measurement, 15,* 237–261.

Glass, G. V. (1979, May 8). The war on incompetence. *Paper presented at the Distinguished Scholars Series,* Oregon College of Education; Monmouth, Oregon.

Gould, S. J. (1981). *The mismeasure of man.* New York: Norton.

Gullickson, A. R. (1983). Personal communication.

Gulliksen, H. O. (1950). *Theory of mental tests.* New York: Wiley.

Hawkins, D. (1977). *The science and ethics of equality.* New York: Basic Books.

Hogan, D. B. (1983). The effectiveness of licensing. *Law and Human Behavior, 7,* 117–138.

Hook, S. (Ed.) (1959). *Philosophy, scientific method and psychoanalysis.* New York: New York University Press.

House, E. R., Glass, G. V., McLean, L. D., & Walker, D. (1978). No simple answer: A critique of the Follow-Through evaluation. *Harvard Educational Review, 48,* 128–160.

James, W. (1890). *Principles of psychology.* New York: Holt. 2 volumes.

Kamin, L. J. (1974). *The science and politics of IQ.* Hillsdale, NJ: Lawrence Erlbaum Associates.

Laplanche, J., & Pontalis, J. B. (1973). *The language of psychoanalysis.* New York: Norton.

Loevinger, J., & Wessler, R. (1970). *Measuring ego development, I.* San Francisco, CA: Jossey-Bass.

Marsh, H. W., Smith, I. D., & Barnes, J. (1983). Multitrait-multimethod analyses of the self-description questionnaire: Student-teacher agreement on multidimensional ratings of student self-concept. *American Educational Research Journal, 20,* 333–357.

Meehl, P. (1954). *Clinical vs. statistical prediction.* Minneapolis: University of Minnesota Press.

Meehl, P. (1978). Theoretical risks and tabular asterisks: Sir Karl, Sir Ronald and the slow progress of soft psychology. *Journal of Consulting and Clinical Psychology, 48,* 808–834.

Merton, R. K. (1957). *Social theory and social structure.* New York: Macmillan.

Mills, C. W. (1959). *The sociological imagination.* New York: Oxford University Press.

Mischel, W. (1977). On the future of personality measurement. *American Psychologist, 32,* 246–254.

Oakeshott, M. (1972). Education: The engagement and its frustration (p. 19–49). In R. F. Dearden, P. H. Hirst, & R. S. Peters (Eds.), *Education and the development of reason.* London: Routledge & Kegan Paul.

Olson, P. A. (1983). Credentialism as monopoly, class war, and socialization scheme. *Law and Human behavior, 7,* 291–299.

Selznick, P. (1953). *The theory of social and economic organization.* New York: Oxford University Press.

Shepard, L. A., Smith, M. L., & Vojir, C. P. (1983). Characteristics of pupils identified as learning disabled. *American Educational Research Journal, 20,* 309–331.

Siegler, R. S. (1976). Three aspects of cognitive development. *Cognitive Science, 8,* 481–520.

Simon, H. A., & Chase, W. G. (1973). Skill in chess. *American Scientist, 61,* 394–403.

Sizer, T. R. (1970). Testing: America's comfortable panacea (p. 14–21). In G. V Glass (Ed.), *Proceedings of the 1970 invitational conference on testing problems.* Princeton, NJ: Educational Testing Service.

Smith, M. L. (1982). *How educators decide who is learning disabled.* Springfield, IL: C. C. Thomas.

Smith, M. L., Glass, G. V., & Miller, T. I. (1980). *The benefits of psycho-therapy.* Baltimore, MD: John Hopkins University Press.

Snyderman, M., & Herrnstein, R. J. (1983). Intelligence tests and the Immigration Act of 1924. *American Psychologist, 38,* 986–995.

Thurstone, L. L. (1935). *The vectors of mind.* IL: University of Chicago Press.

Wiley, R. C. (1961). *The self concept: A critical survey of pertinent research literature.* Lincoln: University of Nebraska Press.

3 Computer Technology in Testing

Gale H. Roid
Western Psychological Services

COMPUTER TECHNOLOGY IN TESTING

Although computers have had an important role in educational and psychological testing for decades, the widespread availability of personal computers has focused interest on the appropriate role of computerization in the development, administration, scoring, and interpretation of tests. Although the early decades of computer usage found hardware and services concentrated in large computer installations, future decades will find hardware and services distributed more widely among individual users. With this rapid diffusion of technology and the lightning speed with which technology is changing, it will be increasingly difficult to predict the directions that computerized testing will take. Therefore, the purpose of this review is to discuss some broad themes in the future of computer technology as applied to testing, but, at the same time, restricting the discussion to methodology and usages that appear feasible for application in the near future.

Although it is always tempting in the discussion of any innovation, this review will resist the urge to view computerized testing or computerized interpretation of tests as a panacea for all of the limitations of nonautomated procedures. In fact, computerized test interpretation raises several new ethical issues and complications that magnify the latent problems of inexperienced test users (Zachary & Pope, 1983). It remains clear that skillful and imaginative clinical use of tests and assessment will always require the reasoned guidance of the experienced professional. Computers remain a tool to the professional, admittedly a more complex tool than previously available to the individual user.

This chapter has two time perspectives: (1) current status, and (2) future directions. Within each of these perspectives, four areas of computerization are discussed: (a) aides to test development, (b) test administration, (c) scoring, and (d) the interpretation of test results.

CURRENT STATUS OF COMPUTER TECHNOLOGY IN TESTING

Test Development

Perhaps it may seem odd for the reader interested in computerized administration or scoring of tests to begin with a discussion of the seemingly obvious role of computers in the development of tests. However, there are a number of steps in test development that are often invisible to the consumer of tests, and many of these steps have involved the use of extensive computer analyses, particularly during the last 2 decades. One need only read the descriptions by Terman and Merrill (1937) of the use of Hollerith machines in the processing of standardization data for the *Stanford-Binet Tests of Intelligence* to appreciate how far test development has progressed and how much computing power now lies in the hands of the individual owner of even the most basic home computer.

Although the home computer shows promise in contributing to the efforts of test developers, there remain some difficult technical problems and a time lag in the adaption of packaged statistical and data analysis programs to the microcomputer. Perhaps by stimulating further interest in the role of computers in test development, this chapter can contribute to an awareness of the need to bridge the existing gap between the software available to test developers at large computer installations and that available to the owner of a personal computer. Building such an awareness is seen as important because of the beneficial role that objective procedures and data-based design can have on the generation of test items (Roid & Haladyna, 1982), the calibration and banking of items and scales (Bock & Mislevy, 1982; Choppin, 1968; Gorth, Allen, & Grayson, 1971), and documentation of the psychometric properties of tests, to name only three elements of test development.

Item Writing. Although extensive reviews of item writing methodology are provided elsewhere (Roid, 1984; Roid & Haladyna, 1980), a brief overview of the current status of computerized item writing is given. The majority of computer applications in test item generation have been in the areas of achievement testing and instructionally based testing systems. The distinctive feature of such applications is that *item programs* that direct the computer to assemble a related set of unique items are stored, not the items themselves. Examples include the early work by Suppes, Jerman and Groen (1966) and Atkinson and Wilson (1969) in computer-assisted instruction, the implementation of test generators in university science courses (Johnson, 1973; Millman, 1980; Olympia, 1975), military-training applications (e.g., Braby, Parrish, Guitard, & Aagard, 1978), and the assessment of specific skills such as spelling (Fremer & Anastasio, 1969) and computer programming (Vickers, 1973).

The work of Suppes, Atkinson, and their colleagues at Stanford (Atkinson & Wilson, 1969) was of historical importance to the field of computerized instruction and testing because it demonstrated three concepts: (a) that individual student-computer interaction was feasible and cost-effective, (b) that sophisticated hardware and software could be designed for the specilized functions of instruction and testing, and (c) that psychological theories of learning and cognition could be integrated into daily lessons and tests in complex and experimentally meaningful ways. Their heavily funded projects were a stimulus for the development of the IBM 1500 computer-assisted instruction system, the COURSEWRITER II author language, and a lower-cost PDP-1 system that delivered drill-and-practice instruction on teletypes to as many as 3000 students per day. They developed a series of COURSEWRITER *macros* (computer commands that call up more detailed segments of computer programming) that allowed the individual test-like events in instruction to be varied at will. For example, sentences like "Dan saw the (tan, fat, man, run) hat," would appear on the computer screen with the tape-recorded message "Touch and say the word that belongs in the sentence." The segment had been programmed by a series of COURSEWRITER commands that could be varied by a macro command listing the sentence text, alternatives, response time-limit, and other parameters. The selection of sentences and alternatives was guided by a psycholinguistic theory based on *vocalic center groups* which are words containing a vowel nucleus with zero to four preceding or following consonants (e.g., at, cat, scat). Experiments could then be designed to verify rules such as "Vocalic center groups with zero preceeding consonants should be introduced to the student before those having initial consonant clusters (e.g., "at" before "gnat").

The contribution of the work of Vickers (1973), Olympia (1975), Millman (1980), and others is the development of computer software for computerized item generation without the need for extensive item banks containing prewritten items. Vickers, for example, used a large university computer system to generate test items for a course in FORTRAN programming with an enrollment of 400 students. Random number generators were used to select item types, distractors, and the letters or numbers used to compose the names of variables in FORTRAN statements (e.g., XY2 = JCEQ5 + N3). Fremer and Anastasio (1969) contributed methods for programming computers to implement the erroneous rules students use in misspelling words, as part of the computerized generation of spelling test items. Hively, Patterson, and Page (1968) and Osburn (1968) advanced the theory underlying computerized item generation by describing the formal properties of "item forms." Item forms are sets of specifications that provide a fixed syntactical structure for items and variable elements that are systematically replaced to create unique items (e.g., "What is the standard error of measurement of a scale with a standard deviation of X and reliability equal to Y?"). Millman (1980) and Millman and Outlaw (1978) described an extension of the BASIC computer programming language that made possible the programming of item forms with greater ease that would be required if every test were generated by its own unique computer program. All of these computerized testing projects were important in demonstrating the feasibility and methodology of systems that did not require item banks containing prewritten items.

The work of Braby, Parrish, Gitard, and Aagard (1978) moved the technology of computerized achievement testing a step further by developing systems that generated both instructional sequences and the mastery tests used to assess learning from each sequence. They discovered that most of the training programs designed to teach symbols or codes (e.g., Morse code, weather-report codes) had a generic structure that could be computerized. Sequences of teaching materials, followed by practice examples, followed by unit mastery tests were all programmable on a computer. The computer could generate and print not only the tests but the entire training manual as well.

A new and growing area for the application of computers to the generation of items (or, more precisely, test-like events) is in the area of perceptual, cognitive, and memory assessment (e.g., Barrett, Alexander, Doverspike, Cellar, & Thomas, 1982). Posner and his colleagues (e.g., Posner & Osgood, 1980) developed many sophisticated computer-control programs for laboratory computers used in assessing perceptual and memory functions. Mapou (1982) recently implemented tachistoscopic functions on a microcomputer. Kornbrot (1981) developed a specialized computer language and system for creating and running psychological experiments. Recently, increased attention to the assessment of memory functions in the aged has included the development of sophisticated microcomputer programs that involve the dynamic generation of graphic patterns on the computer screen (e.g., Gilmore, Royer, Tobias, & Ruffing, 1983; Hertzog, 1983).

For reasons that are unclear, certain areas of testing such as personality assessment have not pursued any of the technologies of test-item writing (Roid & Haladyna, 1980) that lead to automated item generation as is done in achievement or aptitude measurement. Although computerized versions of standard psychlogical tests have proliferated, little is currently available that can be truly described as generative of items in the same sense as is done in the composition of a mathematics problem using random number generators. The early work of Colby and associates (e.g., Colby, Watt, & Gilbert, 1966) on the generation of counseling or psychotherapuetic conversations has not been followed by a widespread application to psychological assessment. Perhaps this is because of the extremely complex nature of human dialogue and natural language.

Although it may be difficult to adjust the natural language sequences that occur in personality inventory items, certain key words can be rather easily adjusted to match the characteristics of examinees who are completing psychological inventories. Johnson, Giannetti, and Williams (1979) have developed some response-contingent systems that adapt to the demographic characteristics of the examinee. For example, if the examinee has only an older sibling, certain items would be reworded to refer to "your older brother" or "your older sister."

Item Calibration and Item Banking. Whether test items are generated by the computer, or written offline and simply stored in computer files, it is possible to collect and store them in extensive "item banks" or "item pools." Again, educational measurement has led the way in the development of numerous and extensive item banks at regional or university centers (e.g., Gorth, Allen, & Grayson, 1971), school districts (e.g., Forster & Doherty, 1978), or at the national level (e.g., Popham, 1980; Wood & Skurnik, 1969). Also, recent years have seen a proliferation of published, criterion-referenced achievement tests that feature the possibility for school districts to adapt test content to their particular curricular emphases. These developments are part of the movement toward a closer linking of testing and instruction in educational program evaluation (e.g., Airasian & Madaus, 1983).

Another arena in which educational measurement has broken new ground is in the development of theories and statistical models that are an alternative to classical test theory. The development of these *item response theory* (IRT) models such as the Rasch model (Rasch, 1980; Wright & Stone, 1979) and the 2- and 3-parameter models (Lord, 1980; Hambleton & Swaminathan, 1985) have been the cornerstone of new methods of computerized adaptive testing (Urry, 1977; Weiss, 1979). Implementation of these methodologies to microcomupters (e.g., Bock & Mislevy, 1982) are just now appearing on the horizon. The promise of such methods is that they will save up to 50% on the number of items required for equivalent precision in comparison to longer paper-and-pencil instruments, and will provide more precise measurement by matching the difficulty of items to the functioning level of the examinee (Haladyna & Roid, 1983; Hansen, 1969).

A number of important applications of the Rasch and 3-parameter IRT models to educational achievement tests and aptitude or intelligence batteries have appeared in recent years. Woodcock (1973, 1978) has shown, through the application of Rasch scaling to widely used tests in reading achievement and psychoeducational assessment, that IRT models can help to provide accurate diagnostic information. By showing the positioning of test items (and the skill-content of the items) along the underlying Rasch scale of a test, Woodcock has advanced the interpretability of IRT applications in the practical world of assessment in school psychology and special education. Of similar impact is the recent work by Elliott (1983a) in applying the Rasch model to the development and interpretability of a complex and comprehensive intelligence and achievement battery that includes 23 subtests that span the ages of 2 1/2 to 17. Because each item in each subtest has a calibrated Rasch difficulty, it is possible to assemble short forms of each subtest, and to link new items to an existing subtest (Elliot, 1983a, pp. 25–29). Also, Elliott (1983a) has argued that Rasch scaling allows subjects tested on different subsets of items to be compared on a common scale, so that younger and older subjects can be compared even though they are given different item sets, and change over time can be measured on a common scale. Numerous studies have proliferated in recent years claiming the inaccuracy of Rasch scaling for vertical equating (e.g., Slinde & Linn, 1979) and other purposes. In a recent rejoiner to such studies, however, Elliott (1983b) expressed concern that comparative research on IRT models has tended to apply the 1-parameter model to preexisting item and test data that were not specifically developed to fit that model. For example, Rasch himself expressed concern about multiple-choice items in describing the military intelligence tests he used to develop his most widely applied model (Rasch, 1980, p. 62)—"V is a test of verbal analogies, formally a multiple choice test, but with so many answers offered that the deficiencies of a multiple choice test are practically eliminated." The contributions of Woodcock and Elliott to computer-based item calibration for widely used tests rests on their development of items and tests targeted to the Rasch model, with rigorous attention to model-data fit during the development and field testing of items. In contrast, the indiscriminant application of IRT models to any tests, especially those not specifically designed to fit the model, would seem to be "technology gone wild." The challenge of computerized item calibration in the future will be to ensure its appropriate application, not as an appendage, but as a central part of an item and test development effort.

Another major example of the application of IRT models to widely used achievement tests is the recent development of the *Comprehensive Tests of Basic Skills* (CTBS), Forms U and V, (CTB/McGraw-Hill, 1982; Yen, 1982) using the 3-parameter model developed by Lord (1980) and others. By calibrating difficulty, discrimination, and guessing parameters of individual items, it is claimed that the response patterns of each student are treated differently (because each item is weighted in a formula score on the basis of difficulty, discrimination, and guessing parameter weights). Clearly, without computerized item calibration and computerized scoring, such complex weighted scores would not be feasible for users who wish to survey the achievement of large groups of students.

A few isolated examples of the calibration of personality inventory items via item-response theory models have been reported such as the Rasch analysis of the *Tennessee Self Concept Scale* (Stanwyck & Garrison, 1982), and the Rasch analysis of the *Central Life Interest* measure (Schmitt, 1981) used to assess an employee's degree of job orientation. A good deal of conceptual work has been done on the application of latent-trait or item-response theories to personality and attitude scales, but there have been few actual implementations of the item banking concept or computerized-adaptive testing outside of achievement or ability testing.

Damarin (1970) constructed a rather elaborate theory including specification of the probability that a subject responds ''true'' to an item as a function of the individual's position on one or more latent dimensions. However, Damarin concentrated his applications on the problem of acquiescence and response bias on the MMPI, without extension to the general problem of calibrating items on *content* dimensions such as depression, anxiety, etc. A more recent theoretical contribution has been made by Thissen, Steinberg, Pyszczynski, and Greenberg (1983) using LISREL analyses (Joreskog & Sorbom, 1981) in attitude scale construction. Perhaps the most extensive discussion of item-response theory analysis of rating scales or questionnaire items is that of Wright and Masters (1982) using the Rasch model. All of these are important theoretical contributions, but there has been a lag of several years in the distribution of computer programs to analyze test items so that item banking and computerized adaptive testing can be implemented. Examples of actual item calibration, followed by computerized-adaptive administration of personality inventories or scales are rare. In a study of the *California Psychological Inventory,* Sapinkopf (1978) used the adaptive methodologies of Weiss (1979) to tailor items to subjects. A considerable savings in the number of items adminstered was achieved (67% fewer items), but with some reduction in scale reliability. Although lower reliability may be expected if fewer items are used, adaptive methods draw items from a large pool of potential items, presumably selecting the most precise items for each subject. Thus, Hansen (1969), Weiss (1979), and Hambleton and Swaminathan (1985), have shown empirically that sequential adaptive tests can provide greater reliability and precision with fewer items than conventional tests in the area of achievement assessment.

Developing the Psychometric Properties of Tests. Perhaps more than any other contribution of computer technology to testing, the use of large scale computers in the sophisticated, multivariate analysis of test data has contributed importantly to the overall improvement in the precision and accuracy of educational and psychological tests during this century. Although somewhat invisible to the average user, the monumental contributions of computers to the development of college entrance exams, professional licensure tests, standardized achievement batteries, and objectively scored clinical instruments is staggering. In fact, by the very nature of the *Joint Technical Standards for Educational and Psychological Testing* (American Psychological Association, 1985) and their requirement for quantitative indexes of reliability and validity, it is difficult to imagine a test that could be developed and published without some form of computerized analysis.

A new emphasis in the psychometric development of tests is the establishment of the validity of computerized interpretations and computer-generated reports. The development of specifications for a fully interpretive computer report forces one to pose detailed questions relating the research base of a test to specific numerical rules of score interpretation. For example, research on MMPI profile elevations (e.g., identifying the 2 or 3 highest scores on the standard profile) provides a basis for generating descriptive statements when such elevations are found in an individual's profile. On the *Barclay Classroom Assessment System* (Barclay, 1983), a specific range of low scores on peer and teacher support in the classroom signals a student at risk for psychosocial stress that may effect learning and achievement. On other interpretive reports (see following section on Test Interpretation) narrative material has been generated based on correlations between clinicians personal observations and profile patterns. Such computerized interpretations require particular kinds of research targeted for eventual use in documenting computerized decision rules. A rigorously valid interpretive report forces the developer to plan verification studies (or the adaption of existing research into the framework of computerized decision rules) that might otherwise not have been so obviously needed. Thus, the computer has an important role in both facilitating the completion of decision-rule verification studies, and in encouraging the validation of specific interpretations of tests. As the new *Joint Technical Standards* (APA, 1985) have emphasized, there are as many "validities" or kinds of validity evidence for a test as there are interpretations or uses of the test. Computerized reporting may have inadvertently heightened the attention to this important principle. Further discussion of the controversies surrounding computerized test interpretation is discussed under the heading "Test Interpretation."

Despite the importance of computer analysis to the development of tests, there remains a surprising void in the availability of computer programs specifically designed in an integrated package for the development and analysis of tests and test items. Perhaps because so few centers of test development exist and because the developers of questionnaires and other informal scales are often not simultaneously skilled in their content area, computer programming, and measurement statistics, there has previously been no integrated and widely used set of computer programs for test analysis similar to the statistical packages currently in worldwide use. The computer-assisted data analysis (CADA) package of Novick and colleagues (Novick et al., 1983) has recently been expanded and released in a new package that includes a component on psychometric methods containing score equating and item analysis (norm-referenced and criterion-referenced) modules. The CADA system is probably the most comprehensive test-analysis package available, but it also emphasizes Bayesian statistical method with which some users may be unfamiliar. Consequently, the test developer must use considerable ingenuity in putting together a sequence of computer programs to develop a high-quality test or scale. If measurement is to further develop as a science, it would seem to be crucial to have objective methods of field-test refinement, reliability and validity estimation, item- and test-bias determination, and derivation of methods of test interpretation that exist as an integrated whole. Such an integration would be possible for the test developer if easily usable sets of computer aides were readily available and tailored to the specific needs of test development.

Test Administration

One of the assumptions of this chapter, which is widely shared by many measurement professionals, is that computer applications are ill-fated unless they provide a new dimension to testing not possible with nonautomated techniques. For example, use of the computer as a "page turning" device to present the same questions appearing on a printed form is clearly degenerative, unless another function such as automatic storage and interpretation of data is coexistant with it. Even the use of a microcomputer time clock may be a frill unless each response is timed and interpreted or some other dimension of timing is implemented that would be too costly or difficult with a hand-held stopwatch.

The best of the current offerings in computerized test administration do add considerable benefits over and above those afforded conventional assessment techniques. Klingler, Johnson, and Williams (1976) have shown that savings in staff time, and increased acceptance by patients, have justified the use of a comprehensive system for the computerized intake assessment of mental health patients in large facilities. Urry (1977), Schmidt, Urry, and Gugel (1978), Weiss (1979), Croll (1981), Weitzman (1982) and others have shown the advantages of computerized adaptive testing in achievement and aptitude assessment in government, military, and university settings.

Adaptive Testing. What are the proported advantages and the possible limitations of adaptive, computerized testing? A review of the literature in adaptive, computerized test design reveals at least five major advantages claimed for the technique:

(a) increased precision of measurement,

(b) improved efficiency and time savings for the examinee,

(c) increased breadth of trait or achievement levels assessable, resulting in more accurate decision making (e.g., identification of students who have mastered a unit of instruction),

(d) improved examinee motivation due to the perceived objectivity of computers or fairness in the choice of items selected for a given examinee, and

(e) possible technical improvements in the selection of test items having desirable statistical properties for certain subjects.

In terms of increased measurement precision for adaptive testing, Hambleton and Swaminathan (1985), Lord (1980), and Weiss (1979) have shown that information curves of adaptive tests are superior to conventional tests, particularly for examinees who are of lower or higher ability than the average difficulty level of the conventional test. Haladyna and Roid (1983) recently showed that adaptive criterion-referenced tests had lower errors of measurement than conventional tests composed of items randomly sampled from achievement-item domains.

Studies of adaptive testing can be traced back to Cowdon (1946) and Fiske and Jones (1954) who argued for the cost-efficiency of sequential testing in which costly assessment items (e.g., medical or physiological measures each of which is time consuming or intrusive to the examinee) can be evaluated in sequence, with testing terminated when an underlying parameter (such as proportion correct or probability of illness) can be estimated accurately. Hansen (1969) presented through evidence that an adaptive, science-achievement test could be administered via computer with 50% fewer items (but equivalent or better information precision) than conventional tests. Weiss (1979) similarly found a cost-efficiency in which 50% of the length of achievement tests in college biology and military technical exams were saved by computerized adaptive testing strategies.

Another advantage of adaptive testing is in the breadth of coverage possible when sequential levels of tests can be designed. In the statewide assessment of special-education students (Brodsky & Roid, 1977), for example, adaptive tests have been designed to span the broad range of functioning characterized by mildly to severely retarded children (e.g., in dressing or eating skills as well as academic achievment). Adaptive assessment saves time for the examiner who has many students to individually test and yet provides a scaling such that each level of the test can be related to a common numerical scale. Cleary, Linn, and Rock (1968) argued that the breadth of coverage possible with adaptive achievement tests would prevent the *topping out* or *bottoming out* that can occur when examinees represent a wide range of skill or achievement. Clearly, a test which *tops out* has little variance from which differential evaluation of individuals can be completed. In addition, students or applicants for employment may perceive adaptive tests as more fair and less punishing because tests are tailored to their level of competence (Schmidt, Urry, & Gugel, 1978).

Finally, adaptive testing makes possible the implementation of various technical improvements in the selection of test items. Tests can be specifically designed to be highly sensitive to abilities or achievement in a narrow range such as that defined by a mastery criterion or *cutting score* level (Lord, 1980). In addition to selecting items based on their difficulty level, the discriminating power of items or their relationship to external criteria can be used for composing tests having differential validity for specific individuals. Hansen (1969), Fossum (1973), and Roid (1969) presented experimental adaptive methods of selecting items to increase the correlation between a test and external criteria. Small but important improvements in validity were found by these methods. However, the potential of such methods is limited to cases in which the cost of testing or the importance of decision making (e.g., in the assessment of suicide potential) is significantly high.

With all of the potential advantages of adaptive testing, a very perplexing question for many enthusiasts of tailored tests is, "Why has adaptive testing been so slow to appear as a widely-used method?" McArthur (1984) has pointed out some important reasons for the resistence to adaptive, computerized methods: (a) that the American tendency is not to accept packaged curricula or test-item banks that do not allow local control of content in academic subject matter, (b) educators and other professionals may be concerned that adaptive tests place too much faith in individual items, and (c) there has been an information gap and an absence of widely implemented software that has prevented the development of the knowledge and skill needed by small institutions and individuals to apply the Rasch, 3-parameter, or other IRT models to practical testing problems. Other factors that have contributed to the slow growth in adaptive testing may be the unavailability of large sample sizes required for item calibration (although some Rasch devotees would argue that samples as small as 50–200 may suffice), the complexity of score interpretation in comparison to simple number-correct scoring, the enormous work required to make an IRT-based scale curriculum-referenced (Haladyna & Roid, 1983), and the cost concerns surrounding the hardware and software needs of an individualized, computerized testing system for large numbers of examinees.

The promise of increased precision, breadth of coverage, and efficiency remain, but widespread dissemination of adaptive testing may require future technological developments such as lower-cost respondent keyboards with built-in storage for remote data acquisition. But, more importantly, there remains a need for more practical demonstrations and more time for professionals to learn the terminology and inner workings of IRT methods. These and other future perspectives will be discussed in a later section, "Future Directions."

Research on Computerized Test Administration. Numerous studies over the last decade have contrasted computer-administered and paper-and-pencil versions of widely used tests. Some of these studies have involved straightforward administration of items, without adaption to the examinee, but nevertheless have been important in documenting the similarities and the small number of differences between computer and conventional test administration. Results of these studies are briefly presented in Table 3.1.

The majority of studies reviewed and presented in Table 3.1 showed nonsignificant differences between computer administered and conventional test administrations. This is somewhat surprising for performance vocabulary tests such as the *Peabody Picture Vocabulary Test.* Elwood (1969) found slightly lower WAIS IQ estimates from an automated administration which could be explained as an elimination of examiner subjectivity as easily as an interference from automated procedures. One study found differences in state anxiety under computerized administration (Lushene, O'Neil, & Dunn, 1974) but another found no differences (Katz & Dalby, 1981). One of the studies (O'Brien & Dugdale,

TABLE 3.1
Studies Comparing Computer Versus Conventional Test Administration

References	Test Name or Type	Examples of Results
Elwood (1969)	WAIS (Wechsler Adult Intelligence Scale)	High correlation between computer and face-to-face testing with performance IQ lower on automated WAIS. (N = 35)
Hedl (1971)	Slosson Intelligence Test	Correlation of .75 between computerized and conventional, but higher state anxiety on computer. (N = 48)
Lushene, O'Neil, and Dunn (1974)	MMPI	Correlations between computer and booklet modes comparable to booklet and card form correlation. (N =63)
Scissons (1976)	CPI	Differences between subscale scores between modes particularly for males. (N = 20)
Klinge & Rodziewicz (1976)	Peabody Picture Vocabulary Test	No difference in IQ or test-retest reliability between modes of testing. (N = 52)
Elwood & Clark (1978)	Peabody Picture Vocabulary Test	Nonsignificant differences between testing modes on IQ or test-retest practice effects. (N = 76)
Biskin & Kolotkin (1977)	MMPI	No differences except computer slightly higher on Paranoia scale and lower on "Cannot Say" scale. (N = 165)
O'Brien & Dugdale (1978)	A questionnaire on personal bathing habits	Tendency for computer responses to be nearer the "honest" end of each scale. (N = 126)
Johnson & White (1980)	Wonderlic Personnel Inventory	Elderly subjects given training on the computer prior to testing performed significantly higher than those not treated. (N = 20)
Katz & Dalby (1981)	FIRO Scales and State-Trait Anxiety Inventory for Children	No differences between testing modes on test-retest correlations. Time savings for computer version. No anxiety differences. (N = 80)
Harrell & Lombardo (1984)	16PF	Multivariate analyses revealed no significant differences between Apple II computer and standard booklet modes of presentation. (N = 80)

1978) suggested increased honesty or openness for the computer administration. Biskin and Kolotkin (1977) also found the expected decrease in number of "Cannot Say" responses on the MMPI, indicating increased willingness to make a response committment. With computers, the examinee's responses can also be screened during testing to catch any unexpected responses or the double marks common to paper-and-pencil answer forms. However, the differences in personality scale scores found in some studies listed in Table 3.1 indicate the highly specific effects that computerized administration can have.

Test Scoring

Clearly, computer technology has had an enormous effect on the scoring and reporting of results for standardized achievement tests used in the nation's schools. Although school personnel occasionally regret the inevitable delays between testing and receipt of reports, it is truly remarkable that literally millions of answer sheets are processed each year, with detailed reports generated for individual students, classrooms, school buildings, districts, and States. Also remarkable is the proliferation of types of derived scores for educational tests—percentiles, stanines, NCEs, grade-equivalents, and normalized standard scores. With the development of computer algorithms for the direct calculation of normalized standard scores from percentages (Beasley & Springer, 1977), some of the common derived scores can be obtained directly rather than retrieved from lengthy tables.

In psychological testing, computerized scoring allows the derivation of complex scores such as factor scores, Bayesian-derived probability scores for low base-rate behaviors such as suicide (Greist, Gustafson, Strauss, Rowse, Laughren, & Chiles, 1973; Vanderplas & Vanderplas, 1979), item-option weighted scores (Cull & Gill, 1982; Roid, 1983a), profile similarity indexes for the test scores of married couples (Krug, 1983), weighted scores from tailored, adaptive, or multilevel tests calibrated with the 3-parameter model (Lord, 1980; Weiss, 1979), and sociometric ratings from entire classrooms contrasted with self and teacher ratings from individual students (Barclay, 1983). For psychological and vocational tests having complex and numerous scores, such as the MMPI, the *Strong-Campbell Interest Inventory,* the *Tennessee Self Concept Scale,* or the *16 Personality Factor Questionnaire,* computer scoring provides a richness of interpretive data that could not otherwise be obtained without enormous effort.

The studies listed in Table 3.1 also suggest that, with few exceptions, the published norms of tests may be applicable without adjustment to computerized version of these tests. However, more research is needed on the effects of computerization. It may be that effects on score distributions are slight but highly specific to each instrument, hence, the need for comparability studies similar to those required for alternative forms of tests. The availability of inexpensive home computers and methods of field testing in which a computer is left unattended but responsive to each examinee who voluntarily steps forward in locations such as community centers (McArdle & Kirson, 1983) show promise for the automated collection of norms specific to computerized test administration.

Test Interpretation

Although the interpretation of test results for individuals continues to require first-hand knowledge of the examinee by experienced clinicians, educators, or other professionals, the computer has an increasing role in the processing of test results using statistical and actuarial methods that are complex or time consuming for the professional. When test results are aggregated for classrooms, schools, or other groupings computerized summarization of results seems natural and widely accepted particularly in educational measurement. With the recent advent of computerized interpretation of individual psychological, vocational, employment screening, medical-psychological, special education, and counseling instruments, concern the for bounds of acceptable computerization has increased. The important ethical issues in the clinical use and interpretation of computerized test results are complex, and thorough reviews are available elsewhere (e.g., Bersoff, 1981; Zachary & Pope, 1983).

The best of available test-interpretive computer programs for major vocational and psychological tests have been designed on the basis of empirically validated decision rules and intended for the use of the trained professional who is otherwise experienced with the instrument and its supportive research. Unfortunately, the worst of available programs include the private and subjective narratives of individuals who developed the programs without benefit of empirical studies. The problem facing the field of measurement is to provide criteria for distinguishing between the objective and subjective programs.

Much of the controversy over computerized psychological testing (e.g., Matarazzo, 1983) is based on at least four prominent concerns:

1. that it is questionable whether there are real advantages to computerized interpretation of tests as compared to the clinician working without a computer,
2. that computerized interpretive reports will reach the hands of inexperienced or unqualified individuals who will respond to the halo-effect of objectivity projected by a computerized report,
3. that publishers or developers of computerized interpretive programs may not openly reveal their decision rules for professional review, and
4. that computerized reports (particularly those that cannot be evaluated closely) will not be sufficiently validated.

Each of these concerns are discussed in turn.

Advantages of Computerized Interpretation. The controversy seems to spring from a combination of true belief in the richness of clinical interpretation, philosophical differences in approach to clinical prediction (Meehl, 1954), as well as an information gap that separates those who are familiar with the inner workings of interpretive programs and the professional consumer. Obviously, the long-standing debate on clinical versus actuarial prediction will continue, with proponents of clinical prediction citing examples of poorly constructed computer reports and proponents of actuarial prediction citing examples of rigorously validated, empirically developed reports. Devotees of empirically based interpretive reports who never intended such reports to *replace* all clinical judgment do not understand the arguments that say all computerized reports attempt to replace human interpretation. An analogy with statistical computer programs is useful in pointing out that because some users of multivariate analysis of variance, for example, may misapply the underlying general linear model, can the statistical program therefore be condemned? And, more pointly, should a misused multivariate analysis program be outlawed because some users may reach false research conclusions (by violating the assumptions of the program), and all multivariate analyses of variance be conducted by hand? Developers of sophisticated computerized reports would argue that their validated programs include complex calculations and decision rules that approach the complexity of some of the packaged statistical programs in wide use today.

Several practical advantages to computerized interpretive programs seem clear to most developers of such programs. First, accuracy of scoring and retrieval of norms from complex norm tables should provide a measure of quality control (many of those who supervise interns would readily attest to the frequent error rates in psychological test scoring and norms retrieval). Second, the time saved by clinicians who are relieved of hand-scoring and profiling could be invested in additional testing or personal interviews that would supplement and add fidelity to the computerized report. Third, because the decision rules for interpreting multiscale tests are often complex and numerous, it seems illogical to argue that human memory can retain and access all such rules in the same fractions of seconds required by a computer—therefore the computer acts as a memory aid. Fourth, when there is research showing moderator effects on test interpretation (e.g., that certain age groups or ethnic groups have different ranges or patterns of scores), a computerized interpretive report again provides a memory aid which reminds the clinician of such moderator effects. Finally, as is detailed below, there are numerous technical advances in profile analysis and statistical processing of scores that would be impossible to implement in a hand-scoring system without complex calculations by each clinician (and a resulting complexity in the published profile sheets or test manuals).

Unauthorized Use. The concern that unqualified users will be attracted to computerized interpretations seems to be an issue of *controlling access* to such programs or reports, and the ethical responsibility of test distributors and users. Clearly, any system of controlling access will be imperfect to some degree if an unqualified user is determined to bend the rules to obtain a copy. The screening currently done by test publishing companies is more extensive than may be apparent to the professional consumer (e.g., qualifications questionnaires, approvals by supervisors, registry procedures that assign user-numbers only to qualified applicants), but, even so, it is not perfect if applicants exaggerate qualifications or lend copies of software to those who were not screened. Also, the frequent practice of honoring an institutional order for test materials from an approved clinic, hospital, or school can result in the placement of test materials in a location where both qualified and unqualified users may gain unauthorized access to computerized reports. Therefore, the *Joint Technical Standards* (APA, 1985) emphasize the individual responsibility of the *users* of computerized interpretive reports to be familiar with the research base of such reports, to have test manuals available for reference, and to use appropriate caution in making decisions from these reports. Clearly, both distributors and users (individuals and institutions) need to continue to examine their procedures for allowing access to computerized interpretive reports with an eye to the problems of potential unqualified use.

Documentation of Decision Rules. The ethical responsiblity of the developer of computerized interpretive reports is to carry out and document validity studies of the underlying decision rules for each program. However, a controversy has developed over that extent to which the inner workings of a commercially-distributed computer report should be exposed. From the view of the pure scientist and individual helping professional, it must seem that all such aids to assessment should be part of the public domain, contributing to the advancement of science and the health and welfare of people. For programs developed by public funds, this view seems entirely appropriate. Also, if a researcher wishes to donate his or her efforts to the social good, it becomes a matter of individual choice to make such a contribution. However, if a private individual or organization has invested years of study, research, computer programming, and other resources into an inventive program, and is not in a position to donate such efforts to free public use there is legitimate concern over protection of one's proprietary rights. From the developer's view, there are at least two important issues: (a) the legal and ethical rights of developers who wish to retain their rights to an inventive creation, and (b) the economic realities of producing and distributing computerized reports.

The issue of rights to an inventive creation are clarified by the wordings of laws and regulations for patents and copyrights, and the many formalized procedures established by universities and research centers. Also, case law which develops from the successive decisions of court actions in representative cases (e.g., in disputes over copying of computer software), is relevant here. A discussion of legal issues goes beyond the present review, and the reader would need to consult with recognized legal experts.

The importance of the economic aspects of computerized program development was discussed several years ago by Campbell (1976) when the *Strong-*

Campbell Interest Inventory became the first major computer-scored test to withhold its scoring weights from public release. Campbell argued that the "pragmatic research scientist" must recognize that research funding to expand and improve an inventory, and to "cover all new issues" (e.g., the new concerns about the ethnic bias of tests for the growing subpopulation of hispanics comes to mind) is not usually supported by public or nonprofit foundation grants. Therefore, if a widely used commercial instrument is to be improved over time, revenue must be protected from the errosion created by copying of tests and the proliferation of competing commercial scoring services. Unfortunately, years of experience have shown publishers that one of the only effective controls for certain tests is to withhold keys, norms, or portions of the interpretive decision rules. The present author would argue that documentation of the validity of decision rules is not incompatible with securing the rights to a program. It should be possible in nearly every case to withhold some central element of scoring- or interpretive-program logic and still document in detail the validity of the resulting report. For example, the research base (including all references to published articles) and even most of the numerical decision rules can be revealed in the documentation of a program without having to publish the entire operating specifications of a scoring/interpretive program.

Another approach to documenting computerized interpretive programs and discussing the validation of such programs is to provide a typology of different categories of programs. A typology of computerized programs in testing would allow developers and users to have appropriate descriptive labels to distinguish one program from another. A typology with examples of how each category of program might be validated is given in the following section.

A Typology of Computerized Interpretive Programs

A proposed typology for programs, which may be useful in labeling and distinguishing among the various commercially available programs, uses four categories (Roid & Gorsuch, 1983): (a) scoring only, (b) descriptive, (c) clinician-modeled, and (d) clinical actuarial. Proper categorization and labeling of commercially offered programs, particularly those for microcomputers, would contribute to informed usage. Each category of program beyond the scoring-only level will be briefly reviewed (the reader interested in scoring programs is referred to the previous section on "Test Scoring").

Descriptive Programs. Once the subtest or scale scores for a test are available, and perhaps presented on a profile, quantitative criteria such as cut-off scores are often applied to describe the test results. Because the computer can store literally thousands of quantitative criteria and descriptive words attached to give the criteria meaning, computerized descriptive programs can be useful to the trained and experienced test user. The simplest of descriptive programs provide phrases such as "above average," "in the gifted range," "indicates mastery of this objective,'' ''in the disabling range,'' and ''possible organic problems,'' to list only a few of hundreds of examples. The more sophisticated descriptive programs use alternative modifiers within the same score range (''average,'' ''at the mean,'' or ''typical performance'') so that the report does not become overly redundant and repetitive with the same descriptors used line after line. Some other attributes of the better descriptive programs include: (a) selection of descriptive words based on empirical studies of language, (b) narrative paragraph composition, and (c) statistical description of differences among subtest scores, each of which is briefly described.

Empirical research on the scaling properties of words, modifiers, adverbs, and other verbal phrases have been used to design Likert-type rating scale questions, but may also have a role in the design of descriptive computer reports of tests (Gorsuch, 1982). Hakel (1968) and Lichtenstein and Newman (1967) studied the scaling properties of words and phrases such as ''often,'' ''seldom,'' ''very likely,'' ''unlikely,'' and ''highly improbable.'' Altemeyer (1970) studied the equal-interval scaling of sets of adverbs such as ''completely, substantially, somewhat'' and identified several with good interval scaling properties. More recently, Pohl (1981) analyzed 39 expressions such as ''frequently,'' ''occasionally,'' and ''seldom'' in relation to the anchor referent ''sometimes.'' These studies provide empirically based standards for composing computer-assembled descriptors of test score intervals.

Some of the currently available microcomputer programs for test analysis have descriptive reports that are highly redundant with wording such as ''He is above average on Scale 1... average on Scale 2... average on Scale 3....'' Using more of the power of the computer, it is possible to combine sentences into paragraphs, using different modifiers and sentence forms to create more readable reports. Extensive use of this method was recently implemented by Barclay (1983) in a comprehensive computer report for students in elementary classrooms as a means of summarizing multiple indicators of social competence based on self, peer, and teacher ratings.

Some of the true power of the computer comes into play when test scores can be analyzed statistically, such as is done to describe the ''scatter'' of profile scores on tests such as the *Weschler Intelligence Scale for Children—Revised* (Kaufman, 1979, pp. 195–209). Figure 3.1 shows a sample of a computerized report for the *Luria-Nebraska Neuropsychological Battery* (Golden, Hammeke, & Purisch, 1980) that implements for the type of analysis of profile scores suggested by Reynolds (1982). Figure 3.1 is intended as a statistically rigorous study of strengths and weaknesses within an individual score profile. The mean of all 14 profile scores in Fig. 3.1 is calculated and printed at the top of the display (mean = 51.89. Then, each profile score is subtracted from the mean and the difference is plotted (e.g., scale M1 was 52.30, and has a difference of +0.41 from the mean of all profile scales). Thus, positive differences are potential ''weaknesses'' in the profile, (because all scales are keyed in the clinical

MEAN T-SCORE FOR CLINICAL AND SUMMARY SCALES

SCALES	DIFFERENCE FROM MEAN	SEM
M1	.41	6.24
M2	19.32***	3.46
M3	-1.61	3.61
M4	-6.34	4.90
M5	-5.23	8.66
RH1	-4.75	4.58
T1	11.44	5.57
T2	.25	4.69
V1	4.61	5.83
V2	21.95**	6.63
R1	-11.89*	3.87
R2	-9.26	5.10
R3	-6.18	8.49
R4	-4.39	7.42

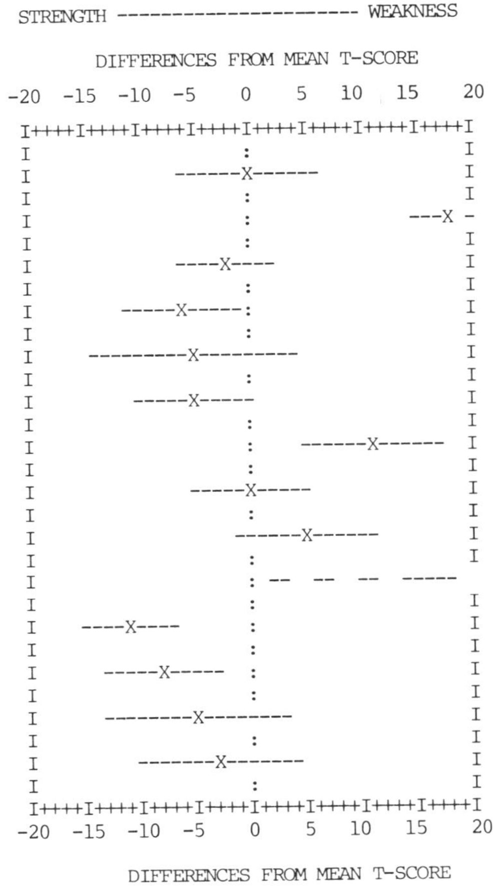

*p < .10. **p < .05. ***p < .01.

FIG. 3.1. Example of a Scatter Analysis Profile Generated by Computer. Reprinted from the WPS TEST REPORT for the Luria-Nebraska Neuropsychological Battery, Copyright © 1983 by Western Psychological Services. Reprinted by permission of Western Psychological Services, 12031 Wilshire Blvd., Los Angeles, CA 90025.

direction), and negative differences are potential "strengths." What remains is to test the significance of these differences using information about the reliability and standard error of measurement of each scale. Also, a correction for multiple comparisons is included. As shown in Fig. 3.1, scales M2 and V2 are significant weaknesses and scale R1 is a significant strength for this patient.

Profile analysis such as that shown in Fig. 3.1 accomplishes several objectives. Specifically, it: (a) displays confidence intervals for observed scores based on the standard error of measurement of each scale, (b) provides a statistical test of the differences between each scale value and the mean of all profile scale values as a rigorous test of "scatter," and (c) corrects for the multiple comparisons using a Bonferroni-type correction to the *t*-statistic used in testing hypotheses of significant scatter. The methodology for implementing these procedures is given by Davis (1959), Dunn (1961), and Bailey (1977), and most recently Stoline (1983) has provided even more complex statistical procedures with increasingly fine accuracy. Clearly, such statistical operations would be virtually impossible for the clinician or educator who approaches profile-score analysis *bare handed* without the benefit of either extensive tables of pairwise score differences or some form of calculator or computer aid. The practical benefit of such methods is that they replace the subjective *reading* of profiles which may be characterized by overinterpretation of small differences between profile scales.

Clinician Modeled. Another type of computerized interpretation of individual test results can take two general forms: (a) where the program simulates the interpretive decisions of a renowned clinician, and (b) where a statistical model is constructed from studies of groups of expert clinicians and programmed into the computer (e.g., Goldberg, 1970).

Work in progress by the current author is aimed at modeling the clinical interpretations of the *Louisville Behavior Checklist* (Miller, 1981) as provided by its developer, Dr. Lovick Miller. Tape recordings of his actual case interpretations are being studied, and objective decision rules extracted from this rich clinical source. Several cycles of development will be used to produce trial interpretive programs, apply them to actual results on the *Checklist,* present them for reinterpretation by Dr. Miller (under blind conditions), and validate the fit between the objectively programmed rules and those actually used by this experienced clinician and test developer.

Wiggins (1973), in summarizing the work of Goldberg (1968, 1970) and his colleagues, noted that a statistical model of clinician judgments often can be more accurate than individual clinicians working in isolation. In a classic series of studies (Goldberg, 1970) examined the ways in which clinicians diagnosed psychosis versus neurosis from the standard profile scales of the MMPI. Although clinicians reported that they were combining information in complex *interactive* ways, the data showed that simple linear regression models were effective in describing how they made diagnoses from the test-score information. When the methodology reflected in these classic experiments are applied to other psychological tests, it becomes an empirical question whether or not a model with acceptable accuracy can be derived.

Clinical Actuarial. Following the rationale proposed by Meehl (1954), numerous computerized actuarial systems have been developed for educational (McDermott, 1980, 1982) and psychological tests (e.g., Lachar, 1974). The term *actuarial* is perhaps an unfortunate choice of wording, because it brings to mind the statistical tables developed by insurance actuaries in the assessment of the likelihood of death or accident, but the term has an historical tradition particularly in MMPI prediction research. McDermott (1982, p. 248–249) recently expanded the definition of actuarial assessment to include a broad array of multivariate statistical procedures useful in making decisions about people on the basis of test and nontest information.

In educational measurement, the work of McDermott (1980) in the identification of students in a variety of special education categories using computer algorithms is an example of sophisticated methodology. McDermott implemented a computer program that examines intelligence, achievement, and adaptive-behavior scores of individual students in order to make quantitative judgments about diagnoses such as learning disability status. For example, test scores from the WISC-R, WRAT-R, and *Adaptive Behavior Scales* might be input for a student who had particularly low mathematics achievment. The computer program would have stored information on the reliability, standard error of measurement, and intercorrelation of these tests and would calculate the statistical significance of the descrepancy between the WISC-R and WRAT-R scores. The program would then print out the numerical estimates and description of the result. Obviously, such calculations are possible by hand, but with the numerous comparisons possible, the busy evaluator equipped with a computer could make more such comparisons in less time than an evaluator using a hand-calculator.

The exceedingly complex computer analysis provided for the *Barclay Classroom Assessment System* (Barclay, 1983) is another example. Barclay distilled 25 years of multivariate statistical studies of self, peer, and teacher ratings of elementary students into a computerized interpretive program that provides a narrative, diagnostic, and prescriptive report (up to 100 pages in length for a given classroom) useful for teachers, school psychologists, and other personnel or resource professionals. The logistics of attempting to analyze just the peer data (sociometric choices by each member of a classroom), by hand, would be challenging enough, let alone the integration of self and teacher ratings, and achievment test scores. Clearly, the computer has an inherent value in such applications that cannot be dismissed as extravagant technologizing.

In psychological testing, the most widely used and discussed actuarial programs are probably those of the MMPI. Although some investigators (e.g., Matarazzo, 1983) may wish to see additional evidence, there have been a considerable number of research studies aimed at assessing the validity of the narrative reports generated by computers for the MMPI and other psychological tests. Lushene and Gilberstadt (1972) used independent judges to rate the accuracy of 3,926 statements in 355 computerized reports and found 79% of the statements judged correct and 93% of the reports rated favorable overall. Lachar (1974) studied computerized reports for 1,410 adult patients and found that clinicians rated 107 frequently occurring paragraphs (which appeared 7,555 times in the reports) as accurate 90.3% of the time. In a study of the use of adolescent norms in computerized MMPI reports, Lachar, Klinge, and Grisell (1976) found that clinicians rated 20% of narratives based on standard adult norms as inaccurate for 100 patients aged 12–17, whereas narratives based on adolescent norms were judged inaccurate in only 10% of the cases.

Another approach to increasing the validity and accuracy of computerized reports based on statistical/actuarial methods, is that of tailoring reports of results on a comprehensive personality inventory to findings from studies with particular patients or applications (Krug, 1982). For example, a computerized report for the *16 PF* (Dee-Burnett, Johns, & Krug, 1982) was developed specifically from validity studies in law-enforcement settings. Another report (Krug, 1983) is specifically designed for use in marriage counseling.

Another approach to assuring the accuracy of descriptions generated by computerized reports is the method of *replicated correlates* used extensively by Lachar and Alexander (1978) on the MMPI, and by Lachar and Gdowski (1979) on the *Personality Inventory for Children* (Wirt, Lachar, Klinedinst, & Seat, 1977). In this method, clinicians who personally interview each patient are asked to provide detailed ratings using a behavioral and symptom checklist. All patients are then given the inventory which is to be computer scored, and the profile scales from the inventory are plotted for each subject on a standard T-score profile. Each profile scale is divided into *elevations* or segments such as 80T+, 70-79T, 60-69T, 41-59T, and 40T-. The frequency of each checklist description of the patients is then calculated for each elevation on each scale. High frequency checklist items are called correlates of a given scale. A new sample of subjects is used to replicate the findings, and only replicated checklist descriptors are used in the final computerized report to describe the potential behavior and symptoms of the patient.

Even though there remain examples of undocumented computer programs that provide narrative reports similar to the clinical-actuarial programs described above, the best of the reports provide extensive documentation. Certainly, as a means of combating the aura of objectivity projected by computerized reports, it is essential that detailed documentation of the empirical bases for decision rules and narratives be provided, such as is done in the manual for the *Strong-Campbell Interest Inventory* (Campbell, 1977) and the book-length monograph by Lachar and Gdowski (1979).

FUTURE DIRECTIONS IN THE APPLICATIONS OF COMPUTERS IN TESTING

In an interesting book on the early history of computer programming languages, Sammet (1969), one of the codevelopers of the COBOL language, used the analogy of the biblical story of Babel to describe the proliferation of languages in computer technology. Similarly, the future of computerized testing and test interpretation will undoubtedly be characterized by incredible diversity and lack of standardization of procedures. The current proliferation of brands of microcomputers is but one example.

Amidst the diversity that seems to be inevitable in the fast-changing world of computer technology, there are several important forces that may help to unify some of efforts of independent researchers and developers. The developments of CP/M (Kildall, 1982) and the highly acclaimed UNIX operating system (Christian, 1983) promise to increase the machine-independence of computer software and systems, thus making wider distribution of computer programs possible. The developments of memory devices such as the inexpensive *floppy disk* which will become increasingly miniaturized and higher in capacity, will allow the distribution of testing programs (and translators required to adapt them to particular hardware) that feature exceedingly large item banks or narrative-interpretive material. However, it is likely that a continuing problem into the future will be the compatibility of various types of hardware and software.

This segment of the chapter proceeds with a review of the four major areas introduced earlier: (a) test development, (b) test administration, (c) scoring of tests, and (d) test interpretation. Emphasis is on trends that are in the experimental stage now that seem feasible for wide-spread application in the near future.

Test Development

Item Writing. The availability of microcomputers and the software developed especially for them, has brought some new advantages to the developer of tests and test items. Many of the microcomputer word-processing packages are more sophisticated than those previously available on large mainframe computers. For example, page-oriented editing systems are more efficient than line-by-line editors and widely distributed software is available for correcting spelling, grammar, and the general readability of test items and other material needed for computerized test reports. The future promises more and more aides for the item writer and test developer.

Roid (1984) called for the development of software that would be useful for the automated development of reading comprehension tests keyed to textbooks. Some merging of the methods used in library science to access the keywords in text, and the methods of transforming text into test questions (Bormuth, 1970; Roid & Haladyna, 1982, chapter 6) would seem to be helpful to the publishers or widely used school textbooks. As mentioned earlier, such automated methods may improve the match between teacher and testing as reviewed by Airasian and Madaus (1983). Computerized versions of word lists, such as that compiled by Carroll, Davies, and Richman (1971) for American textbooks would play an important role in such methods.

For years, the field of computer-assisted instruction has been experimenting with methods of helping authors create computerized lessons. Systems called "authoring aids" have been developed for the PLATO system at the University of Illinois (Alpert & Bitzer, 1970) and at several military CAI installations (Schulz, 1979). These aids help the developer to create common forms of questions such as multiple choice, constructed response, and matching items. For reasons that are not entirely clear, there has been little cross-fertilization between the fields of CAI and standardized testing, but the future promises to see more sharing of ideas between these fields as more and more testing is implemented on small computers used in schools.

Another important area of cross-fertilization between CAI and testing will hopefully be in the definition of programming languages useful for constructing items and implementing computerized testing. Currently, the most widely used programming languages for computerized testing appear to be BASIC and PASCAL, which are general purpose languages. Instead, great advantages would accompany the usage of application languages such as the TUTOR language for the PLATO CAI system. Cory, Rimland, and Bryson (1977) used an IBM 1500 CAI system, which includes the COURSEWRITER language, to develop a battery of information-processing tests. These tests were used by Cory (1977) for predicting job performance.

In the area of perceptual and cognitive experimental psychology, which should contribute more and more to the assessment of memory and cognitive functioning as evaluated by computerized testing in the future, several high-level programming languages have been developed. The LAB-TALK language (Maxwell & Schvaneveldt, 1983) is useful in presenting stimuli, collecting responses, and recording data. Two other examples are the EXPERIMENT WRITER language of Posner and Osgood (1980) and the ARTIST system of Kornbrot (1981). These higher-level languages should serve as a model for the development of applications languages in computerized adaptive testing, particularly in cognitive/perceptual evaluation where complex graphic and multiple-trial stimulus events are used. Perhaps the future implementation of such languages on widely distributed microcomputers will play a key role in making computerized testing more feasible.

Item Banking. The future development of sophisticated and miniaturized memory devices, as mentioned earlier should encourage the increasing use of item banking, particularly in the area of criterion-referenced achievement testing. There are recently developed examples of such item banks published for use on microcomputers. One large-scale project (Forster & Doherty, 1978) in the Portland (Oregon) Public Schools has included the development of Rasch-calibrated items numbering 1000 or more in each of the three basic achievement domains, reading, mathematics, and language arts. Work is currently in progress to implement these large item banks on microcomputers tailored to the needs of individual school districts. Haladyna and Roid (1983) showed that adapting the difficulty of mastery tests to the functioning achievement level of students provided greater measurement precision than random sampling of items from domains, and this is the approach being implemented in the Forster and Doherty (1978) system. Furthermore, computer tailored tests such as these promise to provide a "curriculum-referenced" interpretation to achievement tests (Rentz, 1982; Woodcock, 1982). Since achievement test items can be calibrated for difficulty along the same scale as the estimates of achievment for each student (Wright & Stone, 1979), it is possible to draw a curriculum continuum that maps the specific skills achieved by students at various score levels on a test. For example, in basic arithmetic, items of long division would be more difficult than items of addition. If a test includes, subtraction, multiplication, and division, it is possible to draw a continuum showing where specific skills (e.g., 2-place addition with carrying) lie, and then to reference the total test score of a student to this curriculum-continuum. Woodcock (1982) uses this system for his finely detailed profiles such as those available with computer scoring on the *KeyMath Diagnostic Arithmetic Test* (Connolly, Nachtman, & Pritchett, 1976).

Developing the Psychometric Properties of Tests. Among the many new multivariate analysis methods, only possible on computers, that will help to shape the tests of the future, are two that deserve special attention in the next decade: (a) increasing use of linear structural relationship (LISREL) analyses in the development of evidence for the construct validity of tests, and (b) the use of factor analyses specifically designed for dichotomous items, including the important exploration of the unidimensionality assumption for achievement and ability tests.

A number of multivariate methods known as causal modeling (Bentler, 1980), structural equation modeling (Joreskog & Sorbom, 1981), or path analysis (Wolfle, 1980) show promise for the study of construct validity of tests. Because the study of construct validity (Cronbach & Meehl, 1955; Messick, 1980) involves the comparison between empirical findings and a theoretical *nomological network* that posits the expected relationships between variables measured by tests and variables in the *real world*, models of multivariate relationships apply. As LISREL-type programs become more available to a wider circle of test developers, it will be possible to use them increasingly in test development research (e.g., Marsh & O'Neil, 1984). These programs will be useful in demonstrating that the latent variables underlying a test battery are related to the latent variables underlying a series of observations collected by means other than the test being examined. For example, a test battery measuring teacher ratings of students could be examined in relation to known correlates of teacher ratings, such as achievement, parent behavioral ratings, and readiness tests among children entering the first grade.

Another area of new interest requires considerable computer power. With growing interest in item-response theory for achievement and ability tests, has come the increasing concern that some tests do not fit the unidimensionality assumptions of IRT models. Some investigators have simultaneously questioned the value of traditional indexes of test homogeneity, such as the *alpha* internal-consistency coefficient, as indicative of unidimensionality (Green, Lissitz, & Muliak, 1977; McDonald, 1981; Smith, 1980). In response to these concerns, several new statistical models allowing for multidimensional tests to be analyzed with IRT-like models have emerged (Reckase, 1979; Stegelmann, 1983; Whitely, 1981). Considering that Damarin (1970) called for multiple latent-variable models for psychological test item analysis, these developments have great importance for psychologists as well as for educators and other social scientists.

Why should test developers be interested in the debate on unidimensionality? There is a reason perhaps even more important than the concern that an IRT model may not fit the data for an educational test in wide usage. Just as factor analysis has often motivated psychologists to add or subtract certain items or scales from personality and ability tests, because of a desire to measure documented factors, perhaps more frequently, educational test developers will examine achievement tests in order to add items that measure secondary factors above and beyond the single-factors that may have been assumed in the past.

The new developments that make the investigation of multidimensionality possible are the new methods of factor analysis specifically designed for dichotomous items (Green, 1983). Gorsuch and Yagel (1982) recommended two types of factor analysis: (a) the factor of small groups of items called "parcels" as used by Cattell (1956, 1974), and (b) hierarchical factor analysis (Gorsuch & Dreger, 1979) which extracts higher-order factors from the first-level of factors extracted (including potential spurious factors due to binary items). Examples of the application of hierarchical factor analysis are provided by Gorsuch (1983a, chapter 11) and in studies such as Wallbrown, Blaha, and Wherry (1973).

Another approach to factor analyzing dichotomous items is described by Muthen (1978, 1981) who is developing a comprehensive computer program that will be an important addition to the test-developers software collection in the future. Roid (1984) has emphasized the value of these new factor analysis programs in the development of criterion-referenced test items to measure potential multiple dimensions in achievement tests. However, because so many psychological tests and checklists have dichotomous items, new vistas in the explorations of test dimensionality remain for psychological and clinical tests as well as achievement testing. And, clearly, the factor analyses of data matrices having 100 items or more is not possible without increasingly sophisticated computer technology and software, coupled with advances in computer-memory technology.

Test Administration

Perhaps more than any of the other areas discussed, test administration by computer will be the area of most tangible and observable progress in the future. The massive effort to computerize many of the tests that have been developed and published during this century has begun. For those tests that are not amenable to computerized test administration (such as individually administered tests for young children), efforts are underway to implement computerized scoring and interpretive programs.

Computers promise to expand the range of human responses that can be recorded automatically during test administration. Whereas the technology of automated test scoring was limited by the medium of optical-scanning answer sheets marked by pencil, the new wave of computerized tests will include a wide array of input media. These will include touch-sensitive screens, light-pens and toggle-levers, physiological receptors sensitive to changes in skin response and heart-rate, etc., and even voice-pattern recognizers. After years of development, touch-screens will now be more easily obtained for adaption to the responses of young children and anyone who cannot easily communicate via a typewriter keyboard. Inexpensive touch devices are also available for adaption to existing equipment (Cumming, 1983). Richards, Fine, Wilson, and Rogers (1983) recently reported success with a voice-operated microcomputer system that allows the patient to respond *True* or *False* vocally to MMPI items presented on a computer screen, by using voice-pattern analysis methods in the computer.

Similarly, the future will see increasing use of sophisticated test-stimulus displays, following price reductions on complex multimedia equipment. The work of Elwood and Griffin (1972) to administer the full-battery WAIS via tapedecks and complex equipment, for example, may give way to computer/video-disc systems such as used by Morf, Alexander, and Fuerth (1981) to administer a picture-preference test. Again, as mentioned earlier in this review, the two worlds of computer-assisted instruction and standardized testing will hopefully meet in the future, to the benefit of both. It would seem that a complex battery such as might be used to diagnose a learning disability would benefit from the sophisticated branching, response-time recording, and graphic/multimedia nature of CAI systems such as PLATO, the IBM 1500 system, or other CAI facilities currently in wide use in military and industrial training.

As discussed in the first section of this chapter on test administration, there has been a surprising lag in the implementation of computerized adaptive administration of tests, which have been possible since the late 1960s (e.g., Hansen, 1969). Perhaps the recent distribution of new item-analysis programs (Wingersky, Barton, & Lord, 1982; Masters, Wright, & Ludlow, 1980) will contribute future development of calibrated item collections needed for adaptive on-line testing. The work of Jensema (1976, 1977) may be very important in the implementation of the 3-parameter model. Jensema (1976) provided estimates of the 3-parameter model that are easily programmed, and easily understood by most psychologists without detailed training in IRT models. Jensema (1977) also provided guidelines for building a good, tailored item bank. As more and more test developers become skilled in the use of such programs, perhaps more adaptively-administered tests will appear such as those being released for use in government agencies and the military (e.g., Schmidt, Urry, & Gugel, 1978). Perhaps also, computerized adaptive testing will emerge more strongly when general-purpose computer software is available to help test authors prepare an adaptive test for a particular computer. A general software package could connect files of items with files of item-calibrations (e.g., item difficulty estimates) to present some standard methods of testing using adaptive branching. Development of such software may break the log-jam created by the effort required for a test developer to create not only the items (and the field-test statistics) but the test-administration program as well.

There remain numerous technical issues that must be resolved in adaptive computerized testing. First, many educators and psychologists will question the nature of normative comparisons that might be made with adaptive testing. Because each subject may be given a different set of items, there appears to be a certain statistical *wizardry* in calculating a total score for normative purposes. Only through numerous practical applications of adaptive testing will test users begin to see examples of scoring and methods of interpreting total scores. When IRT models underlie adaptive testing, total scores are estimates of the trait, ability, or achievement continuum assessed by an item pool. As discussed earlier in this chapter, in a process similar to "curriculum referencing" (Haladyna & Roid, 1983; Rentz, 1982), and IRT-based test score can be made interpretable by defining various points along the latent continuum. If such definition can be achieved, the resulting scale can provide both normative and criterion-referenced interpretations. The interpretation is normative if all items have been calibrated on representative samples in which all users have confidence and studies have been conducted to determine the relative number of students expected to score at successive points on the continuum (from which some new type of percentile can be derived). Total scores can be given criterion-referenced interpretations by the anchoring of specific items, skills, or meaningful trait levels along the continuum (e.g., as in the *Woodcock-Johnson* or *KeyMath* tests, Woodcock, 1982). Second, several technical issues surround the problem of optimal ways to determine the starting place for adaptive testing. The most sophisticated solution proposed to date for this problem involves the storage of longitudinal records of examinee or student performance on previous tests which can be used to begin subsequent testing. Some adaptive testing systems (noncomputerized), rely on the judgments of teachers in placing the student at an approximate level for beginning testing, followed by readministration of scales that prove to be improperly tailored (e.g., the use of special education teachers to estimate functional levels of retarded students assessed by the statewide assessment survey of Brodsky & Roid, 1977). Clearly, there is much that needs to be done to develop viable solutions to these technical problems with adaptive testing.

Test Scoring

Computer technology opens several new avenues for test scoring. In the past, the research finding that item weighting was usually not necessary (e.g., Stanley & Wang, 1970) gave us little reason to search beyond the basic *total score* (sum of a series of item scores) method used for most tests in both education and psychology. Also, for tests that are hand-scored, it is very difficult for users to calculate anything but integer scores for items (zero/one for dichotomous items, 1-5 for five-point Likert items, etc.). With new research emerging on item *option* weighting (Downey, 1979; Roid, 1983a; Stanley & Wang, 1970), and with continuing interest in factor scales for tests, there is increasing likelihood that computers will play an important role in providing more complex scoring systems for educational and psychological tests.

In educational measurement, the extensive research by Wilcox (e.g., Wilcox, 1981) on answer-until-correct scoring for achievement or ability tests is very promising and could be implemented in sophisticated ways using computer technology. New work on diagnostic scoring for achievement tests (Birenbaum & Tatsuoka, 1982, 1983) promises to allow for the diagnosis of erroneous problem-solving rules used by students. The new multicomponent models (e.g., Sternberg, 1977, 1979, 1981; Whitely, 1977, 1981) for achievement and ability tests of the problem-solving type would require complex scoring procedures because each item performance may entail several cognitive steps each of which may be scored separately.

In both psychological and educational measurement, a promising new method of computerized scoring for norm-referenced tests may prove useful in the future. Because the computer can ask the examinee to give exact demographic facts such as age in months (or this can be retrieved by processing a birth date in relation to the current date of testing), it may be possible to calculate what are called "continuous norms" (Gorsuch, 1983b; Roid, 1983b; Wendler, 1983; Zachary & Gorsuch, 1985). In continuous norming, one or more continuous variables such as age are examined in extensive computer analyses of field-test or normative-data results to discover whether or not a statistical formula can be derived to "fit" the pattern of test parameters (e.g., means, standard deviations) observed across the range of the variable. For example, it is often found with cognitive or skill tests that the mean score on a test increases steadily from ages 5 to 10. The traditional way of norming such tests is to provide separate norm tables for each year or 6-month increments of age. However, as Zachary and Gorsuch (1985) showed on an intelligence battery, the traditional norm table may inaccurately estimate the examinee's score if the age of the examinee is on the borderline between two adjacent norm tables. In continuous norming, values of test means and standard deviations are smoothed across a full range of age groups, so that estimates can be made at each and every continuous age level rather than in the graded steps implied by the use of printed norm tables.

An example of continuous norming taken from Roid (1983b), for a test in a learning disabilities battery, is presented in Fig. 3.2. Figure 3.2 shows the fitting of a polynomial regression equation to the progression of mean test scores across the age of students (in months). The vertical axis of Fig. 3.2 is the mean test score of an auditory memory test for school children. The horizontal axis of Fig. 3.2 is age in months (from 66 months to 162 months, i.e., 5.5 to 13.5 years).

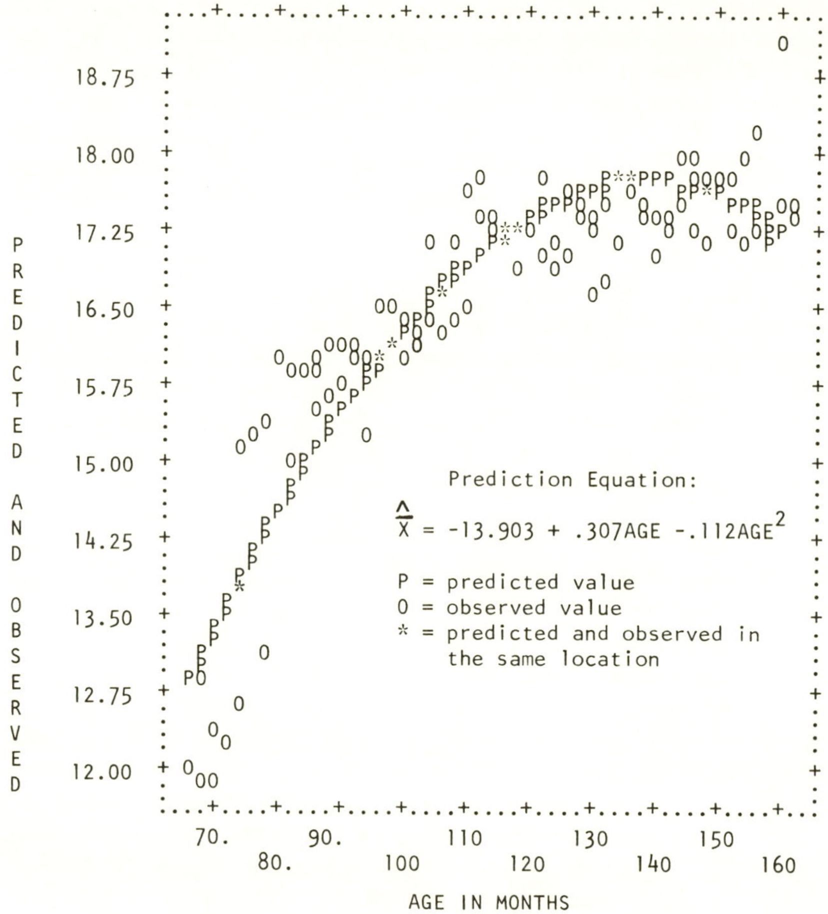

FIG. 3.2. Example of Continuous Norming: Plot of Predicted (P) and Observed (0) Age-Group Mean Scores on an Auditory-Memory Test by Age of the Subjects. Adapted from Roid (1983b).

The plotted points in Fig. 3.2 are either observed (O) or predicted (P) values of mean test score for groups of students at each age level. By drawing a best-fitting regression line through the predicted (P) points one can see that the mean test score steadily increases up to about 140 months, at which point it decreases slightly (due to a possible flaw in the sampling of older students, or some factor related to ''topping out'' on the test among older students). Because the fit of the regression line is adequate (R = .92 between mean score and age), test score means can be estimated for intermediate values of age (e.g., 120.5 months). Also, standard-score norms employing mean estimates derived from the prediction equation shown in Fig. 3.2 can be smoothed across the age span shown in the Figure. Operationally, the prediction equation would be programmed into a computer, the student's age requested for input, and a predicted mean test score for his or her age group calculated from the equation. In a similar fashion, estimates of standard deviation can be obtained so that normative standard scores are derived for each individual subject.

To determine the applicability of continuous norming, it is necessary to investigate each score on a test across a wide ranging sample and to discover whether or not a statistical formula (e.g., polynomial regression equation or cubic splines) is significantly accurate as an estimation device. Then, the formula or equation would be programmed for a computer-scoring routine. Error of estimation would also be evaluated to display the accuracy of continuous norming (Gorsuch, 1983b) along with the scoring output. Additional research is needed to extend the concept and methodology of continuous norming to linear (e.g., standardized T-scores) and nonlinear (e.g., normalized NCE scores) scoring of tests.

Test Interpretation

Some sophisticated methods of test interpretation become possible when the great memory and logic power of the computer can be carefully used following empirical studies of the links between test score patterns and verifiable behaviors or characteristics of examinees. For example, (R. L. Gorsuch, personal communication, 1983) following a study that demonstrates the discriminating power of a test to distinguish between examinees in various criterion groups (e.g., different psychiatric classifications, various types of dyslexia, etc.), the discriminant function equations derivable from such a study could be programmed into a test interpretive program so as to calculate the probability that an examinee belongs to a given criterion group. This involves only a linear equation with weights for each test score, but is often too complex of a calculation to do by hand, particularly if there are numerous criterion groups to assess.

A promising new technique has been developed by Huba (1985) for the matching of psychological test profiles to prototype profiles of criterion groups. Using multivariate techniques, an individual's vector of profile scores can be matched to a vector of criterion-group means using a *chi*-square test of goodness of fit. Huba's method is one of the first to take into account, explicitly, the correlations among profile scales.

Another realm in which computerized interpretation of tests may be important is in the establishment of links between two or more tests. For example, a brief test may be used to predict performance on a lengthier test, such as when full-scale WAIS IQ is predicted from a brief intelligence test (Zachary, Crumpton, & Spiegel, 1985). Typically, empirical studies of the brief tests have included a regression analysis in which scores on the longer test are predicted from the brief test. If the regression changes for different subgroups of examinees, as it often does for different age or ethnic groups, then the prediction of scores on the longer test involves a lengthy series of equations which are difficult to implement by hand. The computer, however, easily calculates any number of such predictions, and can also print confidence intervals and appropriate cautions to consider in evaluating the accuracy of such predictions.

In general, the future should see computer technology assisting professionals in integrating results of diverse test data. In McDermott's (1982) actuarial system for helping school psychologists diagnose learning-related problems, diverse tests of achievement, cognitive potential, and adaptive behavior are interrelated using statistical information such as the reliabilities of each test and their intercorrelations. Barclay's (1983) system for analyzing self, peer, and teacher ratings is another example of integrating diverse test results for practical diagnosis and prescription in the schools. As modes of inputting data from diverse tests progresses, and as more and more schools and clinics have their own computers with large memory capacities, more integration of test results and intertest interpretations will be possible.

Barclay's classroom assessment system (1983) also suggests another important advantage of computer technology in the interpretation of test results—the use of measures of the environment or situational factors in the assessment of individual differences. As Mischel (1979) Moos and Trickett (1974), and Walberg (1980) among others have been emphasizing for a number of years, the environment and changing situations of the individual must be taken into account in educational and psychological assessment. The computer's ability to analyze patterns of data collected across situations, across time periods, and from diverse sources such as self and observer perspectives, should make the evaluation of person-environment fit (or the lack thereof) more feasible.

SUMMARY

The field of measurement and testing, with its affinity for objective scores and statistical sophistication is a natural breeding ground for the emergence of computer technology in education and psychology. Dedication and attention to detail will be required of those who attempt to implement computerized testing and interpretation, if the facade of objectivity created by such systems is to be backed by empirically-validated procedures. Extensive and clear documentation has always been a somewhat difficult challenge in computer science, and the temptation will continue to be great to create novel programs that are undocumented.

Constant reminders of the probabilistic nature of computerized interpretations, and the errors of prediction inherent in them, will need to be woven into computer-printed reports that otherwise appear to have an aura of authoritative objectivity, especially for users who are less clinically experienced or knowledgeable of the limitations of all imperfect measuring instruments. As with any sophisticated tool, the professional must learn the limits of misinterpretation possible with computerized tests and interpretations. Clearly, technology cannot replace the complex evaluations that the experienced educator, researcher, or clinician can bring to the assessment of a child or adult when the individual is observed functioning in his or her environment over a period of time. An area of difficult assessment comes to mind that emphasizes the limitations of technology. It is in the assessment of complex learning diabilities, particularly in the case of the child who may appear to be retarded but whose inconsistent performance includes obtaining a very low or borderline score on an intelligence test and yet the child obtains isolated high scores on various diagnostic tests including adaptive behavior measures. The parents may also report inconsistent behavior, including "flashes of brilliance" which come and go. This child does not match the classical picture of the learning disability student who has average or above-average intelligence with specific achievement delays or underachievement (Gaddes, 1980). A complex pattern of perceptual, communication, and perhaps neuropsychological disorders may interfere with the child's performance on many tests, computerized or conventional. The response-timing features of computerized testing may be helpful in the diagnosis of conditions such as word-finding disorder (German, 1979, 1983), in which the child knows the answer but cannot retrieve it fast enough for a timed test, but the larger problem is that word-retrieval is only one of many potential disorders that interact with standardized test performance. Such disorders can interfere with assessment so that a false picture of the child's true potential is given. Clearly, skilled clinical observation, input from parents or other observers, and recognition of the limitations of testing must be considered in the interpretation of test results in such cases.

Despite its inevitable limitations, computerized testing and interpretation shows interesting promise for the future. The cost-effectiveness of relieving the professional from tedious hours of hand-scoring tests or in calculating various statistical indexes derived from test scores is clearly apparent. The value of computers in the multivariate studies that lead to test refinement and development has been clear for decades. If the fields of measurement and computer-assisted instruction can become even more closely related, there is great promise for the development of tests that use multiple input and output media for presenting test-items (or test-like experimental events such as in the assessment of memory and perception) and for displaying results. As the equipment used in applications such as speech and hearing therapy and biofeedback treatment become linked more and more to computers, new forms of auditory and physiological data will become amenable to computerized interpretation along with more conventional test items and scales.

ACKNOWLEDGMENT

This chapter is dedicated to the memory of Bruce Choppin who encouraged the author to pursue some of the ideas presented here during informal meetings and symposia at the UCLA Center for the Study of Evaluation. Acknowledgments are extended to Manson

Western Corporation for their support of research and development activities at Western Psychological Services. Many of the major ideas presented in this chapter originated in conversations with Richard L. Gorsuch, Fuller Graduate School of Psychology, Pasadena, California, whose advice is based on a breadth of experience spanning the history of computerization in psychology. Dr. Gorsuch's contribution is gratefully acknowledged. However, any omissions or errors remain the responsibility of the author. Opinions expressed or future predictions made are those of the author, not the official policy of any corporation or institution.

REFERENCES

American Psychological Association. (1985). *Joint technical standards for educational and psychological testing*. Washington, D.C.: Author.

Airasian, P. W., & Madaus, G. F. (1983). Linking testing and instruction: Policy issues. *Journal of Educational Measurement, 20,* 103–118.

Alpert, D., & Bitzer, D. L. (1970). Advances in computer-based education. *Science, 167,* 1582–1590.

Altemeyer, R. A. (1970). Adverbs and intervals: A study of Likert scales. *Proceedings 78th Annual Convention of APA,* pp. 397–398.

Atkinson, R. C., & Wilson, H. A. (1969). *Computer-assisted instruction: A book of readings*. New York: Academic Press.

Bailey, J. R. (1977). Tables of the Bonferroni *t* statistic. *Journal of the American Statistical Association, 72,* 469–478.

Barclay, J. R. (1983). *Barclay Classroom Assessment System manual*. Los Angeles: Western Psychological Services.

Barrett, G. V., Alexander, R. A., Doverspike, D., Cellar, D., & Thomas, J. C. (1982). The development and application of a computerized information-processing test battery. *Applied Psychological Measurement, 6,* 13–29.

Beasley, J. D., & Springer, S. G. (1977). The percentage points of the normal distribution. *Applied Statistics, 26,* 118–121.

Bentler, P. M. (1980). Multivariate analysis with latent variables: Causal modeling. *Annual Review of Psychology, 31,* 419–456.

Bersoff, D. N. (1981). Testing and the law. *American Psychologist, 36,* 1047–1056.

Birenbaum, M., & Tatsuoka, K. (1982). On the dimensionality of achievement test data. *Journal of Educational Measurement, 19,* 259–266.

Birenbaum, M., & Tatsuoka, K. (1983). The effect of a scoring system based on the algorithm underlying the students' response patterns on the dimensionality of achievement test data of the problem solving type. *Journal of Educational Measurement, 20,* 17–26.

Biskin, B. H., & Kolotkin, R. L. (1977). Effects of computerized administration on scores on the Minnesota Multiphasic Personality Inventory. *Applied Psychological Measurement, 1,* 543–549.

Bock, R. D., & Mislevy, R. J. (1982). Adaptive EAP estimation of ability in a microcomputer environment. *Applied Psychological Measurement, 6,* 431–444.

Bormuth, J. R. (1970). *On the theory of achievement test items*. Chicago: University of Chicago Press.

Braby, R., Parrish, W. F., Guitard, C. R., & Aagard, J. A. (1978). *Computer-aided authoring of programmed instruction for teaching symbol recognition*. (TAEG Report No. 58) Orlando, FL: U.S. Navy Training Analysis and Evaluation Group.

Brodsky, M., & Roid, G. H. (1977). *Examiner's manual for the Student Progress Record*. Salem, OR: Oregon Mental Health Division.

Campbell, D. P. (1976). Author's reaction to Johnson's review. *Measurement and Evaluation in Guidance, 9,* 45–56.

Campbell, D. P. (1977). *Manual for the Strong-Campbell Interest Inventory.* CA: Stanford University Press.

Carroll, J. B., Davies, P., & Richman, B. (1971). *Word frequency book.* Boston: Houghton-Mifflin.

Cattell, R. B. (1956). Validation and intensification of the Sixteen Personality Factor Questionnaire. *Journal of Clinical Psychology, 12,* 205–214.

Cattell, R. B. (1974). Radial parcel factoring versus item factoring in defining personality structure in questionnaires: Theory and experimental checks. *Australian Journal of Psychology, 26,* 103–119.

Choppin, B. (1968). An item bank using sample-free calibration. *Nature, 219,* 870–872.

Christian, K. (1983). *The UNIX operating system.* New York: Wiley.

Cleary, T. A., Linn, R. L., & Rock, D. A. (1968). An exploratory study of programmed tests. *Educational and Psychological Measurement, 28,* 345–360.

Colby, K. M., Watt, J., & Gilbert, J. A. (1966). A computer method of psychotherapy. *Journal of Nervous and Mental Disease, 142,* 148–152.

Connolly, A. J., Nachtman, W., & Pritchett, E. M. (1976). *KeyMath Diagnostic Arithmetic Test.* Circle Pines, MN: American Guidance Service.

Cory, C. H. (1977). Relative utility of computerized versus paper-and-pencil tests for predicting job performance. *Applied Psychological Measurement, 1,* 551–564.

Cory, C. H., Rimland, B., & Bryson, R. A. (1977). Using computerized tests to measure new dimensions of abilities: An exploratory study. *Applied Psychological Measurement, 1,* 101–110.

Cowden, D. J. (1946). An application of sequential analysis to testing students. *Journal of the American Statistical Association, 41,* 547–556.

Croll, P. R. (1981). *Computerized adaptive testing: From psychometrics to systems.* Paper presented at the annual meeting of the Military Testing Association, Arlington, Virginia, October.

Cronbach, L. J., & Meehl, P. E. (1955). Construct validity in psychological tests. *Psychological Bulletin, 52,* 281–302.

CTB/McGraw-Hill. (1982). *Comprehensive Tests of Basic Skills, Forms U and V.* Monterey, CA: Author.

Cull, J. G., & Gill, W. S. (1982). *Suicide Probability Scale (SPS) manual.* Los Angeles: Western Psychological Services.

Cumming, G. A. (1983). A rudimentary touch input device for visual displays. *Behavioral Research Methods and Instrumentation, 15,* 72–74.

Damarin, F. (1970). A latent-structure model for answering personal questions. *Psychological Bulletin, 73,* 23–40.

Davis, F. B. (1959). Interpretation of differences among averages and individual test scores. *Journal of Educational Psychology, 50,* 162–170.

Dee-Burnett, R., Jones, E. F., & Krug, S. (1982). *Law enforcement assessment and development report manual.* Champaign, IL: Institute for Personality and Ability Testing.

Downey, R. G. (1979). Item-option weighting of achievement tests: Comparative study of methods. *Applied Psychological Measurement, 3,* 453–461.

Dunn, O. J. (1961). Multiple comparisons among means. *American Statistical Association Journal, 56,* 52–64.

Elliott, C. D. (1983a). *British Ability Scales Technical Handbook.* Windsor, England: NFER-Nelson.

Elliott, C. D. (1983b). *Vertical equating of unidimensional tests using the Rasch model.* Unpublished paper, Department of Education, University of Manchester, England.

Elwood, D. L. (1969). Automation of psychological testing. *American Psychologist, 24,* 287–289.

Elwood, D. L., & Clark, C. L. (1978). Computer administration of the Peabody Picture Vocabulary Test to young children. *Behavior Research Methods and Instrumentation, 10,* 43–46.

Elwood, D. L., & Griffin, H. R. (1972). Individual intelligence testing without the examiner: Reliability of an automated method. *Journal of Consulting and Clinical Psychology, 38,* 9–14.

Evan, W. M., & Miller, J. R. (1969). Differential effects on response bias of computer vs. conventional administration of a social science questionnaire: An exploratory methodological experiment. *Behavioral Science, 14,* 216–227.

Fiske, D. W., & Jones, L. V. (1954). Sequential analysis in psychological research. *Psychological Bulletin, 51,* 264–275.

Forster, F., & Doherty, V. (1978). *Using the Rasch model to improve a district's testing program.* Paper presented at the meeting of the American Educational Research Association, Toronto, March.

Fossum, J. A. (1973). An application of techniques to shorten tests and increase validity. *Journal of Applied Psychology, 57,* 90–92.

Fremer, J., & Anastasio, E. J. (1969). Computer-assisted test writing—I (Spelling items). *Journal of Educational Measurement, 6,* 69–74.

Gaddes, W. H. (1980). *Learning disabilities and brain function: A neuropsychological approach.* New York: Springer-Verlag.

German, D. J. (1979). Word finding skills in children with learning disabilities. *Journal of Learning Disabilities, 12*(3), 43–48.

German, D. J. (1983). I know it but I can't think of it: Word retrieval difficulties. *Academic Therapy, 18,* 539–545.

Gilmore, G. C., Royer, F. L., Tobias, T. R., & Ruffing, A. G. (1983). *Software approaches to the analysis of global form processing.* Paper presented at the annual meeting of the American Psychological Association, Anaheim, California, August.

Goldberg, L. R. (1968). Simple models or simple processes? Some research on clinical judgments. *American Psychologist, 23,* 483–496.

Goldberg, L. R. (1970). Man vs. model of man: A rationale, plus some evidence for a method of improving on clinical inferences. *Psychological Bulletin, 73,* 422–432.

Golden, C. J., Hammeke, T. A., & Purisch, A. D. (1980). *The Luria-Nebraska Neuropsychological Battery manual.* Los Angeles: Western Psychological Services.

Gorsuch, R. L. (1982). *Standard adjectives and modifiers for computerized interpretive programs.* Unpublished paper, Western Psychological Services, Los Angeles, CA.

Gorsuch, R. L. (1983a). *Factor Analysis* (2nd ed.). Hillsdale, NJ: Lawrence Erlbaum Associates.

Gorsuch, R. L. (1983b). *Continuous norming: A procedure for more accurate estimation of normative test statistics as function of a conditioning variable.* Paper presented at the annual meeting of the American Psychological Association, Anaheim, California, August.

Gorsuch, R. L., & Yagel, J. C. (1982). *Exploratory item factor analysis.* Unpublished paper, Fuller Graduate School of Psychology, Pasadena, California.

Gorsuch, R. L., & Dreger, R. M. (1979). "Big jiffy": A more sophisticated factor analysis and rotation program. *Educational and Psychological Measurement, 39,* 209–214.

Gorth, W. P., Allen, D. W., & Grayson, A. (1971). Computer programs for test objective and item banking. *Educational and Psychological Measurement, 31,* 245–250.

Green, S. B. (1983). Identifiability of spurious factors using linear factor analysis with binary items. *Applied Psychological Measurement, 7,* 139–147.

Green, S. B., Lissitz, R. W., & Muliak, S. A. (1977). Limitations of coefficient alpha as an index of test unidimensionality. *Educational and Psychological Measurement, 37,* 827–838.

Greist, J. H., Gustafson, D. H., Strauss, F. F., Rowse, G. L., Laughren, T. P., & Chiles, J. A. (1973). A computer interview for suicide risk prediction. *American Journal of Psychiatry, 130,* 1327–1332.

Hakel, M. D. (1968). How often is often? *American Psychologist, 23,* 533–534.

Haladyna, T. M., & Roid, G. H. (1983). A comparison of two approaches to criterion-referenced test construction. *Journal of Educational Measurement, 20,* 271–282.

Hambleton, R. K., & Swaminathan, H. (1985). *Item response theory: Principles and applications.* Boston: Kluwer-Nijhoff.

Hansen, D. N. (1969). An investigation of computer-based science testing. In R. C. Atkinson & H. A. Wilson (Eds.), *Computer-assisted instruction: A book of readings.* New York: Academic Press.

Harrell, T. H., & Lombardo, T. A. (1984). Validation of an automated 16PF administration procedure. *Journal of Personality Assessment, 48,* 638–642.

Hedl, J. J. (1971, June). An evaluation of a computer-based intelligence test. *CAI Technical Report,* Computer-Assisted Instruction Laboratory, Florida State University, No. 21.

Hertzog, C. (1983, August). *Cognitive assessment of adults using microcomputer techniques.* Paper presented at the annual meeting of the American Psychological Association, Anaheim, California.

Hively, W., Patterson, H. L., & Page, S. A. (1968). A 'universe-defined' system of arithmetic achievement tests. *Journal of Educational Measurement, 5,* 275–290.

Huba, G. J. (1985). *New psychometric directions for computer-based test interpretations.* Unpublished paper, Western Psychological Services, Los Angeles, California.

Jensema, C. (1976). A simple technique for estimating latent trait mental test parameters. *Educational and Psychological Measurement, 36,* 705–715.

Jensema, C. (1977). Bayesian tailored testing and the influence of item bank characteristics. *Applied Psychological Measurement, 1,* 111–120.

Johnson, J. H., Giannetti, R. A., & Williams, T. A. (1979). Psychologial systems questionnaire: An objective personality test designed for on-line computer presentation, scoring, and interpretation. *Behavior Research Methods & Instrumentation, 11,* 257–260.

Johnson, K. J. (1973). Pitt's computer-generated chemistry exam. *Proceedings of the Conference on Computers in Undergraduate Curricula,* 199–204.

Johnson, D. F., & White, C. B. (1980). Effects of training on computerized test performance in the elderly. *Journal of Applied Psychology, 65,* 357–358.

Joreskog, K. G., & Sorbom, D. (1981). *LISREL V.* Chicago: International Educational Services.

Katz, L., & Dalby, J. T. (1981). Computer-assisted and traditional psychological assessment of elementary school aged children. *Contemporary Educational Psychology, 6,* 314–322.

Kaufman, A. S. (1979). *Intelligent testing with the WISC-R.* New York: Wiley.

Kildall, G. (1982). CP/M: A family of 8- and 16-bit computer operating systems. *Education, 102,* 211–219.

Klinge, V., & Rodziewicz, T. (1976). Automated and manual intelligence testing of the Peabody Picture Vocabulary Test on a psychiatric adolescent population. *International Journal of Man-Machine Studies, 8,* 243–246.

Klinger, D. E., Johnson, J. H., & Williams, T. A. (1976). Strategies in the evaluation of an on-line computer-assisted unit for intake assessment of mental health patients. *Behavior Research Methods and Instrumentation, 8,* 95–100.

Kornbrot, D. E. (1981). ARTIST: A computer system for creating and running psychological experiments. *Behavior Research Methods and Instrumentation, 13,* 351–359.

Krug, S. E. (1982, August). *Tailoring interpretive programs to specific users.* Paper presented at the annual meeting of the American Psychological Association Washington, D.C.

Krug, S. E. (1983). *Marriage Counseling Report manual.* Champaign, IL: Institute for Personality and Ability Testing.

Lachar, D. (1974). Accuracy and generalizability of an automated MMPI interpretation system. *Journal of Clinical and Consulting Psychology, 42,* 267–273.

Lachar, D., & Alexander, R. S. (1978). Veridicality of self report: Replicated correlates of the Wiggins MMPI content scales. *Journal of Consulting and Clinical Psychology, 46,* 1349–1356.

Lachar, D., & Gdowski, C. L. (1979). *Actuarial assessment of child and adolescent personality: An interpretive guide for the Personality Inventory for Children Profile.* Los Angeles: Western Psychological Services.

Lachar, D., Klinge, V., & Grisell, J. L. (1976). Relative accuracy of automated MMPI narratives generated from adult norm and adolescent norm profiles. *Journal of Consulting and Clinical Psychology, 44,* 20–24.

Lichtenstein, S., & Newman, J. R. (1967). Empirical scaling of common verbal phrases associated with numerical probabilities. *Psychonomic Science, 9,* 563–564.

Lord, F. M. (1980). *Applications of item response theory to practical testing problems.* Hillsdale, NJ: Lawrence Erlbaum Associates.

Lushene, R. E., & Gilberstadt, H. (1972, March). Validation of VA MMPI computer-generated reports. Paper presented at the Veterans Administration Cooperative Studies Conference, St. Louis.

Lushene, R. E., O'Neil, H. F., & Dunn, T. (1974). Equivalent validity of a completely computerized MMPI. *Journal of Personality Assessment, 38,* 353–361.

Mapou, R. L. (1982). Tachistoscopic timing on the TRS-80. *Behavior Research Methods and Instrumentation, 14,* 534–538.

Marsh, H. W., & O'Neil, R. (1984). Self Description Questionnaire III: The construct validity of multidimensional self-concept ratings by late adolescents. *Journal of Educational Measurement, 21,* 153–174.

Masters, G. N., Wright, B. D., & Ludlow, L. H. (1980). *Item calibration and person measurement with ordered response categories: The CREDIT program.* Chicago: Department of Education, University of Chicago.

Matarazzo, J. D. (1983). Computerized psychological testing. *Science, 221*(4608), (Editorial comment).

Maxwell, K, J., & Schvaneveldt, R. W. (1983). The PSYCHLAB programming system: A tool for developing experimental control programs. *Behavior Research Methods and Instrumentation, 15,* 49–56.

McArdle, J. J., & Kirson, D. (1983, August). *Individualized psychological measurement via microcomputer.* Paper presented at the meeting of the American Psychological Association, Anaheim, California.

McArthur, D. C. (1984, April). *Emerging models for assessing intellective functioning interactively.* Paper presented at the National Conference on the Development and Assessment of Human Competence, Howard University, Washington, DC.

McDermott, P. A. (1980). A systems-actuarial method for differential diagnosis of handicapped children. *Journal of Special Education, 50,* 223–228.

McDermott, P. A. (1982). Actuarial assessment systems for the grouping and classification of schoolchildren. In C. R. Reynolds & T. B. Gutkin (Eds.), *The handbook of school psychology.* New York: Wiley.

McDonald, R. P. (1981). The dimensionality of tests and items. *British Journal of Mathematical and Statistical Psychology, 34,* 100–117.

Meehl, P. E. (1954). *Clinical vs. statistical prediction: A theoretical analysis and review of the evidence.* Minneapolis: University of Minnesota Press.

Messick, S. (1980). Test validity and the ethics of assessment. *American Psychologist, 35,* 1012–1027.

Miller, L. C. (1981). *Louisville Behavior Checklist Manual.* Los Angeles: Western Psychological Services.

Millman, J. (1980). Computer-based item generation. In R. A. Berk (Ed.), *Criterion-referenced measurement: The state of the art.* Baltimore, MD: Johns Hopkins University Press.

Millman, J., & Outlaw, W. S. (1978). Testing by computer. *Association for Educational Data Systems (AEDS) Journal, 11,* 57–72.

Mischel, W. (1979). On the interface of cognition and personality: Beyond the person-situation debate. *American Psychologist, 34,* 740–754.

Moos, R. H., & Trickett, E. J. (1974). *Classroom Environment Scale manual.* Palo Alto, CA: Consulting Psychologists Press.

Morf, M., Alexander, P., & Fuerth, T. (1981). Fully automated psychiatric diagnosis: Some new possibilities. *Behavior Research Methods and Instrumentation, 13,* 413–416.

Muthen, B. (1978). Contributions to factor analysis of dichotomous variables. *Psychometrika, 43,* 551–560.

Muthen, B. (1981). Factor analysis of dichotomous variables: American attitudes toward abortion. In E. Borgatta & D. J. Jackson (Eds.), *Factor analysis and measurement in sociological research: A multidimensional perspective.* Beverly Hills: Sage Publications.

Novick, M. R., Chen, J. J., Mayekawa, S., Woodworth, G., Libby, D., Cook, J., & Novick, R. L. (1983). *The computer-assisted data analysis (CADA) monitor.* Iowa City: Lindquist Center for Measurement, University of Iowa.

O'Brien, T., & Dugdale, V. (1978). Questionnaire administration by computer. *Journal of the Market Research Society, 20,* 228–237.

Olympia, P. L., Jr. (1975). Computer generation of truly repeatable examinations. *Educational Technology, 14*(6), 53–55.

Osburn, H. G. (1968). Item sampling for achievement testing. *Educational and Psychological Measurement, 28,* 95–104.

Pohl, N. F. (1981). Scale considerations in using vague quantifiers. *Journal of Experimental Education, 49,* 235–240.

Popham, W. J. (1980). Domain specification strategies. In R. A. Berk (Ed.), *Criterion-referenced measurement: The state of the art.* Baltimore, MD: Johns Hopkins University Press.

Posner, M. I., & Osgood, G. W. (1980). Computers in the training of inquiry. *Behavior Research Methods and Instrumentation, 12,* 87–95.

Rasch, G. (1980). *Probabilistic models for some intelligence and attainment tests.* Chicago: University of Chicago Press. (Originally published by Denmarks Paedagogiske Institut, Copenhagen, 1960).

Reckase, M. D. (1979). Unifactor latent trait models applied to multifactor tests: Results and implications. *Journal of Educational Statistics, 4,* 207–230.

Rentz, R. R. (1982, March). *Using the Rasch model to develop a curriculum-referenced norm.* Paper presented at the meeting of the National Council on Measurement in Education, New York.

Reynolds, C. R. (1982). Determining statistically reliable strengths and weaknesses in the performance of individuals on the Luria-Nebraska Neuropsychological Battery. *Journal of Consulting and Clinical Psychology, 50,* 525–529.

Richards, J. S., Fine, P. R., Wilson, T. L., & Rogers, J. T. (1983). A voice-operated method for administering the MMPI. *Journal of Personality Assessment, 47,* 167–170.

Roid, G. H. (1969). *Branching methods for constructing psychological test scales.* Unpublished doctoral dissertation, University of Oregon.

Roid, G. H. (1983a, April). *A successful case of using item-option weighted scores to increase the validity of a suicide-risk scale.* Paper presented at the meeting of the Western Psychological Association, San Francisco.

Roid, G. H. (1983b, August). *Generalization of continuous norming: Cross-validation of test-score mean estimates.* Paper presented at the meeting of the American Psychological Association, Anaheim, California.

Roid, G. H. (1984). Generating test items. In R. A. Berk (Ed.) *A guide to criterion-referenced testing.* Baltimore, MD: Johns Hopkins University Press.

Roid, G. H., & Gorsuch, R. L. (1983). Development and use of test interpretive programs on microcomputers. In M. Schwartz (Ed.), *Guide to clinical use of computers.* New York: Haworth Press.

Roid, G. H., & Haladyna, T. M. (1980). The emergence of an item-writing technology. *Review of Educational Research, 50,* 293–314.

Roid, G. H., & Haladyna, T. M. (1982). *A technology for test-item writing.* New York: Academic Press.

Sammet, J. E. (1969). *Programming languages: History and fundamentals.* Englewood Cliffs, NJ: Prentice-Hall.

Sapinkopf, R. G. (1978). A computer adaptive testing approach to the measurement of personality variables. *Dissertation Abstracts International, 38*(10-B), 4993.

Schmitt, N. (1981). Rasch analysis of the Central Life Interest measure. *Applied Psychological Measurement, 5,* 3–10.

Schmidt, F. L., Urry, V. W., & Gugel, J. F. (1978). Computer assisted tailored testing: Examinee reactions and evaluations. *Educational and Psychological Measurement, 38,* 265–273.

Schulz, R. E. (1979). Computer aids for developing tests and instruction. In H. F. O'Neil (Ed.), *Procedures for instructional systems development.* New York: Academic Press.

Scissons, E. H. (1976). Computer administration of the California Psychological Inventory. *Measurement and Evaluation in Guidance, 91,* 22–25.

Slinde, J. A., & Linn, R. L. (1979). A note on vertical equating via the Rasch model for groups of quite different ability and tests of quite different difficulty. *Journal of Educational Measurement, 16,* 159–165.

Smith, J. K. (1980). On the examination of test unidimensionality. *Educational and Psychological Measurement, 40,* 885–889.

Stanley, J. C., & Wang, M. D. (1970). Weighting test items and test-item options: An overview of the analytical and empirical literature. *Educational and Psychological Measurement, 30,* 21–35.

Stanwyck, D. J., & Garrison, W. M. (1982). Detection of faking on the Tennessee Self Concept Scale. *Journal of Personality Assessment, 46,* 426–431.

Stegelmann, W. (1983). Expanding the Rasch model to a general model having more than one dimension. *Psychometrika, 48,* 259–267.

Sternberg, R. J. (1977). *Intelligence, information processing, and analogical reasoning: The componential analysis of human abilities.* Hillsdale, NJ: Lawrence Erlbaum Associates.

Sternberg, R. J. (1979). The nature of mental abilities. *American Psychologist, 34,* 214–230.

Sternberg, R. J. (1981). Testing and cognitive psychology. *American Psychologist, 36,* 1181–1189.

Stoline, M. R. (1983). The Hunter method of simultaneous inference and its recommended use for applications having large known correlation structures. *Journal of the American Statistical Association, 78,* 366–370.

Suppes, P., Jerman, M., & Groen, G. J. (1966, April). Arithmetic drills and review on a computer-based teletype. *Arithmetic Teacher,* pp. 303–308.

Terman, L. M., & Merrill, M. A. (1937). *Measuring intelligence: A guide to the administration of the new revised Stanford-Binet Tests of Intelligence.* Boston: Houghton Mifflin.

Thissen, D., Steinberg, L., Pyszczynski, T., & Greenberg, J. (1983). An item response theory for personality and attitude scales: Item analysis using restricted factor analysis. *Applied Psychological Measurement, 7,* 211–226.

Urry, V. W. (1977). Tailored testing: A successful application of latent trait theory. *Journal of Educational Measurement, 14,* 181–186.

Vanderplas, J. M., & Vanderplas, J. H. (1979). Multiple versus single-index predictors of dangerousness, suicide, and other rare behaviors. *Psychological Reports, 45,* 343–349.

Vickers, F. D. (1973). Creative test generators. *Educational Technology, 13*(3), 43–44.

Walberg, H. J. (1980). A psychological theory of educational productivity. In F. H. Farley & N. Gordon (Eds.), *Perspectives on educational psychology.* Berkeley, CA: McCutchan Publishing.

Wallbrown, F., Blaha, J., & Wherry, R. J. (1973). The hierarchical factor structure of the Wechsler Preschool and Primary Scale of Intelligence. *Journal of Consulting and Clinical Psychology, 41,* 356–362.

Weiss, D. J. (1979). Computerized adaptive achievement testing. In H. F. O'Neil, Jr. (Ed.), *Procedures for instructional systems development*. New York: Academic Press.

Weitzman, R. A. (1982). Sequential testing for selection. *Applied Psychological Measurement, 6,* 337–351.

Wendler, C. L. W. (1983, August). *Continuous norming of a projective test*. Paper presented at the meeting of the American Psychological Association Anaheim, California.

Whitely, S. E. (1977). Some information-processing components of intelligence test items. *Applied Psychological Measurement, 1,* 465–476.

Whitely, S. E. (1981). Measuring aptitude processes with multicomponent latent trait models. *Journal of Educational Measurement, 18,* 67–83.

Wiggins, J. S. (1973). *Personality and prediction: Principles of personality assessment*. Reading, MA: Addison-Wesley.

Wilcox, R. R. (1981). Solving measurement problems with an answer-until-correct scoring procedure. *Applied Psychological Measurement, 5,* 399–414.

Wingersky, M. S., Barton, M. A., & Lord, F. M. (1982). *LOGIST user's guide: LOGIST 5, Version 1.0*. Princeton, NJ: Educational Testing Service.

Wirt, R. D., Lachar, D., Klinedinst, J. K., & Seat, P. D. (1977). *Multidimensional description of child personality: A manual for the Personality Inventory for Children*. Los Angeles: Western Psychological Services.

Wolfle, L. M. (1980). Strategies of path analysis. *American Educational Research Journal, 17,* 183–209.

Wood, R., & Skurnik, L. S. (1969). *Item banking*. London, England: National Foundation for Educational Research in England and Wales.

Woodcock, R. W. (1973). *Woodcock Reading Mastery Tests manual*. Circle Pines, MN: American Guidance Service.

Woodcock, R. W. (1978). *Development and standardization of the Woodcock-Johnson Psycho-Educational Battery*. Hingham, MA: Teaching Resources.

Woodcock, R. W. (1982, March). *Interpretation of the Rasch ability and difficulty scales for educational purposes*. Paper presented at the meeting of the National Council on Measurement in Education, New York.

Wright, B. D., & Masters, G. N. (1982). *Rating scale analysis: Rasch measurement*. Chicago: MESA Press.

Wright, B. D., & Stone, M. H. (1979). *Best test design: Rasch measurement*. Chicago: MESA Press.

Yen, W. M. (1982, March). *Use of three-parameter item response theory in the development of CTBS, Form U, and TCS*. Paper presented at the meeting of the National Council of Measurement in Education, New York.

Zachary, R. A., Crumpton, E., & Spiegel, D. F. (1985). Estimating WAIS-R IQ from the Shipley Institute of Living Scale. *Journal of Clinical Psychology*.

Zachary, R. A., & Gorsuch, R. L. (1985). Continuous norming: Implications for the WAIS-R. *Journal of Clinical Psychology, 41,* 86–94.

Zachary, R. A., & Pope, K. S. (1983). Legal and ethical issues in the clinical use of computerized testing. In M. D. Schwartz (Ed.), *Using computers in clinical practice*. New York: Haworth Press.

II EDUCATIONAL AND ACADEMIC PROFESSIONAL DIRECTIONS

4 Future Directions for Educational Achievement and Ability Testing

Nancy S. Cole
University of Pittsburgh

Forecasting future directions is at best a risky business. Few can claim to see with confidence the shape of things to come. In addition, the issues related to the future directions of educational achievement and ability testing are broad and numerous. They range from such important, but difficult to forecast, areas as the national mood toward education and educational accountability, legislation and the political arena, and court rulings to theoretical and technological advances. However, instead of providing a survey of possible futures resulting from these diverse and numerous potential influences, this chapter focuses on two areas of overriding importance in considering these future directions. The two areas to be addressed are: (a) the evolving conceptual understanding of the nature of the achievement and ability constructs, and (b) the opportunities afforded by advances in computer technology.

The chapter begins with a brief review of the current status of such testing to establish a context for the consideration of these two central issues and their implications for the future. It then turns, first, to the conceptual directions likely to evolve from changes in the familiar ability and achievement constructs and, second, to some of the implications of computer technology for testing in the future.

CURRENT STATUS OF ACHIEVEMENT AND ABILITY TESTING

A Recent Period of Criticism

To examine the current status of educational testing, it is necessary to recognize that the period from the late 1960s through the 1970s was a time of intense criticism of standardized testing. During this period of increased social awareness, test score differences between groups were attributed quickly to bias in tests, particularly when the major alternative explanation under debate was hereditary differences in intelligence between groups (Jensen, 1969). Access to education and appropriate employment for minority applicants was one major objective of the civil rights movement and often tests were viewed as tools which stood in the way of such access. The courts were asked to judge the appropriateness of tests and test use and tests became central in several court rulings. When the concern with citizen rights made its way into the market place and the consumer movement began in earnest, the result was calls for "truth in testing," which was implemented in some states basically as a requirement for periodic public disclosure of the content of major tests.

As this time of criticism blossomed, testing specialists were in the forefront both criticizing and defending tests in the public and professional literature and seeking ways to improve them. Measurement specialists began to examine with a new seriousness the possibility of test bias. The professional literature contains hundreds of studies conducted during this period (e.g., see Cole, 1981, 1983) which address technical approaches and substantive issues of bias. It became common for manuals of educational tests to routinely address the issue of bias. More measurement professionals began to disassociate themselves from arguments that test score differences necessarily reflect hereditary differences and in so doing emphasized with new vigor the effects of educationally related experience on educational ability tests. In addition, many test developers began to adopt a more critical view of their tests in preparation for disclosure and other forms of public scrutiny—to view them as an interested public might—and in so doing discovered a number of small ways of improving test content.

This period of criticism perhaps ended about the time of the 1982 National Academy of Sciences's Committee on the Ability Testing report (Wigdor & Garner, 1982), which provided a broad survey of the social context of testing and concluded with both criticisms and endorsements of ability test use. In addition to a number of specific recommendations, the Committee noted several broad limitations of tests:

> Although a well-developed test can be a reasonable good predictor of the performance of people in the aggregate, it may be a poor predictor of the performance of any particular individual.
>
> Ability tests do not measure many things that are important to performance in school and at work.
>
> The relative immaturity of theories of cognition places signifcant limits on the explanation of abilities that can be derived from test results (p. 237).

The Committee described its general view as a "call for balance":

> By emphasizing the limitations of tests we mean to counteract the widespread tendency to look to ability tests as a panacea for deep-seated social ills; and by discussing testing in the context of social developments that far transcend it in

importance or effect, we hope to counter the equally prevalent tendency to use tests as a scapegoat for society's ills (p. 236).

Renewed Demands for Tests

Long before the period of criticism had run its course, there were trends in counter directions. "Back to the basics" became a rallying cry, particularly for elementary education, both as the focus of federally financed compensatory programs and in response to a growing public concern about educational quality. The more test scores declined, the more demand there was for testing as a method of educational accountability. The move to evaluate programs encouraged the development of tests designed for such purposes, and there arose a special concern to have tests directed more specifically to the particular curriculum in question. Criterion referenced tests (CRTs) became popular basically as tests referenced to a set of specific educational objectives. The minimum competency testing movement gained force with several states adopting statewide tests prior to high school graduation; although among testing professionals the movement had its critics (e.g., Madaus & McDonagh, 1979) as well as its defenders (e.g., Lerner, 1981). Test use in the schools continued at a high level apparently little affected by the period of criticism noted above. Houts (1975) estimated that students received from six to twelve full batteries of achievement tests during their school years. Anderson (1982) noted that such a high level of testing continues and is only a part of the total school testing program of most school districts.

Current Practice

Although educational achievement and ability tests appear to be on the brink of major types of change, current practice is notably similar to testing in previous decades in terms of the types of items and the methods of administration. By far the largest amount of commercially prepared standardized educational testing is in a paper-and-pencil, multiple choice, group-administered form. School achievement batteries continue to cover the same general areas (reading, mathematics, language, social studies, science) and report norm-referenced scores. A major alteration from past decades is the addition of score reporting in terms of item clusters tied to particular educational objectives.

School group ability tests continue to focus on verbal and quantitative areas, featuring changes in name in an effort to discourage the hereditary interpretations often associated with terms such as intelligence and aptitude. Individual intelligence tests are being updated after being challenged in the courts for out-of-date and biased content but focus on the same features of language, reasoning, and general knowledge as before.

Admissions testing continues to feature verbal and quantitative skills in the Scholastic Aptitude Test (SAT) of the College Board, the Graduate Record

Examination (GRE), the Law School Admissions Test (LSAT), and the Graduate Management Admissions Test (GMAT). The American College Test (ACT) which is organized instead around high school subject (i.e., English, mathematics, social studies, and natural science) remains an exception as do several attempts to expand and redefine in directions of analytical reasoning (GRE, LSAT)and more realistic contexts (Medical College Admissions Test).

Two areas currently receiving renewed attention are writing and reasoning/problem-solving. Concern among the lay public with the writing skill of students has arisen anew. Professional study of writing and how to teach it is active. Development of writing assessment procedures has been spurred by work on the National Assessment of Educational Progress and reports from the results of NAEP assessments has further stimulated concern. An optional writing sample has been returned to the SAT; many states mention writing skills much more directly as educational goals with some requiring written products from students in statewide assessments.

Renewed concern with reasoning and problem solving has also become prevalent. College, graduate, and professional school faculty frequently name reasoning skill as an area of deficiency in students and are sympathetic to attempts to include such skills in admissions assessment. Cognitive psychologists have been centrally interested in the cognitive processes of reasoning and problem solving and their attention has furthered this area as well.

Thus, although technological advances and concerns about the nature of the constructs being mesured are both increasing, current practice continues primarily as in the past. The economy and efficiency of multiple-choice group testing with machine scorable answer sheets makes it quite resistant to change. To replace it, any new assessment procedure must not only be educationally useful but also be practical, economic, and efficient. The time may be near when new computer technology will make possible those practical, economic, and efficient replacement methods. Although the tendency may be to transfer old constructs to the new methodology, this period of pending change begs as well for a reexamination of the familiar constructs. The remainder of this paper is devoted to a discussion of future conceptual and technical directions and some of the aspects that need to be reexamined.

CONCEPTUAL DIRECTIONS—THE NATURE OF ACHIEVEMENT AND ABILITY

The Traditional Achievement Construct

Traditionally the notion of achievement has been linked closely to the school curriculum. Achievement is expected in areas in which there has been instruction and, commonly, the content of instruction is used to determine areas of achievement to be assessed. Also, the names of areas of study are used to name the achievement tests. Teacher-made tests in schools are prototypical achievement tests and practically all standardized achievement tests justify their content on the basis of addressing common elements in a school curriculum.

Historically, achievement testing has been the concern primarily of measurement specialists with backgrounds in education, as opposed to psychology, and the definition of achievement has been closely linked, as noted, to educational goals rather than to psychological constructs. E. F. Lindquist and his colleagues and students, such as Robert Ebel, advanced one line of educational thought in which achievement of understanding of content areas and the application of that understanding to new contexts and to the solution of problems was the primary educational goal as well as the primary target of educational measurement (Ebel, 1974; Lindquist, 1951). Similar lines, still from the educational perspective, produced schemas such as Bloom's (1956) taxonomy of educational objectives as a blueprint for both educational goals and educational tests.

When achievement assessment is tied so closely to educational goals, its nature is subject to change as society's educational goals change. With the back-to-the-basics philosophy of the 1970s and with public attention primarily at the elementary school level, emphasis on shorter-term and often behavioral objectives in the classroom and in achievement testing became prominent. Criterion-referenced testing has been the popular term associated with such testing. During this period both the educational goals and the assessment became more specific and more directly tied to what was being taught in relatively short instructional periods. However, both the broader view of achievement with concern for understanding and application of knowledge and the more recent narrow one in terms of very specific instructional objectives have been viewed as extensions of educational goals (Haertel & Calfee, 1983) and have been largely unencumbered by elaborate theory.

The Traditional Ability Construct

In contrast to achievement, the notion of ability has a more complex history involving psychologists more than educators and closely tied to the concept of intelligence. The popular early notion of intelligence was as an innate, general cognitive ability. The development of the construct has centered around two primary issues: (1) the issue of intelligence as one general cognitive ability versus several general abilities versus many highly specific abilities, and (2) the issue of the extent to which intelligence is inherited.

Modern theories of intelligence began in the late-19th century with British writers who tended to view intelligence as predominantly unitary—a single or a very few general abilities. Later, theorists elaborated separate factors. Thurstone (1938) identified seven primary mental abilities: verbal comprehension, word fluency, number, reasoning, spatial visualization, perception speed, and memory. Guilford (1967), perhaps the most extreme of the multiple factor theorists, identified 120 separate ability factors in his structure of the intellect by crossing five operations by six products by four contents. Horn (1968) and Cattell (1971) proposed subdividing the general construct into crystallized intelligence (information-based and developed through environmental influences) and fluid intelligence (abstract and general and thought to be innate).

From a more educational perspective came definitions such as Thorndike's (1911) definition of intelligence as the ability to learn. Early stimulus-response (S-R) theorists saw intelligence as resulting from a build up of S-R bonds. From this trend Gagne (1970) proposed a hierarchical theory of eight kinds of learning from simple Pavlovian conditioning to rule learning and problem solving.

Many recent theories of intelligence have arisen from an information processing perspective. Sternberg's (1977) componential theory of intelligence is one example. A component, in this conception, is an elementary information process that may be classified by its function and level of generality. Higher order processing similar to general factor notions of intelligence are referred to as metacomponents in Sternberg's system. Other authors approaching the concept of intelligence from an information processing view include Hunt (1978), Pellegrino and Glaser (1979), and Snow (1979).

The issue of the extent to which intelligence is inherited began as well with early British theorists (Galton, 1883) who strongly supported the view that intelligence represented an innate ability. Relying heavily on the evidence from twin studies, the work of Burt (1940) and others supported a strong inherited component. More recently, some writers (e.g., Anastasi, 1976) have pointed to difficulties in twin studies and the many environmental influences on cognitive performance have received more emphasis.

Early notions of intelligence as largely perceptual and sensory (e.g., Galton, 1883) led to tests, measuring perceptual skills (e.g., Cattell, 1890). However, the more successful efforts to assess intelligence often arose in more practical than theoretical contexts. For example, Binet (Binet & Simon, 1905) tried to assess skills that would identify children who would have difficulty in a regular French school context. His efforts and other later assessments were often successful in the sense of identifying performances that related to other similar performances and predicted important subsequent school behaviors. Analyses of the results of most educationally related ability tests produced empirical evidence for general factors across types of performances consistent with many theories of intelligence. However, the evidence did not provide any resolution to the theoretical debates and, in fact, there was often little connection between the theoretical concerns and the development of intelligence and ability tests designed for practical use. An exception is the recent Kaufman Assessment Battery for Children (Kaufman & Kaufman, 1983) in which the theoretical notions of fluid and crystallized intelligence are applied.

Today we view the results of intelligence tests as showing what a person can now do based on learning and experience (i.e., a developed ability) without the implication of permanence or heredity it once held (Anastasi, 1981). We know such scores relate to many significant events from educational to job performance. We know that what is effective, competent functioning differs in some ways from culture to culture and may differ qualitatively at different developmental levels. We know that some theorists still focus on a single, general concept of intelligence but many other consider instead more specific abilities. And we know that measures we now have, though quite efficient and predictive, only approximate the construct we conceptually describe.

Distinctions Between Achievement and Ability

These descriptions of achievement and ability, though suggesting the educational content relatedness of the former and the more psychologically theory-based character of the latter, still give only hints of the similarities and differences inherent in measures of the two concepts. Yet there have been reasons to distinguish ability (intelligence, aptitude) from achievement for as long as there have been standardized tests, and conceptually the distinctions go back much farther. The constructs have been distinguished historically in terms of the relationship of each to heredity (i.e., the measurement of innate capacity versus the measurement of learned accomplishment). Currently, however, most writers recognize now that this distinction is largely inappropriate because both types of measure reflect current attainments learned though a variety of experiences. Another distinction often made concerns the use to be made of the information (i.e., to predict future performance versus to assess past learning). Yet another distinction involves the nature of the skills measured (i.e., generalized skills usually involving reasoning versus specific knowledge of information in particular content areas). Finally, distinctions have been made about the length of time required to learn the skill and the site of the learning with ability tests measuring skills learned over longer periods of time from many types of learning experiences with achievement tests measuring skills learned in shorter time periods, primarily in school. Anastasi (1976) and Cronbach (1970) described aptitude and achievement tests as falling at different locations on a continuum with the continuum reflecting the nature of the skills, source of the learning, and time of the learning period.

Many (e.g., Cooley & Lohnes, 1976) have noted that one cannot rely on a label on current tests to distinguish the types of skills, because many current ability tests ask for specific factual knowledge or specific instructed skills whereas many current achievement tests include questions requiring general reasoning skills and application of knowledge to new domains. The distinctions have become increasingly fuzzy as cognitive psychologists have directed attention to knowledge structures in specific content fields such as mathematics, reading, and science as typified by Glaser's (1981) examples of potential contributions of cognitive psychology to psychometrics. Thus, although there are many conceptual distinctions that can be made between achievement and ability, there remains considerable confusion, especially in relation to current test content.

Messick (1984) suggested that we begin directly with the constructs not with the possibly contaminated content of current tests. What do we and should we mean by the concept of educational achievement and that of educational ability? According to Messick, (1984),

> Educational achievement refers to what an individual *knows* and *can do* in a specified subject area. At issue is not merely the amount of knowledge accumulated but its organization or structure as a functional system for productive thinking, problem solving, and creative invention in the subject area as well as for further learning (p. 1).

Messick goes on to develop the idea of the organization or structure of knowledge as follows:

> A person's structure of knowledge in a subject area includes not only declarative knowledge about substance (or information about *what*), but also procedural knowledge about methods (or information about *how*), and strategic knowledge about alternatives for goal-setting and planning (or information about *which, when,* and possibly *why*).

Achievement is thus viewed as a knowledge structure and contrasts with ability which may be conceptualized as a process structure, a relatively stable constellation of psychological processes (Messick, 1982b) developed over time through learning into "a coherent set of habit skills, knowledge, conceptual developments, and tactical and strategic 'know how' " (Cattell, 1971, p. 319) or, in information processing terms, as assembly and control processes (Snow, 1980), functioning much like subroutines or prior assemblies in computer terms.

Not all writers agree with this dual formulation of achievement as knowledge structure and ability as process structure. Ebel (1969, 1974, 1982) described the knowledge structure similarly to Messick's definition, but did not distinguish the ability or process structures from the knowledge structure. In fact, Ebel seemed to argue that we have tended to call too many of the higher level achievements abilities when they are, in fact, a crucial part of achievement. He was especially concerned that we not leave to the achievement construct only lower level factual acquisition. Anastasi (1976, 1980, 1981) treated achievement within the notion of developed abilities and described achievement and ability within a continuum from recent, more factual, school-based performance (achievement) to longer term, high level, more generalized performance (ability).

In spite of the complexities in distinguishing abilities and achievements in practice, at this point it seems useful to accept the viability of aspects of both constructs with several reservations: (a) that we not relegate achievement to acquisition of facts and leave the complex accomplishments to the ability domain; (b) that we retain the learned nature of each; and (c) that we remain willing to mix the two constructs in our measurement efforts because we cannot measure achievement without ability being reflected nor ability without achievement in one or many domains.

Ability and achievement can be conceptualized as a matrix, with the knowledge structure in a content area (i.e., achievement) crossed with more general process structures (i.e., abilities). Such a matrix structure illustrates the conceptual distinction but also their interrelated nature when measured. Measurement occurs then as tasks are presented involving one or more than one cell in the matrix. When we assess factual recall we are at an information level on the achievement dimension and perhaps the memory level on the ability dimension. For high levels of achievement a person is presented with tasks requiring the higher levels of the knowledge structure called achievement and using advanced processes called abilities. Any educational performance can then be viewed as a combination of exercising a portion of a knowledge structure (i.e., achievement) and certain portions of a process structure (i.e., ability).

Research is occurring in many content areas directed toward defining the knowledge structure in each area. Although Bloom's (1956) taxonomy provides one type of general terminology that could be used to define the knowledge structure in a variety of subjects, most recent work seems quite specialized for the particular content domain and suggests a unique knowledge structure definition for each subject. For example, structures have been proposed for several subject areas (Glaser, 1981).

The ability structure is the focal interest of many current cognitive and information processing theorists. It may be defined by elemental to higher order information processes or components or in traditional terms such as memory, reasoning, etc. The ability dimension of the matrix is less content specific and one definition of the ability structure is likely to be useful in a number of cognitive content areas.

Even with such a cursory look at this conceptually complex area several features are clear:

1. Both achievement and ability as described here have tremendous educational significance and can even be characterized as the ultimate goals of education in terms of developing knowledge structures (achievement) and process structures (abilities) in students.
2. Both structures as described include high level cognitive activities and build upon lower level ones.
3. Both arise through learning often over a substantial period of time.
4. The two structures are intricately interrelated and each relates to the accomplishment of the other.

Implications of the Constructs for Education. A first implication in the consideration of these constructs for education is the importance of high level cognitive accomplishments in an adequate description of educational attainment. Although educators have long endorsed the importance of higher levels of learning such as reflected in Bloom's taxonomy, there have sometimes been demands to teach and test at lower levels of information and skill acquisition. With requirements for evaluation, people have tried to measure those things that can be taught successfully and to demonstrate that success in a short time by focusing on precisely definable and immediately observable instructional objectives.

There are indications of renewed concern for higher levels of learning. For example, public attention is now directed to concerns with excellence at the secondary education level (National Commission on Excellence in Education, 1983; Boyer, 1983). With concern for excellence at the high school level, it seems likely that concern with higher cognitive structures will occur.

A desirable direction for education is to move toward greater concern with the instruction and the assessment of higher level educational goals whether we call them comprehension, analysis, and evaluation as Bloom did, or knowledge structures with terms specific to each of the content domains. With such higher level goals, it will be necessary to address process structures as well as knowledge structures. It would likely be possible to address the development of process structures more directly educationally with such a focus. At these higher levels there may be diagnostic educational purposes for which we will wish to distinguish achievement from ability in order to assist student development or to predict future behaviors in which more general process structure accomplishments will be useful.

Finally, we would profit educationally from a better understanding of ability and achievement constructs, how they are distinguished, and how they work together. As one examines Messick's (1973, 1982, 1984) theoretical notions of knowledge structures and process structures, the interrelationship of the two in the learning process and their intermixing on many present day tests becomes easier to understand. However, the distinction is likely not an essential one for all time and all places. In education, particularly, as we better understand knowledge structures in various subject areas and study such things as word decoding, semantic access, sentence processing, and discourse analysis in reading or error patterns in mathematics and common misunderstandings of physical principles (Glaser, 1981), we may find achievement and abilities mixed together without loss in the mixture. The proper goal here is not clear and complete separation—likely an impossible goal—but better understanding of the interrelation and how tests reflect both. This understanding may help us concentrate more explicitly in the educational process on both the high level knowledge structures indicative of important achievement in a field and on the process structures that are developing through learning experiences as well.

TECHNOLOGICAL DIRECTIONS—USING COMPUTER TECHNOLOGY

In this section the focus is on one major technical development that holds the promise of a revolution in educational testing—computerized testing. Of course, the notion of computerized testing is not new. What makes it potentially revolutionary today is that it has become practically feasible with small computers linked to TV screens in essentially every school in the nation.

In this section, three major areas in which computerized testing may revolutionize testing are addressed. The revolution seems more certain to occur in the first of the three areas and seems desirable, if not certain, in the latter two. The three areas are:

1. Use of computers to handle clerical functions of administering tests (e.g., presenting stimuli, recording answers), scoring tests, and reporting results back immediately to students and teachers.
2. Use of computers for adaptive testing so that each individual is tested with different questions depending on previous responses.
3. Use of computers to develop entirely new types of questions involving more complex stimuli and responses which affect additional stimuli.

Computers As Clerks

Clerical functions have always been one of the strengths of the computer. Today, with computers in nearly every classroom, it seems virtually certain that clerical functions of managing instruction, keeping track of student progress, presenting and scoring practice and drill, and presenting and scoring tests will become commonplace within 10 to 20 years. Of course, there are complexities in accomplishing this goal (e.g., compatibility of different machines in the same school, sales and ownership of computer-based instruction and testing) but the complexities seem small compared to the opportunities for more efficient classroom operation in a form likely to be accepted by schools.

Adaptive Testing Via Computer

Throughout its history, standardized testing has been locked into fixed tests in which all individuals took all items regardless of their performance. Because one gets more information about an individual level of performance by testing near that person's performance level it was necessary for tests to have a wide spread of item difficulty when used with groups with broad spreads of performance levels. Hence, the better performers had to answer many questions too easy to provide much information on them and the poorer performers had to answer many questions too hard to help much in locating their performance level. Thus, in a 100 item test, maybe only 30 to 50 items were giving much information with respect to each individual.

Two things were needed to give each test taker the *right* 30 to 50 items: (a) a statistical procedure to guide the selection of which items to give each individual and to produce a final score on a scale comparable to the scores others taking different items would receive, and (b) hardware to record the individual's response to items, perform the analyses for the next item choice, and present the next item to the test taker. Item response theory has provided a needed statistical procedure and today's minicomputers and microcomputers provide the hardware.

Adaptive testing provides the opportunity not only to assess more efficiently the global achievement and ability accomplishments we now assess but to probe adaptively into the misconceptions underlying errors made to previous questions. The opportunities of fitting in such information directly into instruction are exciting indeed.

New Item Types by Computer

In this area we have only begun to tap the surface of the possibilities for developing entirely new item types. However, some possibilities can already be envisioned in the types of complex stimuli that are being presented in the arcade and home video games and in the possibilities for studying problem solving strategies or perceptual processes in learning.

A VIEW OF THE FUTURE

One of the rewards of preparing a paper on future directions is the license it gives for sharing one's own images of a future time. I am taking that license here to describe to you what I think educational achievement and ability testing might be like by about the beginning of the 21st century. What I do here, however, is to combine those events which I feel confident will occur with some about which I have considerably less confidence. Although to some this may be an uncomfortable vision, I view it positively with many opportunities for better education.

We begin this fantasy by entering an elementary school classroom. What strikes us first is that on every child's desk is a small computer and screen and at several locations in the room are large screens visible to the whole class. The teacher is talking to a group of 10 youngsters about social studies and uses the computer at hand to call up a short film strip on one of the large screens to illustrate a point. As the lesson ends, each child activates his or her own computer and is given questions over the material covered that day. When a mistake is made, a hint is given and the child tries again and each student may receive a different set of questions depending on earlier answers. The student also types in short answers to some questions. While this group of students moves on to lessons on science or art, the teacher calls up a display of the results of the children's performance on the questions asked, which includes an error analyses identifying particular children with incorrect conceptions or misunderstandings of the content. The results are summarized by various levels of knowledge from acquisition of facts to higher levels of learning with categories appropriate to the content area—the type of learning we used to call comprehension and application. Today's lesson was near the end of a unit and had more questions in the comprehension and application categories although several weeks ago the emphasis had been more on facts. For students who perform poorly on these higher levels, several special activities were suggested. The teacher notes which topics or activities to review the next day. Tomorrow the students will first review today's material, then hear from the teacher, then answer some questions over the entire unit.

While this lesson was going on, three students sat with headphones on, operating computers at their desks. These students were drilling on multiplication facts and since each differed in the way they learned these facts from other children (i.e., perviously they were called learning disabled), they were receiving auditory as well as visual stimuli as well as hints when they made mistakes and words of encouragement for correct answers. These three students had been identified as having difficulty learning the math facts in the usual way and had been given a special diagnostic procedure to identify process structure difficulties. Problems in visual memory had been identified and these students were receiving training to use their more effective auditory memory systems to compensate for the visual memory problems. After 10 minutes of this drill, the teacher called up the results for the three students separately. The computer report indicated how many facts they attempted, answered correctly, and the prominent errors.

That evening at home, each child would contact the school computer to get a report on that day's activities for the parents. These reports indicated each student's progress and areas needing special work and described the child's own individual homework assignment based on the day's performance. The parents could receive a special message from the teacher or leave one for the teacher for the next day.

Meanwhile the principal, with several classes at this grade, wishes to check on the children's progress in science in each classroom and calls up a summary report by classroom. Several times during the year the principal must report to the superintendent on the students' progress in all major content areas for sharing with the school board. Those reports will be due in 2 weeks and this check is to insure the students are on track. That report to the school board will include norm-referenced information on the students' standing in relation to students nationally but will require no special testing. Instead it uses particular questions from the student's regular daily and end-of-unit review questions administered by the computer. This report is organized as well by categories within the knowledge structure showing both information acquisition and higher levels as well. In addition, however, the test report from the computer notes that certain higher levels are not covered well by the computerized procedures and refers the principal to projects and reports evaluated by the teacher as a supplement to the computerized segment.

A mile away in the high school building, two juniors were preparing to begin their college admission tests. The school principal connected one to Princeton for the College Boards and one to Iowa City for the ACT using a voice identification procedure to identify the caller and begin the process. Both students took adaptive tests in the major areas of high school achievement required for college as defined slightly differently by two major commissions. Both would have been called achievement tests in the 1890s although there was a mix of process structures in the content areas of concern. Because of their adaptive nature, the basic survey tests required on 1½ hours to complete compared to 3 hours in years past. The remaining time was spent by the student preparing short written responses to questions requiring high levels of cognitive functioning in the content areas. Immediately, the survey scores would be transmitted along with the short answers to the colleges of the students' choice.

CONCLUDING REMARKS

Fantasy aside, it is an exciting time for educational achievement and ability testing with tremendous opportunity through technology to become *not* an add-on process done exclusively for evaluation or accountability reasons but an integral part of the instructional process much as homework assignments and ditto sheets now are. Once computerized we can then select the relevant information about students' performance for those evaluation and accountability purposes too, choosing the part of the information best suited to those uses. But even those functions become an ongoing monitoring process that can support the instructional enterprise, not a separate once a year event.

Even with these advances all our problems will not, however, be solved. It seems likely that classrooms of the future will involve primarily the familiar multiple choice format or open-ended versions which produce machine scorable answers even though we would wish to see many new item types as well. There will continue to be difficulty assessing many important educational goals. Perhaps the ultimate advantage of the technology is that, if correctly programmed, it can remind us of those areas not measured well in the system and requiring special teacher instructional attention and evaluation through other means.

In conclusion, it should be noted what the successful implementation of such a system will involve. As an integral part of the instructional system, its programming will require as much educational theory as psychometric theory. We will have to spend as much or more time learning how to ask questions to help children learn as in assessing their learning. The largest focus of our attention will be on achievement testing even though we may explicitly include under that label those ability processes involved in learning and accomplishing content domains. Educational testing with a predominant ability focus will occur on a smaller scale in a diagnostic mode and will result in educational prescriptions. Teachers will study testing, not as a separate subject, but as part of instructional methods. Teachers will be trained to spend their time considerably less on classroom management and drill and considerably more on activities related to higher levels of learning.

Let's hope we're up to the tasks ahead.

REFERENCES

Anastasi, A. (1976). *Psychological testing* (4th ed.). New York: Macmillan.

Anastasi, A. (1980). Abilities and the measurement of achievement. In W. B. Schrader (Ed.), *New directions for measurement: Measuring achievement: Progress over a decade—Proceedings of the 1979 ETS Invitational Conference.* San Francisco: Jossey-Bass.

Anastasi, A. (1981). Diverse effects of training on tests of academic intelligence. In B. F. Green (Ed.), *New Directions for testing and measurement: Issues in testing-coaching, disclosure, and ethnic bias,* (no. 11, pp. 5–20). San Francisco: Jossey-Bass.

Anderson, B. (1982). Test use today in elementary and secondary schools. In Wigdor and Garner (Ed.), *Ability testing: Uses, consequences, and controversies.* Part II: pp. 232–285) Documentation Section. Washington, DC: National Academy Press.

Binet, A., & Simon, T. (1905). Methods nouvelles pour le diagnostic du niveau intellectuel des anormaux. *Annee psychologique, 11,* 191–244.

Bloom, B. (Ed.) (1956). *Taxonomy of educational objectives, Handbook I: Cognitive domain.* New York: Longmans, Green & Co.

Boyer, E. L. (1983). *High school.* New York: Harper & Row.

Burt, C. (1940). *The factors of the mind.* London: University of London Press

Cattell, J. M. (1980). Mental tests and measurements. *Mind, 15,* 373–380.

Cattell, R. B. (1971). *Abilities: Their structure, growth, and action.* New York: Houghton Mifflin.

Cole, N. S. (1981). Bias in testing. *American Psychologist, 36,* 10, 1067–1077.

Cole, N. S. (1983, Jan.). Test bias: A review. *NERA Researcher,* pp. 3–12.

Cooley, W. W., & Lohnes, P. R. (1976). *Evaluation research in education.* New York: Irvington.

Cronbach, L. J. (1970). *Essentials of psychological testing.* Third Edition. New York: Harper & Row.

Ebel, R. L. (1969). Knowledge vs. ability in achievement testing. *Proceedings of the 1969 Invitational Conference on Testing Problems: Toward a theory of achievement measurement.* Princeton, NJ: Educational Testing Service.

Ebel, R. L. (1974). The relation of aptitude for learning to achievement in learning. In D. R. Green (Ed.), *The aptitude-achievement distinction.* Monterey, CA: CTB/McGraw-Hill.

Ebel, R. L. (1982). Writing achievement test items: Current issues. *Paper presented at the meeting of the American Psychological Association,* Washington, DC.

Gagne, R. M. (1970). *The conditions of learning.* Second Edition. New York: Holt, Rinehart, & Winston.

Galton, F. (1883). *Inquiries into human faculty and its development.* London: Macmillan.

Guilford, J. P. (1967). *The nature of human intelligence.* New York: McGraw-Hill.

Glaser, R. (1981). The future of testing: A research agenda for cognitive psychology and psychometrics. American Psychologist, *36*, 9, 923–936.

Haertel, E., & Calfee, R. (1983). School achievement: Thinking about what to test. *Journal of Educational Measurement, 20*, 2, 119–132.

Horn, J. L. (1968). Organization of abiliites and the development of intelligence. *Psychological Review, 75*, 242–259.

Houts, P. L. (1975). Standardized testing in America, II. *The National Elementary Principal, 54*, 2–3.

Hunt, E. B. (1978). Mechanics of verbal ability. *Psychological Review, 85*, 109–130

Jensen, A. R. (1980). *Bias in mental testing.* New York: The Free Press.

Kaufman, A. S., & Kaufman, N. L. (1983). *Kaufman Assessment Battery for Children.* Sampler Manual, Circle Pines, MN: American Guidance Services.

Lerner, B. (1981). The minimum competence testing movement: Social, scientific, and legal implications. *American Psychologist, 36*, 1057–1066.

Lindquist, E. F. (1951). Preliminary considerations in objective test construction. In E. F. Lindquist (Ed.) *Educational measurement* (pp. 119–158). Washington, DC: American Council on Education.

Madaus, G. F., & McDonagh, J. T. (1979). Minimum competency testing: Unexamined assumptions and unexplored negative outcomes. In R. T. Lennon (Ed.), *New directions for testing and measurement: Impactive changes on measurement,* (no. 3, pp. 1–14) San Francisco: Jossey-Bass.

Messick, S. (1973). Multivariate models of cognition and personality: The need for both process and structure in psychological theory and measurement. In J. R. Royce (Ed.), *Multivariate analysis and psychological theory.* New York: Academic Press.

Messick, S. (1984). Abilities and knowledge in educational achievement testing: An assessment of dynamic cognitive structures. In B. S. Plake (Ed.), *Buros/Nebraska symposium on measurement and testing. Social and technical issues in testing: Implications for test construction and usage.* Hillsdale, NJ: Lawrence Erlbaum Associates.

Messick, S. (1982). Developing abilities and knowledge: Style in the interplay of structure and process. *Educational Analysis, 4*, 105–121.

National Commission on Excellence in Education. (1983). *A nation at risk: The imperative for educational reform.* Washington, DC: U.S. Department of Education.

Pellegrino, J. W., & Glaser, R. (1979). Cognitive correlates and components in the analysis of individual differences. In R. J. Sternberg & D. K. Detterman (Eds.) *Human intelligence: Perspectives on its theory and measurement.* Norwood NJ: Ablex.

Snow, R. E. (1979). Theory and method for research on aptitude processes. In R. J. Sternberg & D. K. Detterman (Eds.) *Human Intelligence: Perspectives on its theory and measurement* (pp. 27–63). Norwood, NJ: Ablex.

Snow, R. E. (1980). Aptitude processes. In R. E. Snow, P. A. Federico, & W. E. Montague (Eds.), *Aptitude, learning, and instruction* (Vol. 1). Hillsdale, NJ: Lawrence Erlbaum Associates.

Sternberg, R. J. (1977). *Intelligence, information processing, and analogical reasoning: The componential analysis of human abilities.* Hillsdale, NJ: Lawrence Erlbaum Associates.

Thorndike, E. L. (1911). *Animal intelligence: Experimental studies.* New York, Macmillan.

Thurstone, L. L. (1938). *Primary mental abilities.* Chicago: University of Chicago Press.

Wigdor, A. K., & Garner, W. R. (Eds.) (1982). *Ability testing: Uses, consequences, and controversies.* Part 1: Report of the Committee. Washington, DC: National Academy Press.

5 Minimum Competency Testing: Status and Potential

Ronald A. Berk
The Johns Hopkins University

INTRODUCTION

Competency becomes an issue when one seemingly encounters incompetence and its consequences. For example, suppose an automobile is taken to a dealer for brake repair. Once the repair has presumably been completed, the owner drives the car to the first intersection, one block from the dealer, and the brake light in the dashboard appears. This owner would probably begin to question the competence of the attending mechanic. As consumers, employers, or even students, we witness countless other examples of probable incompetence.

It is this questioning of competence that provided the impetus for the minimum competency testing movement in this country. The movement which began in the 1970s developed in two distinct but interrelated fields: education and occupational licensing. In education, the public seriously questioned the meaning of the high school diploma and, in essence, the competence of a high school graduate. Coterminously, thousands of consumer complaints against licensed and certified practitioners brought into question the quality of services rendered and the conditions for relicensure. Although many of the issues in education and licensure are quite similar, especially in regard to assessment, only the competency movement in education will be reviewed here in order to avoid redundancy in this ,chapter and with Kane's chapter in this volume.

Minimum Competence

Despite the recency of the minimum competency testing movement, the concept of minimum competence is not new. It has been an integral part of occupational licensing in the United States for more than 200 years. Licensing is "the process by which an agency of government grants permission to an individual to engage in a given occupation upon finding that the applicant has attained the minimal degree of competency necessary to ensure that the public health, safety, and welfare will be reasonably well protected'' (U.S. Department of Health, Education, and Welfare, 1977, p. 4). This concern for ''minimal competency'' or ''minimum qualifications'' for safe practice underlies the state regulation of more than 800 occupations and professions (Greene & Gay, 1980; Shimberg, 1981). Individuals seeking licensure, for example, physicians, pilots, electricians, lightning rod installers, or horseshoers (see Shimberg, 1982a, chap. 1), are usually required to pass an examination in order to demonstrate their competence. Shimberg (1982a) has listed three responsibilities of licensing boards using such an examination:

1. the examination is a satisfactory measure of competence;
2. it measures the critical and important knowledge, skills, and abilities prerequisite to performance of the job at the minimum level of competence deemed necessary for the public protection; and
3. it is capable of screening out those who lack the requisite level of competence. (p. 56)

The educational analogue of these characteristics will become apparent in subsequent sections of this chapter.

Another perspective on the concept considers minimum competence in the context of United States social policies. Cohen and Haney (1980) have pointed to the longstanding interest in having government promote minimum levels of social welfare. Examples include public health programs, social security, unemployment insurance, welfare programs, and, certainly, a free public education.

Throughout the relatively brief history of the competency testing movement the expression *minimum competence* has engendered a considerable amount of confusion among lay people and educators alike. In practice, it connotes both the type of competence to be measured and the performance standard that is specified to designate attainment of the competencies. A further discussion of this topic is given in the section on ''Definitions.''

Grass Roots Movement

Earlier, it was indicated that the origin of the competency testing movement in education was public questioning of the competence of high school graduates. Public support transformed into legislative action has also been the sustaining force behind its continuation. As Lerner (1981) observed, minimum competency testing is a genuine grass roots movement that is ''clearly being led, or pushed, by noneducators'' (Pipho, 1978, p. 586). To date, the 39 state mandates for minimum competency testing programs were instigated by either legislative action or state board of education action, not by professional educators. Since the first programs were mandated in 1971 (Florida and Georgia), the competency testing movement has been viewed as an outgrowth of the increasing public clamor for accountability in the schools. Beard (1979) stated:

> [Minimum competency testing] has widespread popular appeal to citizens and politicians who see it as a way of holding schools accountable and forcing them "back to basics." These groups are convinced that the quality of public education has been eroded over a period of years and that high schools are graduating significant numbers of students who are unable to read and write, and consequently, unable to support themselves through gainful employment. (p. 9)

Several Gallup polls on public attitudes toward education over more than a decade furnish ample evidence of the massive public support for the movement (Gallup, 1978, 1979, 1980, 1981, 1982, 1983, 1984).

Evidence of Incompetence

One major purpose of minimum competency testing is to restore confidence in the high school diploma by requiring students to satisfy certain standards of basic competence. This focus stems from the accumulating evidence of incompetence in the 1970s. Declining SAT scores (College Entrance Examination Board, 1977) and increasing rates of illiteracy and semiliteracy among American teenagers (National Assessment of Educational Progress, 1976) led to widespread public disillusionment and dissatisfaction with the quality of the entire educational system. Complaints by employers that high school graduates were unable to complete job applications correctly were echoed with complaints by colleges and universities that the reading ability of a substantial number of incoming students was inadequate for college level work (Perkins, 1982), which necessitated the institution of remedial reading classes.

The 1983 report by the National Commission on Excellence in Education, titled "A Nation at Risk: The Imperative for Educational Reform," listed several indicators of risk that convey more dramatically the incompetence of American youth:

1. Some 73 million American adults are functionally illiterate by the simplest tests of everyday reading, writing and comprehension.
2. About 13% of all 17-year-olds in the United States can be considered functionally illiterate. Functional illiteracy among minority youth may run as high as 40%.
3. The College Board's Scholastic Aptitude Tests demonstrate a virtually unbroken decline from 1963 to 1980. Average verbal scores fell over 50 points and average mathematics scores dropped nearly 40 points.
4. Both the number and proportion of students demonstrating superior achievement on the SATs (i.e., those with scores of 650 or higher) have also dramatically declined.

5. Many 17-year-olds do not possess the "higher order" intellectual skills we should expect of them. Nearly 40% cannot draw inferences from written material; only one-fifth can write a persuasive essay; and only one-third can solve a mathematics problem requiring several steps.

6. Between 1975 and 1980, remedial mathematics courses in public four-year colleges increased by 72% and now constitute one-quarter of all mathematics courses taught in those institutions.

7. Business and military leaders complain that they are required to spend millions of dollars on costly remedial education and training programs in such basic skills as reading, writing, spelling, and computation. (pp. 8–9)

From these and many other indicators of risk cited in the report, the Commission concluded that more and more young people emerge from high school ready neither for college nor work. One recommendation was that state and local high school graduation requirements be strengthened and that, at a minimum, all students seeking a diploma be required to lay the foundations in the Five New Basics: English, mathematics, science, social studies, and computer science. The Commission stressed that whatever the student's educational or work objectives, knowledge of the New Basics is the foundation of success for the after-school years.

The pressing need for this re-emphasis on instruction and assessment of basic skills was expressed by the Commission:

> Individuals in our society who do not possess the levels of skill, literacy, and training essential to this new era will be effectively disenfranchised, not simply from the material rewards that accompany competent performance, but also from the chance to participate fully in our national life. (p. 7)

The seriousness of the consequences of incompetent high school graduates and meaningless diplomas was articulated by Lerner (1981):

> Functional literacy and/or numeracy is an essential prerequisite for the competent performance of almost all skilled jobs, blue-collar or white-collar, in the United States or in any other developed nation in the world today. (p. 1059)
>
> With or without diplomas, young Americans who leave school without basic skills face bleak futures. Some will manage to secure unskilled work on at least an intermittant basis. Many others will not, because without those basic skills, they are not just unemployed—they are for most practical purposes in today's economy, unemployable. (p. 1060)

Confronted with these facts, one must decide whether minimum competency testing programs can, at least, partially solve these educational problems or another approach will prove more effective. At present, there are no alternatives with the overwhelming public support accorded competency testing. More than that, however, the testing technology exists and teaching and testing for competence (or mastery) have sound theoretical bases.

This chapter reviews the current status of minimum competency testing and the issues that must be addressed for its future success. Special attention is given to the most thorny technical problems in competency test construction and score analysis and use. Specific recommendations are also offered for the more promising measurement techniques.

CURRENT STATUS OF MINIMUM COMPETENCY TESTING

This section assesses the state of minimum competency testing practices in relation to three major topics: (1) definitions, (2) policy specifications, and, (3) pros and cons.

Definitions

The burgeoning literature on minimum competency testing over the past decade has produced numerous and diverse definitions of *competency, minimum competency, minimum competency test,* and *minimum competency testing program.* Although it is easy to conclude that "there is no consistent terminology for minimum competency testing in use in the testing field" (Gorth and Perkins, 1979a, p. 8), there are certain key characteristics of all of the testing programs in operation that render the differences between the most popular definitions as trivial. Several of these definitions are presented in Table 5.1. A close inspection of the definitions along with analyses of the results of a nationwide survey of 31 states and 20 local district minimum competency testing programs conducted by Gorth and Perkins (1979b) and of a similar survey by Pipho (1983) reveal the following common features:

1. There is an emphasis on the acquisition of minimum skills or competence, usually academic skills (e.g., reading, math, writing) and/or life skills (e.g., following directions, filling out a job application, balancing a checkbook)
2. An explicit performance standard for pass-fail decisions is set so that one can separate the competent from the incompetent
3. The test results are used to make important decisions about individual students such as a promotion to a higher grade (or retention at the same grade), awarding of a high school diploma or a certificate of special recognition (or awarding a certificate of school attendance), or assignment to remedial classes.

These features are reflected in the definition adopted by the widely publicized National Institute of Education sponsored adversary evaluation hearing on the topic held in Washington, D.C., in July, 1981:

TABLE 5.1
Definitions of Minimum Competency Testing (Listed Chronologically)

Source	Definition
Elford (1977)	Minimum competency testing involves: (1) the use of objective, criterion-referenced competency tests; (2) the assessment of reading and computation using "real life" or "life skill" items; (3) the requirement of a specialized mastery level for high school graduation; (4) the early introduction of such testing for purposes of identification and remediation.
American Friends Service Committee (1978)	[Minimum competency testing programs are] organized efforts to make sure public school students are able to demonstrate their mastery of certain *minimum* skills needed to perform tasks they will routinely confront in adult life.
Airasian, Pedulla, and Madaus (1978)	[Minimum competency testing is] a certification mechanism whereby a pupil must demonstrate that he/she has mastered certain minimal (sic) skills in order to receive a high school diploma.
Miller (1978)	Minimum competency tests are constructed to measure the acquisition of competency or skills to or beyond a certain defined standard.
National School Boards Association (1978)	[Minimum competency testing programs are] testing programs which attempt to learn whether each student is at least "minimally competent" by the time the student graduates from public school.
Beard (1979)	Minimum competency testing involves the administration of proficiency tests in order to certify that minimum competency or proficiency exists with regard to a well-defined set of knowledge or skills.
Cohen and Haney (1980)	Nearly all minimum competency testing programs seek to define minimum learning outcomes for students in a variety of academic areas and to insure that these standards are satisfied
Lerner (1981)	[T]he common core of all minimum competence programs is an insistence on the cardinal importance of the basics--reading, writing, and arithmetic--and an equal insistence on hard, objective test data to measure success or failure in the acquisition of those fundamental intellectual tools by school children of all races, classes, and backgrounds.

> Minimum competency testing refers to programs mandated by a state or local body which have the following characteristics: (1) All or almost all students of designated grades are required to take paper-and-pencil tests designed to measure basic academic skills, life or survival skills, or functional literacy; (2) a passing score or standard for acceptable levels of student performance has been established; and (3) test results may be used to certify students for grade promotion, graduation, or diploma award; to classify students in remedial or other special services; to allocate compensatory funds to districts; to evaluate or to certify schools or school districts; or to evaluate teachers.

Obviously the most meaningful and useful strategy for tackling the construct of minimum competency has been to define it operationally in terms of program characteristics. Although such an approach does not help clarify *competency* in contrast to knowledge or performance (see Chickering & Claxton, 1981; Gale & Pol, 1975; Klemp, 1979; Senior, 1976; Shimberg, 1982b), it does direct attention toward the most crucial elements of the problem so that solutions can at least be attempted. The polemics over "what is competence" will probably continue for decades.

Policy Specifications

In most states, the policy of a minimum competency testing program is initiated in one of two ways: legislative action or state board of education action. The legislature may pass one or more laws stipulating the components and requirements of the program and/or the agents responsible for implementation. Alternatively, the state board may be empowered by state law to pass a mandate to establish the program. In a few cases where neither the legislature nor state board instituted the testing program, the local board of education may pass a mandate to initiate a program in its district (e.g., Denver, CO, Gary, IN).

The Gorth and Perkins (1979b) survey of minimum competency testing programs cited previously gathered information on the policy specifications in 31 states. More recently, Pipho (1983) updated that information on 39 states. The type of initial mandate (legislature or state board), the date of that mandate, and the date of completed implementation are identified for each state in Table 5.2. State board action (26 states) was the more frequent source of mandate compared to legislative action (15 states). While most of the mandates were passed by 1979, the implementation of a majority of the programs (22 states) will not be completed until the 1980s.

In addition to the data on state mandates, Table 4.2 displays how the test results are to be used. Only 15 of the states require a student to pass the competency test in order to receive a high school diploma. Arizona, California, Florida, and Maryland also require satisfactory test performance to advance to a higher grade level. In Illinois and New Hampshire both uses of the test scores are optional. For students who pass the test in 20 states, either standard diplomas are

TABLE 5.2
Minimum Competency Testing Policy Specifications in 39 States[a]

State	Mandate: Legis-lature	Mandate: State Board	Mandate: Date	Use of Test Results[b]: Imple.	Use of Test Results[b]: H.S. Diploma	Use of Test Results[b]: Special Recog.	Use of Test Results[b]: Grade Promo-tion	Use of Test Results[b]: Remed.
Alabama		X	1977	1984	X	U		O
Arizona	X	X	1972/75	1976	X		X	
Arkansas	X		1979	1982				M
California	X		1976/81	1980	X		X	M
Colorado	X		1975		O			
Connecticut	X		1978	1980				M
Delaware		X	1978	1981	X			
Florida	X		1971/75/76	1983	X		X	M
Georgia		X	1971/76/77/78	1978				O
Idaho		X	1977	1982	O	X		
Illinois		X	1978	1980	O	O	O	O
Indiana		X	1978					O
Kansas	X		1978/81	1980				O
Kentucky	X		1978	1979				O
Louisiana	X		1979	1980			X	
Maine	X		1978	1980				O
Maryland		X	1976/77/78	1986	X		X	M
Massachusetts		X	1978	1981				O
Michigan		X	1969	1979				O
Missouri		X	1976/78	1979				O
Nebraska		X	1975	1975				O
Nevada	X		1977/79	1979	X			M
New Hampshire		X	1977	1984	O		O	
New Jersey	X	X	1976/79	1978/85	X			M
New Mexico		X	1976	1979		X		M
New York		X	1978/79	1981	X			M
North Carolina	X		1977	1979	X			M
Ohio		X	1982					
Oklahoma		X	1977	1978				
Oregon		X	1972/76/80	1978	X			O
Rhode Island		X	1978	1981				
South Carolina	X		1978	1985				M
Tennessee		X	1977/80/82	1982	X			M
Texas		X	1979	1978				M
Utah		X	1977	1980	X			M
Vermont		X	1977	1979	X			O
Virginia	X		1975	1980	X			M
Wisconsin	X		1982		O			M
Wyoming		X	1977/79	1980				M

[a] Sources: Adapted from Gorth and Perkins (1979b, p.2) and Pipho (1983) with permission of the authors.

[b] The specifications on the use of test results indicate characteristic of program (X), mandatory (M), optional/conditional (O), or undecided (U).

awarded or diplomas with an endorsement or certificate of competency achievement (special recognition) are given. In the 15 states that mandate the test for graduation, the students who fail receive only a certificate signifying completion of all other requirements (attendance or course credits). In Florida, Commissioner Ralph Turlington estimated that 1300 seniors across the state (about 2%) fell into this category in 1983 (Cody, 1983). Remediation for students who fail the test is mandatory in 17 states (including Florida) and optional in 13. Remediation may take the form of providing extra attention and special remedial materials in the regular classroom setting, remedial classes, or other methods to address the skill deficiencies.

Pros and Cons

From the characteristics of the testing programs and the movement described thus far, it should be apparent that minimum competency testing is politically motivated and educationally implemented. As such, it has become a hotly debated topic with growing numbers of proponents (e.g., Beard, 1979; Fisher, 1978; Fremer, 1978; Lerner, 1981; Popham, 1981a) and opponents (e.g., Airasian, Madaus, & Pedulla, 1979; Glass, 1978a, 1978b; Haney & Madaus, 1978; Jaeger, 1982; Lazarus, 1981; Linn, Madaus, & Pedulla, 1982; Madaus, 1981). The proponents argue for its potential benefits and the opponents argue about its potentially harmful effects. Since an extended discussion of pros and cons is beyond the scope of this chapter, only a brief summary of some of the major arguments is given. Interested readers should consult the above references and the transcripts or videotapes of the National Institute of Education hearing on minimum competency testing (National Institute of Education, 1981; Thurston & House, 1981) for a detailed account of the pro (Popham, 1981a) and con (Madaus, 1981) issues.

Perhaps the most up-to-date and comprehensive summary of arguments on both sides is the list of perceived benefits and perceived costs of minimum competency testing completed by Perkins (1982). It is reproduced here in Table 5.3. The benefits seem to be related to five key assertions (Gorth & Perkins, 1979a): "(1) restore confidence in the high school diploma, (2) involve the public in education, (3) improve teaching and learning, (4) serve a diagnostic, remedial function, and (5) provide a mechanism of accountability" (p. 12). The costs tend to concentrate on the harmful effects of the testing on students, teachers and administrators, the curriculum, and the control of education.

Inasmuch as the opposing arguments on minimum competency testing are irreconcilable at this time, although the representatives on each side are convinced that they are right, policy makers should weigh carefully the advantages and disadvantages and then decide for themselves what is the most appropriate course of action. The 50 arguments in Table 5.3 should help guide that informed decision.

TECHNICAL SPECIFICATIONS FOR MINIMUM COMPETENCY TESTS

Another area where "benefits and costs" may be applied is the minimum competence test itself and its technical specifications. There are several strategies now available for constructing a competency test and for assessing the validity and reliability of the scores. The testing technology is derived from the research on criterion-referenced measurement (Berk, 1980c, 1984b, 1984c; Hambleton, Swaminathan, Algina, & Coulson, 1978; Popham, 1978a).

TABLE 5.3
Perceived Benefits and Costs of Minimum Competency Testing[a]

Benefits	Costs
1. restores meaning to a high school diploma	1. emphasis on the practical will lead to an erosion of liberal education
2. reestablishes public confidence in the schools	2. causes less attention to be paid to difficult-to-measure learning outcomes
3. impels us to face squarely the question of "what is a high school education?"	3. promotes teaching to the test
4. sets meaningful standards for diploma award and grade promotion	4. will be the "deathknell for the inquiry approach to education"
5. challenges the validity of using seat time and course credits as basis for certifying student accomplishments	5. oversimplifies issues of defining competencies and standards and of granting credentials to students
6. certifies that students have specific minimum competencies	6. promotes confusion as to the meaning of the high school diploma when competency definition is left to local districts
7. involves the public and local educators in defining educational standards and goals	7. fails to adequately consider community disagreement over the nature and difficulty of competencies
8. focuses the resources of a school district on a clear set of goals	8. will exclude more children from schools and further stigmatize underachievers
9. defines more precisely what skills must be taught and learned for students, parents, and teachers	9. will cause "minimums" to become "maximums," thus failing to provide enough instructional challenge in school
10. promotes carefully organized teaching and carefully designed sequential learning	10. may unfairly label students and cause more of the "less able" to be retained
11. reemphasizes basic skills instruction	11. may cause an increase in dropouts, depending on the minimum that is set
12. helps promote competencies of life after school	12. provides no recognition of the "average" student

13. broadens educational alternatives and options	13. fails to provide alternatives that can "inspire" average students to excel in some areas
14. motivates students to master basic reading, mathematics, and writing skills	14. ignores the special needs of gifted students, giving them less opportunity to be challenged and to expand their horizons
15. stimulates teachers and students to put forth their best efforts	15. may have adverse impact on a student's future career as a result of a withheld diploma
16. identifies students lacking basic skills at an early stage	16. may promote bias against racial, ethnic, and/or special needs groups
17. encourages revision of courses to correct identifies skill deficiencies	17. places the burden of "failure" on the student
18. ensures that schools help those students who have the greatest educational need	18. causes educators to be held unfairly accountable
19. can bring about cohesiveness in teacher training	19. intensifies the conflict for educators between humaneness and accountability
20. can truly individualize instruction	20. increases the record-keeping burden for administrators
21. shifts priorities from process to product	21. does not assure that students will receive effective remediation
22. holds schools accountable for educational process	22. does not assure that all the perceived needs and benefits will be met and realized
23. furnishes information to the public about performance of educational institutions	23. promotes the power of the state at the expense of local district autonomy
24. provides students with an opportunity to remedy the effects of discrimination	24. can be costly, expecially where implementation and remediation are concerned
25. provides greater holding power for students in the senior year	
26. provides for easier allocation of resources	

[a]Source: Reprinted with permission of the author and the National Council on Measurement in Education from Perkins, Marcy. "Minimum Competency Testing: What? Why? Why Not" *Educational Measurement: Issues and Practice*, Winter, 1982, pp. 5-9 (Tables II and III, pp. 7-8). Copyright 1982, National Council on Measurement in Education, Washington, DC.

Before proceeding with an examination of the pertinent technical topics, the following definition is proposed:

> *A minimum competency test is designed to determine a student's performance level with respect to a well-defined domain of competencies and prespecified performance standard.*

Two elements in this definition, competencies and standards, are consistent with most definitions of minimum competency testing (see Table 5.1) and the common features of existing programs which were described previously. In addition, the definition focuses on an individual student's score. The uses of the score for decisions about graduation or promotion are reflected in the types of validity and reliability evidence gathered as the test is developed.

Technical Standards for Competency Test Construction and Score Use

For the construction and use of norm-referenced tests, the *Standards for Educational and Psychological Tests* (APA/AERA/NCME Joint Committee, 1974) has served as the guide and, in essence, the "bible" of acceptable measurement practices. No single set of standards established by a joint committee of national experts is available for minimum competency tests. While it is possible to search through the *Standards* and glean some standards relevant to competency or certification tests, the product of this effort will be far from adequate. The fourth draft (February 1984) of the next edition of the *Standards,* titled the *Joint Technical Standards for Educational and Psychological Testing,* which is to be published in 1985, suggests that a separate section will be devoted exclusively to standards for certification testing in elementary and secondary education.[1] These standards and the collection of standards devised by Hambleton (1982; Hambleton & Eignor, 1978) for criterion-referenced tests will provide the foundation for the technical specifications and issues discussed hereafter.

Among the steps in the development of a minimum competency test (see Hambleton & Eignor, 1980), five are particularly troublesome: (1) defining the domain of competencies, (2) setting the performance standard, (3) gathering appropriate validity evidence, (4) estimating the reliability of the scores and decisions, and (5) equating the scores on different test forms. They are troublesome because there are many methods or statistical procedures one can use at each step and there is no consensus on any best method. Reviews of these technical areas by Hambleton and Eignor (1980), Shepard (1980b), and Jaeger (in press) furnish a few guidelines that may be helpful. Their recommendations will be integrated into this presentation.

[1]There will also be a section on professional and occupational licensure and certification.

Defining the Domain of Competencies

The first step in the construction of a minimum competency test is the specification of what the test is to measure. If one cannot define clearly what the test measures, then the resulting scores will be virtually meaningless. From the standpoint of score interpretation, a score *must* be referenced to the domain of competencies prior to any other type of referencing. For example, a teacher might say that Joanna's score on the test tells that she acquired 80% of the functional reading skills in the areas of survival signs, directional vocabulary, map symbols, and simple forms.

Academic Skills or Life Skills

The types of competencies typically measured by minimum competency tests are academic skills and/or life skills. The academic skills are those which have been traditionally taught in school, usually reading, mathematics, and writing. The life or survival skills often involve the transfer and application of the academic skills to practical "life-like" situations. In the preceding example, a simple form such as a bicycle registration form might be used to test the application of reading skills to a real situation that the student would encounter outside of the school environment.

The Gorth and Perkins (1979b) and Pipho (1983) surveys indicated that among 39 states, 18 states assess both academic and life skills in their minimum competency testing programs. The results are summarized in Table 5.4. All of the other states, with the exception of Georgia, emphasize academic skills only. In other words, almost every state includes academic skills and almost half of the states also test life skills. As shown in the table, the primary subject area coverage of the academic skills is the basic skills or "three Rs." Only six states have measured speaking and/or listening skills. Just what skills comprise the domain of competencies is determined most often by a special state level committee. Parents and citiziens either serve as members of this committee or are surveyed for their reaction to the domain definition.

Traditional Approach

The rigor and precision with which the domain is defined can enhance or diminish the score interpretation. The interpretation may be vague or explicit. Since the 1970s, the leading proponents of criterion-referenced and minimum competency tests have argued that the traditional approach to defining a content domain, which includes a content outline, a list of objectives, and a table of specifications or similar "blueprint," tends to provide an *ambiguous* domain definition. The arguments focus on the subjectivity involved in composing those specifications. That is, the selection of competencies and objectives is quite arbitrary, typically representing only one conceptualization of the domain, that adopted by the state level committee. Such specifications would be open to different interpretations by different teachers, administrators, students, and par-

TABLE 5.4
Types of Competencies Measured by Minimum Competency Testing Programs in 39 States[a]

	Emphasis			Subject Area[b]					
State	Academic Skills	Life Skills	Both	Reading	Math	Writing	Speaking	Listening	Other
Alabama			X	X	X	X			
Arizona	X			X	X	X			
Arkansas	X			X	X				
Calif.			X	X	X	X			X
Colorado			O	O	O	O			
Conn.			X	X	X	X			X
Delaware	X			X	X	X			
Florida			X	X	X	X			
Georgia		X		X	X				X
Idaho	X			X	X	X			X
Illinois			X	X	X	X	X	X	X
Indiana	X			X	X	X			X
Kansas			X	X	X				
Kentucky	X			X	X	X			X
Louisiana	X			X	X	X			X
Maine	X			X	X	X			
Maryland			X	X	X	X			X
Mass.			X	X	X	X	X	X	
Michigan	X			X	X	X		X	X
Missouri			X	X	X				X
Nebraska	X			X	X	X			O
Nevada	X			X	X	X			
New Hamp.	X			X	X	X			X
New Jer.	X			X	X	X			
New Mex.			X	X	X	X			
New York	X			X	X	X			
N. Caro.			X	X	X				
Ohio	X			X	X	X			
Oklahoma			X	X					
Oregon			X	X	X	X	X	X	X
Rhode Is.			X	X	X				
S. Carol.			X	X	X	X			X
Tennessee	X			X	X				X
Texas	X			X	X	X			X
Utah			X	X	X	X	X	X	X
Vermont			X	X	X	X	X	X	X
Virginia			X	X	X				X
Wisconsin	X			X	X				X
Wyoming	X			X	X	X			X

a Sources: Adapted from Gorth and Perkins (1979b, p. 3) and Pipho (1983) with permission of the authors.

b The specifications on the subject area indicate characteristic of program (X) or optional/conditional (O).

ents. Even if the academic and/or life skills are selected and reviewed by these groups, the ambiguity still remains, particularly when one attempts to identify those skills needed by adults to "function" or to "survive" in society. Haney and Madaus (1978) noted:

> People function differently in society, and some do it in ways offensive to others. Are we interested in the "essential skills" of the librarian or the lawyer, the bureaucrat or the baker, the con artist or the congressman? Would prisoners be considered as "functioning" in society? Or pardoned politicians? Even if we could reach agreement on what constitutes success (for example, functioning at a high level of competency) and what constitutes minimum functioning in society, their determinants are simply not very well understood. We do know, for example, that success in school seems not to be a very good predictor of success in later life—at least as measured by social scientists. (p. 465)

For the present, it appears that consensus on what constitutes the domain may be attained by some group which is representative of the lay public and professional educators in a given state, but the inherent ambiguity in an objectives-based definition can not be removed.

Coupled with this criticism is the charge that traditional item construction procedures are also ambiguous inasmuch as they can result in a set of items that manifest the biases and idiosyncrasies of each test maker. Consequently, different test makers would probably develop different items from the same specifications.

Domain Specification Strategies

Although the traditional approach is predominant in minimum competency testing programs, perhaps for practical reasons, several new strategies have been devised in order to overcome the aforementioned deficiencies: (1) amplified objectives, (2) IOX test specifications, (3) item transformations, (4) item forms, (5) algorithms, and (6) mapping sentences (for details, see Millman, 1980; Popham, 1981b, 1984; Roid, 1984; Roid & Haladyna, 1982). The characteristics of these strategies are outlined in Table 5.5. Clearly, the applications have been restricted to reading or mathematics, or both. The first two objectives-based strategies are more adaptable than item transformations, item forms, and algorithms. Although mapping sentences can be applied to most any domain, there are few examples of its utility. As the project applications suggest, life skills is a relatively unexplored domain. Recent developments on other approaches such as Tiemann and Markle's (1983) system derived from the research on concept and rule learning are also rather limited.

One underlying purpose of these strategies is to provide an *unambiguous* definition of a domain by implicitly or explicitly delineating sets of rules for generating test items, such that any two test makers would construct identical items from the same specifications. However, the extent to which the strategies can actually supply an unambiguous link between a domain of competencies and the corresponding test items varies markedly from one strategy to the other (Berk, 1980a). For building minimum competency tests, the ambiguity in defining what are "basic," "essential," "functional," or "survival" skills is still problematic. Regardless of how objective or mechanical the strategies in Table 5.5 operate in producing the items, the subjectivity used to arrive at the competencies remains.

Setting the Performance Standard

Setting the standard for minimum competence is the most important and the stickiest technical topic in minimum competency testing. Since the standard is the point of decision making and the basis for inferences about individual competency, the state department of education must be concerned about whether it provides a foundation for accurate, fair, and equitable decisions. If that founda-

TABLE 5.5
General and Technical Characteristics of Six Strategies for Defining the Domain of Competencies[a]

Strategy	Developer(s)	Project Application/ Domain(s)	Rule Structure	Major Components	Item Domain[c]
Amplified Objectives	Baker (1974) Popham (1974)	IOX Test Development/ Reading, Language, Mathematics, Social Studies	Traditional Item Construction Rules	Objective Sample Item Testing Situation Response Alternatives Criterion for Correctness	Infinite
IOX Test Specifications	Popham (1978a, 1984)	IOX Test Development/ Reading, Language, Mathematics, Social Studies, Affective, Psychomotor	Traditional Item Construction Rules	General Description Sample Item Stimulus Attributes Response Attributes Specification Supplement	Infinite
Item Transformations	Anderson (1972) Bormuth (1970) Conoley & O'Neil (1979) Finn (1975)	UCLA Readability Project/ Reading	Transformational Rules	Base and Derived Sentence	Finite
Item Forms	Hively et al. (1973) Osburn (1968)	MINNEMAST Project/ Mathematics	Generation Rules	Shell Replacement Sets	Finite
Algorithms	Scandura (1973, 1977)	MERG Projects[b]/Mathematics, Reading	Rules of Competence (Higher and lower order)	Equivalence Classes	Finite
Mapping Sentences	Berk (1978) Castro & Jordan (1977) Guttman (1970) Jordan (1978) Schlesinger (1978)	Cross Cultural Research/ Attitudes	Facet Design and Item Construction Rules	Facets Facet Elements Semantic Profiles	Infinite

[a]Source: Adapted from Berk (1980a, p. 51) by permission of Educational Technology Publications, Englewood Cliffs, NJ.
[b]The Mathematics Education Research Group (MERG) Projects provided the bases for most of the empirical research on algorithms.
[c]While an item domain may be viewed theoretically as either infinite or finite regardless of the particular strategy, the distinction between the two types of domain is intended to draw attention to the relative precision of the strategies and the need to consider that characteristic in their application.

tion is shaky, one will inevitably confront the consequences of inaccurate, unfair, or inequitable decisions in the school board room or courtroom.

The performance standard can be expressed as a number (24 out of 30 items), as a percentage (80%), or as a proportion (.80) of the items an individual must answer correctly.[2] The number which is based on the specific item sample measuring a single objective or a cluster of objectives (e.g., total test) is commonly referred to as the cutoff or passing score. It is the score that cuts the score distribution in two mutually exclusive categories: one category containing the scores from which "competency" is inferred and a second category containing the scores from which "incompetency" is inferred. Individuals who are labeled competent must score at or above the cutoff score; those who are labeled incompetent score below the cutoff score.

Although the percentage and proportion correct have been used interchangeably with the term cutoff score, they should be reserved more appropriately for the standard of performance in the item domain.[3] That is, if an individual can answer correctly 24 items in the 30-item sample, it is expected that he or she should be able to answer correctly about 80% of the items in the domain. If the domain happens to consist of 150 items, than 120 items or more should be answered correctly.

The responsibility for setting the standard on a minimum competency test resides with the state in about 80% of the cases (Gorth & Perkins, 1979b). In most other cases, the local districts set the standard. Very often the standard is specified for each subject area measured on the test and for each subset of items comprising a subject area. Only Connecticut and Tennessee are required to set a passing score for the total test.

While the polemics over certain issues in standard setting are far from over, at present there seems to be consensus among the experts on standard setting on at least one issue—*all of the methods involve some form of human judgment.* A completely objective, scientifically precise method does not exist (see Rowley, 1982). Regardless of how complex and statistically sophisticated a method might be, judgment plays a role in the determination of the cutoff score and/or in setting acceptable classification error rates. However, when a legislature sets a standard such as 80% without any foundation or reason, the judgment is capricious. This is the weakest and least defensible approach to standard setting. For its lack of any logical, experiential, or empirical justification, it has been characterized as the "cardiac approach" (Berk, 1979b, 1983), i.e., I know in my heart that she is competent and he is incompetent.

[2]Alternatively, a standard may be expressed as the number or percentage of competencies mastered. Multiple standards or cutoffs may also be used. However, these interpretations are less frequent in minimum competency testing programs than the number of items a student must answer correctly.

[3]The observed percentage correct is not necessarily the best estimate of the domain percentage correct.

Deficiencies of the "Cardiac Approach"

The deficiencies or problems associated with this approach are numerous. In order to appreciate the serious ramifications of decisions made on the basis of that type of standard, a few of the problems are specified below in terms of competency testing practices:

1. *An individual's pass-fail performance on the test has no meaning.* If an individual passes the test, there is no way of knowing whether he or she truly possesses the necessary skills (e.g., academic, life, survival, job-specific). The relationship between the performance standard and competence-incompetence on the actual skills is indeterminable. If an individual does well or poorly, there is no way to explain why.

2. *The percentage of individuals passing the test has no meaning.* This information which is simply an aggregate of individual performance data is supposed to indicate the overall competency of the group (e.g., the percentage that can be certified) and often the effectiveness of the instructional program as well. For example, if 70% of the 11th graders passed a minimum competency test as a requirement for graduation, no explanation of this percentage in terms of competence is possible. Certainly anyone can attach any meanings that they wish; negative inferences would be as unjustified and unfounded as positive ones.

3. *The standard does not reflect the difficulty or complexity of the items* measuring a single objective, a collection of different objectives or the total construct. Given the probable variability in item difficulty levels, an 80% standard may be easily attainable in some objectives or tests and highly unrealistic or unattainable in others.

4. Coupled with this insensitivity to difficulty is *the unavailability of any performance data on how individuals who are judged to be competent* (by their teacher or immediate supervisor) *in their position actually score on the test.* This information is essential to assess whether the standard is too high or too low. It would also provide a means of linking the standard to competency on particular skills.

5. Probably the most unfortunate consequences of using a completely arbitrary standard are the *incorrect, unfair, and inequitable decisions that could be made in individual promotions and graduation certification.* The cardiac approach precludes the estimation of decision accuracy, fairness, and equity. For example, the incorrect decision of denying a high school diploma to an individual who is truly competent (false incompetency error) suggests not only that the individual may be labeled as a failure, but also that the competency test failure may eliminate many potential opportunities and jobs for which that individual might otherwise be qualified. The seriousness of this problem becomes accentuated when one considers that the approach does not permit the decision maker even to estimate how many individuals have been mistakenly promoted or

certified or how many have unjustifiably been denied promotion or certification based on their competency test performance.

Clearly any standard setting method that is recommended as a substitute for the "cardiac approach" must address these problems. Specific criteria by which one can appraise the adequacy of a method will be delineated shortly.

On the spectrum of practicability, ranging from the simplest "cardiac approach" to the most complex Bayesian models, there are more than 30 different standard setting methods (Berk, 1985). Several extensive reviews of these methods have been conducted by Hambleton and his colleagues (Hambleton, 1980; Hambleton & Eignor, 1980; Hambleton & Powell, 1983), Meskauskas (1976), Shepard (1980a, 1980b, 1983, 1984), and Berk (1985). A few summaries, more limited in scope, have also been presented by Berk (1980d), Popham (1978b, 1981b, chap. 16), Livingston & Zieky (1982), and Jaeger (in press). The review of standard setting methods which follows will build on the structure, content, and insights proffered by these earlier works. In order to expedite a more perceptive selection of standard setting methods and to increase the use of the better methods by competency test makers, *criteria* for judging their quality and a *framework* for choosing the most appropriate method need to be developed. The next two sections are devoted to these considerations.

Criteria for a Defensible Standard Setting Method

In view of the aforestated deficiencies of the most popular standard setting method and the requirements of current competency testing programs, a defensible standard setting method should ultimately satisfy the following criteria:

1. Given the variation in the difficulty and complexity of the skills measured by competency tests, the method should be sensitive to the different difficulty levels;
2. Given the variation in the lengths of the tests and their component subtests, the method should be flexible for application to different test lengths:
3. Given the design and overall intent of competency tests, the method should be directly linkable to the performance of individuals who use the skills that are measured by the test in school or on the job;
4. Given the types of decisions for which the competency tests are used, the method should produce classifications of competence and incompetence for the different score continua;
5. Given the need for evidence to defend the accuracy of the decisions based on the standard, the method should provide estimates of probabilities of correct classification decisions and decision errors for any score point in the different score continua;
6. Given the various professional educators and lay people who will need to defend the method and to interpret the results on individuals and programs, the

method should be intuitively sound and conceptually simple, and the results should be easily interpretable;

7. Given the typical practical problems and constraints in educational settings, the method should be practicable in terms of execution and available resources and should be computationally simple.

Recent court decisions pertaining to the choice of a performance standard for a teacher certification test (National Teacher Examination) indicated that in order for the standard to be judged valid, it must be logical and be related to a specific level of job performance (see *Georgia Association of Educators* v. *Nix,* 1976; *United States* v. *State of North Carolina,* 1975, 1977; *United States* v. *State of South Carolina,* 1977). The implications of those decisions for setting minimum competency standards are expressed in criteria 3, 4, and 5. Criteria 1 and 2 focus on the sensitivity of a standard setting method to technical characteristics of competency tests (e.g., difficulty level, test length). The last two criteria stress the utility and practicability of a method. While it may be difficult for any single standard setting method to satisfy all of the criteria, certain criteria should be met so that the method might be defensible legally. Primary emphasis should be placed on criteria 3 and 4 (cf. Bernknopf, Curry, & Bashaw, 1979), and secondary weight should be assigned to criterion 5. The evidence gathered in support of decision accuracy, however, would be highly desirable, where possible.

A Framework for Standard Setting Methods

Numerous classification schemes have been devised to facilitate the study, interpretation, and use of cutoff score methods. From these schemes and the characteristics of the methods, Berk (1980d) derived a rather simple bilevel framework for classifying most available approaches. The first level, adopted from Meskauskas' (1976) review, partitions the methods into two major categories based on their assumptions about the acquisition of the underlying trait or ability: *state models* and *continuum models.* The second level, adopted in part from Hambleton's (1980) review, classifies the methods according to whether they are based purely on judgment or incorporate both judgmental and empirical information: *judgmental methods* and *judgmental-empirical methods/models* (see also Berk, 1985, for an extension of this classification). There are certainly other features that test makers need to consider, such as the definition of the internal or external criterion variable, the type of data, the distribution assumptions, and the specifications of a loss function (utility analysis). However, in the interest of parsimony, the bilevel framework should prove adequate for an analysis of the major methodological issues and to guide the selection of the type of method appropriate for decisions of grade level promotion and high school graduation certification.

The first step toward deducing which standard setting method is best for a particular competency test and decision application is to determine which general standard setting category is most appropriate: state or continuum. The key factor in this determination is the assumption regarding the acquistion of the underlying ability.

State models assume that competence or true-score performance is an all-or-nothing state; the standard is set at 100%. Deviations from this true state are presumed attributable to "intrusion" (false competency) and/or "omission" (false incompetency) errors. After a consideration of these errors, the standard is adjusted to values less than 100%. Glass (1978c) referred to these models as "counting backwards from 100%" (p. 244). Unfortunately, this all-or-nothing assumption is implausible, unrealistic, or difficult to apply to the academic and life skill domains measured by minimum competency tests. Competence is usually conceptualized in "degrees" such that it could be defined at any number of points on a test score continuum.

Continuum models assume that competence is a continuously distributed ability that can be viewed as an interval on a continuum, i.e., an area at the upper end of the continuum circumscribes the boundaries for competence. This conceptualization appears to fit the design and intent of most competency tests.

Given this initial assessment of the two standard setting categories in terms of current practices in competency testing, only a brief description of some state models and a more extensive description of those continuum models with the greatest potential for addressing the standard setting problem will be presented in the succeeding sections.

State Models of Standard Setting

Although a considerable amount of research has accumulated on standard setting, state models have received relatively little attention. Macready and Dayton (1980) have provided the most comprehensive survey of state models to date. The sources for these models are listed in Table 5.6. Although they claim that the models are nonjudgmental in nature, those models possess many of the

TABLE 5.6
Primary Sources for Selected State and Continuum Models of Standard Setting
(Listed Alphabetically by Category)

STATE	CONTINUUM	
Judgmental-Empirical	Judgmental	Judgmental-Empirical
Bergan, Cancelli, and Luiten (1980)	Angoff (1971)	Berk (1976)
Emrick (1971)	Ebel (1979, chap. 17)	Block (1972)
Knapp (1977)	Jaeger (1978)	Huynh (1976b)
Macready and Dayton (1977, 1980)	Nedelsky (1954)	Kriewall (1972)
Roudabush (1974)		Livingston (1975)
Wilcox (1977a, 1977b)		Livingston (1980)
		Livingston and Zieky (1982)
		Novick and Lewis (1974)
		van der Linden and Mellenbergh (1977)
		Wilcox (1979a)

same judgmental and empirical characteristics of the decision-theoretic approaches for continuum competency models. A further discussion of this point follows.

The various models employ decision rules to identify the cutoff score that minimizes expected loss due to classification errors. Examples of these models include Emrick's (1971) mastery testing evaluation model, Roudabush's (1974) true score model, and Macready and Dayton's (1977, 1980) latent state models (see also Bergan, Cancelli, & Luiten, 1980). The decision rules require judgment in designating the loss ratio. The subjectivity involved in this process has been described at length by Shepard (1980a). Macready and Dayton 1980) indicate that all decision making must incorporate implicitly or explicitly a weighting of losses. Yet they also note that this judgmental component can be eliminated by setting the loss ratio equal to 1.0. In addition, they recommend a judgmental assessment of parameter estimates in conjunction with the absolute and relative statistical assessments of model fit. Clearly, judgment is an integral part of the decision-theoretic state models.

There are several specific limitations of the models that render them less compatible with competency testing programs than the continuum models. One limitation is that some of the models (e.g., Knapp, 1977; Roudabush, 1974; Wilcox, 1977a, 1977b) are based on mastery of only one or two items. Decisions at the item level would be appropriate, for example, in the context of algorithmic testing as in Scandura's (1977) structural learning theoretic approach. The use of a single item to measure attainment of an objective, however, is extremely restrictive in view of the structure and imprecision of most domain specifications. Coupled with this limitation is the requisite homogenity of the domain. Only discrete pieces of information (facts, terminology, etc.) or skills where perfection is essential would produce an adequate model fit. This restriction constrains the application of the models to low-level cognitive skills and ultra-specific objectives. The third limitation pertains to the requisite homogeneity of the student population that is tested. The models assume that competents answer all items correctly and they have an equal chance of incurring an inappropriate response (omission error) to an item. The converse assumptions exist for incompetents. Intact classes, schools, and school districts are more heterogeneous than these assumptions would permit. Probably the composition of certain specially formed groups of students would provide the necessary homogeneity. Finally, many of the models are theoretically and statistically complex. This factor alone will limit their application and usefulness.

Continuum Models of Standard Setting

The bulk of the research on standard setting has concentrated on continuum models. In fact, the majority of the cutoff score methods developed within the past decade fall into this category, and consequently the reviews cited previously have focused primarily on these methods. Table 5.6 presents the sources for the methods according to the judgmental and judgmental-empirical classifications.

The *judgmental methods* are based on judgments of the probability that minimally competent persons would select particular distractors in a multiple-choice item (Nedelsky, 1954) or the probability that they would answer the item correctly (Angoff, 1971; Ebel, 1979, chap. 17; Jaeger, 1978). The subjectivity of these item content decisions used to arrive at an overall cutoff score was expressed succinctly by Shepard (1980a): judges have the sense that they are "pulling the probabilities from thin air" (p. 453). This problem is reflected in the variability among judgments within a single method and also across methods (see Berk, 1985; Jaeger, in press). Recent empirical comparisons of the Angoff, Ebel, and Nedelsky methods have found that they produce different cutoff scores and the Nedelsky method yields consistently lower cutoffs than the others (Andrew & Hecht, 1976; Behuniak, Archambault, & Gable, 1982; Brennan & Lockwood, 1980; Colton & Hecht, 1981; Halpin, Sigmon, & Halpin, 1983; Kleinke, 1980; Koffler, 1980; Poggio, Glasnapp, & Eros, 1981; Saunders, Ryan, & Huynh, 1981; Skakun & Kling, 1980). Van der Linden (1982) even identified three possible sources of arbitrariness in the Angoff and Nedelsky techniques: (1) different conceptions of mastery underlying the technique, (2) different interpretations of learning objectives, and (3) intrajudge inconsistency.

This imprecision and the methods' strong dependence on judgments that are relatively unsystematic and arbitrary render these approaches less desirable than the judgmental-empirical methods for use with minimum competency tests. The Angoff method, in fact, does appear to satisfy six of the seven criteria for a defensible standard setting method specified previously; criterion five requires empirical information.

All the remaining standard setting methods not mentioned in the preceding sections can be lumped into the *judgmental-empirical category*. These methods are based on some type of judgment and actual or simulated data, judgmental data, and/or distribution assumptions. To clarify this point and to justify this classification, the specific judgmental and empirical components in 10 continuum methods that have been given wide visibility in the research literature are defined in Table 5.7. They appear to be the primary candidates for resolving the standard setting problem in many competency testing programs. Just how many nominations a method receives will depend largely on how well it meets the seven criteria.

As one examines these methods, the role of judgment should not be underestimated. While the majority of the judgmental-empirical methods are statistically sophisticated, that does not necessarily imply that they are scientifically precise. The judgmental component of each method furnishes the foundation for much of the statistical estimation of probabilities of correct classification decisions and false competency/false incompetency decision errors.

The judgmental-empirical methods differ according to other characteristics as well: (a) overall purpose, (b) type of empirical information, (c) definition of internal or external criterion variable, (d) distribution assumptions, (e) consideration of utilities, (f) statistical sophistication, and (g) practicability. Perhaps the

TABLE 5.7

Judgmental and Empirical Components of Continuum Methods for Setting Cutoff Scores and/or Estimating Error Rates[a]
(Listed in Order of Increasing Overall Complexity)

Method	Source	Judgmental Component	Empirical Component		
			Actual Data	Judgmental Data	Distribution Assumptions
Educational consequences	Block (1972)	Selection of criterion variable	X		
Criterion groups	Berk (1976)	Selection of intact criterion groups	X		
Contrasting groups/Border-line groups	Livingston and Zieky (1982)	Selection of individuals to compose comparison groups	X		
Binomial model	Kriewall (1972)	Setting boundaries for mastery and nonmastery ranges			X
Utility based	Livingston (1975)	Selection of criterion variable; assignment of benefits/costs	X	X	
Linear loss function	van der Linden and Mellenbergh (1977)	Selection of cutoff for latent variable; assignment of losses	X	X	X
Stochastic approximation	Livingston (1980)	Selection of performance criterion	X	X	
Control compar-ison	Wilcox (1979a)	Selection of control by panel of judges	X		X
Beta-binomial model (Empirical Bayesian)	Huynh (1976b), Huynh and Saun-ders (1979), Wilcox (1979b)	Selection of referral task	X		X
Bayesian decision model	Novick and Lewis (1974),Schoon, Gullion, and Ferrara (1979), Swaminathan, Ham-bleton, and Algina (1975)	Setting prior probabilities and loss ratio	X	X	X

[a] Source: Reprinted by permission from Berk (1980d, Table 1, p. 568), *Applied Psychological Measurement, 4*(4), Fall 1980, edited by David J. Weiss, Copyright 1980, West Publishing Company.

most important and basic distinction between these methods, however, pertains to their purposes. Only the Berk (1976) and Livingston and Zieky (1982) approaches are intended to *select a cutoff scores*; all of the remaining methods *presume a standard already exists* on a criterion or latent variable. This standard is then translated into a cutoff score for the test, and decision error rates based on various assumptions are estimated. In some cases those rates can be used to adjust the cutoff. In fact, van der Linden (1980, p. 470) emphasized that even the most complex decision-theoretic models are not techniques for setting standards or optimizing competency decisions; they *are* techniques for minimizing the consequences of measurement and sampling errors once the true cutoff has already been chosen.

Inter alia, the general *unavailability of an acceptable criterion measure* of present or future individual competency makes it extremely difficult to apply the majority of the methods in Table 5.7 to minimum competency tests. Their other deficiencies have been mentioned elsewhere (Glass, 1978c; Hambleton & Eignor, 1980; Shepard, 1984).

Among the remaining methods, Kriewall's (1972) binomial model utilizes an indifference zone instead of a true cutoff score to differentiate between competents and incompetents and has a restricting distribution assumption. While an indifference zone or region of no-decision may be meaningful in sequential mastery testing at the classroom level, an exact point for the dichotomous classification of all individuals is required for most competency test decisions.

The Bayesian decision models permit the incorporation of a loss ratio, prior information on the distribution of domain scores, current information on the person's domain score, and the degree of certainty that a person's domain score exceeds the cutoff score (Schoon, Gullion, & Ferrara, 1979). Unfortunately, those models possess at least three disadvantages pertinent to the seven criteria: (1) they constitute a rather circuitous solution by augmenting as opposed to actually determining a cutoff score; (2) they are theoretically and statistically complex, and (3) their execution would be unwieldy and the results would be difficult to explain given the dimensions and constraints associated with competency test development by school districts and state departments of education.

Recommendations

It would appear as though the original list of potential methods has now been reduced to include only the criterion- and contrasting-groups methods. Despite the fact that no other alternatives are apparent at this time and these two methods are far from perfect (see Berk, 1984e, chap. 6), they do provide a best fit to the criteria for a defensible method. Probably an amalgam of both methods plus some extensions are necessary to address all aspects of the standard setting problem in minimum competency testing.

The method that seems to hold the most promise for competency tests in education can be derived from the construct validation models proposed by Berk

(1976) and Livingston and Zieky (1982) and the variety of statistical techniques suggested by Berk (1976) and Koffler (1980) that can be used in conjunction with those models. The statistical techniques are especially valuable for selecting the optimal cutoff score based upon estimates of correct and incorrect classification probabilities and the weighted cutoff score based upon probabilities that have been adjusted after a cost-benefit utility analysis.

The judgmental component of this approach consists of operationally defining competence in terms of the actual test performance of individuals who have been judged by their teachers, immediate supervisors, or similar persons as competent on an appropriate collection of skills (e.g., Christie & Casey, 1983). Teacher nominations of masters and nonmasters of academic skill objectives in reading, mathematics, and writing could be used effectively. For survival level skills, occupational groups of unskilled and service workers could be compared with unemployed adults or junior high school students. The competency groups are frequently accessible through the coordinators of work-study programs in local districts.

The process of identifying "competent" or "minimally competent" individuals for inclusion in one of the criterion groups represents the Achilles heel of the approach. Regardless of the rigor imposed on the specification of selection criteria and the systematic and standardized procedures used with each teacher or supervisor, there is no known strategy for objectifying the judgments. Interpretations of what is "competent" in relation to a well-defined list of skills may be diverse or comparatively narrow. There is no way to verify either. One must accept this scientific imprecision in the context of the state of the art and proceed to the next steps. If this judgmental process is not credible or intuitively convincing to the decision makers, the empirical component that follows from that premise will be meaningless. The explicit steps for setting the cutoff score have been outlined in the references cited previously (see also Berk, 1984e, chap. 6).

Unless a deliberate and conscientious attempt is made to obtain estimates of how "survivors" in different occupational categories perform on a minimum competency test, decision makers will be hard-pressed to assign meaning to the passing score and to the diploma (Berk, 1983). Only by testing individuals who have been judged competent can one ascertain the validity of the standard and of the decisions based on that standard.

Gathering Appropriate Validity Evidence

Validity is the degree to which a test achieves the purposes for which it was designed. That is, it relates to the intent or purposes of the test. For if a test does not perform its intended functions satisfactorily, why use it? This definition suggests that validity is

1. inferred from the way in which the test scores are used and interpreted;
2. specific to a particular score use;
3. determined ultimately by judgment;
4. expressed by degree.

The three traditional components of validity—content, criterion-related, and construct—are applicable to minimum competency tests. Only the emphases are different from those of norm-referenced tests due to the first three considerations listed above. In fact, the emphases have given rise to some new types of validity which are peculiar to competency testing. There are a few relatively recent discussions of validity for criterion-referenced and minimum competency tests by Hambleton (1980, 1984; Hambleton & Eignor, 1980), Jaeger (in press), Linn (1979b, 1980), Madaus (1983), Millman (1979), and Shepard (1980b). Some of the key issues related to content, curricular, and instructional validity, sex, racial, and ethnic bias, and criterion-related validity are examined here.

Content Validity

Content validity refers to the extent to which the items on a test constitute a representative sample of the domain of items the test is intended to measure. The adequate sampling of the domain of competencies via explicit content specifications is necessary to assure clarity and meaning in test score interpretation. Several procedures for assessing the match between the items and the objectives and the representativeness of the item sample have been suggested by Berk (1984a) and Hambleton (1984).

Unfortunately, the validity evidence gathered by such procedures is not sufficient for minimum competency tests, according to the ruling of the Fifth Circuit Court of Appeals in the trial of *Debra P.* v. *Turlington* (1981). In *Debra P.*, student plaintiffs challenged Florida's functional literacy test as the requirement to receive a standard high school diploma. Functional literacy was defined as "the satisfactory application of basic skills in reading, writing, and arithmetic, to problems and tasks of a practical nature as encountered in everyday life" (p. 259). Experts for the plaintiffs argued that the students should have received instruction on the domain tested if the certification test was to be valid. The Fifth Circuit Court ruled that "the state must demonstrate that the material on the test was actually taught in the state's classrooms in order to establish the requisite 'content validity' " (Citron, 1982, p. 11).

Much of the testimony concentrated on *curricular validity* and *instructional validity,* and the court failed to distinguish between those types of validity and content validity. The confusion in defining these terms is expressed by Madaus (1983): "The court's description of *content* validity—including as it does a reference to *curricular* validity—in fact implicitly incorporates McClung's (1978, 1979) earlier descriptions of *instructional* validity" (p. 25).

Curricular validity refers to the extent to which the items on the minimum competency test measure the content of a local curriculum (cf. McClung, 1979, p. 682). While conceptually similar to content validity (Madaus, 1983; Schmidt, Porter, Schwille, Floden, & Freeman, 1983) and even viewed as synonymous with content validity (Cureton, 1951; Hopkins & Stanley, 1981, chap. 4; Madaus, Airasian, Hambleton, Consalvo, & Orlandi, 1982), curricular validity is operationally very different. In the case of minimum competency tests, it does not always focus on the domain of academic and/or life skills the test was designed to measure; it deals with a specific domain to which the test is applied. The relevance of the test in a specific application is being evaluated. For basic skills, which are typically included in all curricula, this issue of relevance is not a problem. It is the domain of life or survival skills which is not usually part of the curriculum that is troublesome.

Evidence of curricular validity is obtained by determining the degree of incongruence or mismatch. This is based on a systematic, judgmental review of the test against the curricular objectives or materials by content experts. These experts may be classroom teachers or curriculum specialists; they are the only professionals in a position to judge curricular validity. The review can vary as a function of the following: (a) single grade versus cumulative grade content, (b) specificity of objectives or content/process matrix, (c) internal versus external determination, and (d) curricular materials versus actual classroom activities (for details, see Schmidt, 1983a, 1983b; Schmidt et al., 1983). What emerges from this process are several estimates of content overlap, including the amount of content in common, the percentage of the local curriculum measured by the test, and the percentage of items on the test not covered by the curriculum. The second estimate in particular can furnish evidence of the curricular validity of the test.

While curricular validity is an important characteristic, the most crucial legal question deals with whether minimum competency tests measure *what is actually taught* in the schools. Very often it is simply assumed or implied that evidence of curricular validity means that the objectives guided the instruction and the curricular materials were used in the classroom. This does not necessarily follow, as several studies have demonstrated (Hardy, 1983; Leinhardt & Seewald, 1981; Leinhardt, Zigmond, & Cooley, 1981; Poynor, 1978; Schmidt et al., 1983). What is measured by the test is not always the same as what is taught, especially with regard to life or survival skills on minimum competency tests. Hence, a distinction has been made between these different domains to which the test items can be referenced (Schmidt et al., 1983). When the domain is the instruction actually delivered, a "measure of whether schools are providing students with instruction in the knowledge and skills measured by the test" (McClung, 1979, p. 683) is called instructional validity.

Instructional validity refers to the extent to which the items on the test measure the content actually taught to the students. The requirement that minimum competency tests must be instructionally valid strongly suggests that either life skills be taught in the schools as a standard component of the curriculum or those skills should not be tested. If state departments of education tend to choose the latter, in time, the testing programs will probably drift back to the basics and only academic skills may be measured.

Several techniques have been proposed for assessing the overlap between the test and the instruction. Popham (1983) has identified four data-sources for describing whether students have received instruction that would enable them to perform satisfactorily on a test: (1) observation of classroom transactions, (2) analyses of instructional materials, (3) instructor self-reports, and (4) student self-reports. Although he views these sources as methods for determining the adequacy of test preparation (Yalow & Popham, 1983), they can be considered as techniques for gathering evidence of instructional validity. Unfortunately, Popham's (1983) evaluation of those techniques indicates that the process of estimating the percentage of a minimum competency test that has been covered by teaching is fraught with difficulties. Most of these are methodological problems in executing the data-gathering procedures, so as to provide *adequate* evidence (see Leinhardt, 1983; Schmidt et al., 1983). They stem, in large part, from the variability of instructional content, not only among different classes, but within a single classroom.

Despite the conclusion about how instructional validity evidence should be obtained, two recent court rulings revealed that sufficient evidence could be expressed in very different forms. In *Anderson* v. *Banks* (1982) the trial court accepted a Georgia school district's proof of instructional validity based on expert testimony that tested material was covered in their schools' curriculum, and on teacher testimony that that curriculum was actually taught. At the other extreme, in the latest phase of *Debra P.* (1983), Florida conducted an extensive study of instructional validity to amass voluminous evidence that the material covered on the test was indeed taught in the state's classrooms. The study consisted of six components (Fisher, 1983): (1) principals' disemination of the State Student Assessment Test, Part II (SSAT-II) skills, (2) a student remediation study to determine the status of students who failed the test on their first try, (3) a district-by-district analysis of content in the curriculum of the 67 school districts based on self-report, (4) a survey of approximately 65,000 teachers in the state to ascertain whether they taught the SSAT-II skills sufficiently to enable students to master the skills if they applied reasonable effort, (5) on-site visitations of a sample of schools in every district to verify the accuracy of the self-report and to determine if there was evidence of instruction on the SSAT-II skills, and (6) a survey of about 5,000 students asking them whether they had been taught the test material (see also Citron, 1983a). The court concluded that "although the instruction offered in all the classrooms of all the districts might not be ideal, students are nevertheless afforded an adequate opportunity to learn the skills tested on the SSAT-II before it is used as a diploma sanction" (*Debra P.* v. *Turlington,* 1983, p. 186).

Sex, Racial, and Ethnic Bias

Another aspect of validity that must be addressed in the context of minimum competency testing is sex, racial, and ethnic bias. The research and discourse on bias are organized in terms of validity issues and, in fact, reflect the traditional trinary scheme mentioned previously: content, criterion-related, and construct. Bias in the content of the test has been investigated judgmentally and statistically. A *judgmental review* or logical analysis (Shepard, 1982) is intended to detect stereotypic, culture-specific, and offensive language and to assure fair representation in the work roles and life styles of sex, racial, and ethnic groups (Tittle, 1982). The *statistical analysis* based on an appropriate experimental design (Schmeiser, 1982) is conducted to detect discrepancies in item performance between specific groups (e.g., males and females, blacks and whites, Hispanics and whites). When such discrepancies are found, an *a posteriori (judgmental) analysis* is employed to discern whether true item bias is present and, if it is, to deduce explanations for why it occurred and consider procedures for eliminating it (Scheuneman, 1982).

An item is biased if individuals with the same ability have an unequal probability of answering the item correctly as a function of their group membership. This definition is similar to those proposed by Pine (1977) and Scheuneman (1979). Operationally, bias is inferred from differences in performance between groups. The differences are computed using one or more statistical methods (see review by Angoff, 1982, and Ironson, 1982); these methods have been examined in several studies (Burrill, 1982).

Interestingly, item bias has been the predominant form of bias investigation undertaken by publishers of ability and achievement tests, but item bias has not received attention in minimum competency testing until recently (e.g., Christie & Casey, 1983). Initially, the content or behaviors that a test measures is an integral part of all score inferences, and since the item is the most fundamental level of content analysis and the foundation for these inferences, *item bias studies are necessary for all tests.* However, they are not sufficient for all test score inferences and uses. For example, additional studies would be required if the scores are used to make predictions about future performance, which is implied in the construct of life skills. Second, charges of bias from numerous sources frequently include a citation of specific items that are claimed to be biased against a minority population. These sources can be public or professional organizations such as Parents in Action on Special Education (PASE), the National Education Association, and the Association of Black Psychologists (Jackson, 1975; Williams, 1970, 1971), or individual citizens and organizations who take legal action on specific claims of bias (e.g., *Armstead et al.* v. *Starkville, Mississippi Municipal Separate School District,* 1972; *Larry P. et al.* v. *Wilson Riles et al.*, 1979, 1984; *PASE et al.* v. *Joseph P. Hannon et al.*, 1980). Third, the results of bias studies at the subtest and total test levels do not preclude the presence of bias at the item level. For example, a predictive bias study that finds no sex bias does not rule out the possibility that specific items on the test may be biased against females. Fourth, item bias studies can be incorporated into the early stages of test construction and item analysis to minimize the chances of bias accusations arising later. Finally, the elimination of item bias may decrease the likelihood of test bias, although research evidence is needed to verify this relationship.

The *test bias* literature has focused almost exclusively on intelligence and aptitude tests (Jensen, 1980). The studies have dealt with predictive and construct validity issues. *Predictive bias* may be defined as follows:

> Bias exists in regard to predictive validity when there is systematic error in the prediction of the criterion score as a function of group membership.

This definition is a less technical version of the definitions proffered by Cleary (1968), Cleary, Humphreys, Kendrick, and Wesman (1975), and Reynolds (1982a). A slight restatement of Reynolds' (1982b, p. 194). definition of *construct bias* is presented below:

> Bias exists in regard to construct validity when a test measures different psychological constructs as a function of group membership or measures the same construct but with differing degrees of accuracy.

The statistical methods used to detect these two types of bias are no less numerous and varied than those employed in item bias studies (see review by Reynolds, 1982a). The indices which result are intended to signal possible bias and indicate, for example, whether a test predicts the criterion with greater accuracy for whites than for blacks or whether the constructs measured by the test are different for these groups.

Where bias is inferred, the minimum competency test scores for the group in question should be reported by the state. The nature of the bias should be fully explained. Indeed, all pertinent research evidence should accompany any presentation of scores partitioned by sex, racial, or ethnic subpopulations. Test scores may not be validly used without taking account of group differences. In view of the political and social implications of these distinctions, the decision maker should be very cautious in interpreting differential validity evidence.

While the bias literature has concentrated very heavily in the areas of item bias, predictive bias, and construct bias, many other types of bias have been described in relation to minority group populations (Baca & Chinn, 1982; Gonzales, 1982; Oakland, 1980; Oakland & Matuszek, 1977; Reschly, 1979; Samuda, 1975; Sattler, 1982, chap. 19; Ysseldyke, 1979). Examples are atmosphere bias, linguistic bias, examiner bias, and decision-making bias. The descriptions of these various sources of invalidity are usually couched in the context of the litigation involving charges of racial or ethnic bias (see reviews by Bersoff, 1979, 1982a, 1982b; Jensen, 1980, chap. 2; Oakland & Laosa, 1977; Reschly, 1979) or the Public Law 94–142 (1975) mandate for *nondiscriminatory evaluation*.

Criterion-Related Validity

Criterion-related validity refers to the extent to which test performance is related to some criterion measure of performance. For minimum competency tests measuring academic skills, the mastery criterion must be defined operationally in terms of master and nonmaster students. These students can be selected using the criterion- or contrasting-groups procedures described previously. A *concurrent validity* study could then be conducted by correlating competency test performance and the criterion master-nonmaster classification. Alternatively, the test can be correlated with other achievement tests assessing the same content areas (e.g., Christie & Casey, 1983). A *predictive validity* study is appropriate to predict future performance related to life or survival skills. Since it is often impractical to wait several years to obtain criterion performance data on a current group of test takers, one can instead administer the competency test to adults in the community who by their occupation and/or supervisor's evaluation may be judged at a minimum level of survival or higher. Occupational groups of professional, managerial, sales, skilled, and clerical workers can be employed to establish a hierarchy of competency performance. Unskilled and service workers (e.g., cooks, custodians, truck drivers) can comprise a minimum competency (survival) group. Unemployed adults who are actively seeking employment can serve as an incompetent (nonsurvival) group. Correlations between the minimum competency test scores of these adults and their criterion occupational classification can furnish evidence of predictive validity.

One type of criterion-related validity especially important for minimum competency tests is decision validity. *Decision validity* refers to the extent to which a test can yield accurate decisions according to a criterion classification (Hambleton, 1980, 1984). This may be perceived as analogous to concurrent validity. The principal difference lies in what is being studied: the *decisions* reached on the basis of test scores or just the *test scores*. An investigation of decision validity examines the relationship between the decisions made using a specific test and the decisions made using a criterion procedure. In other words, two dichotomous variables are being compared: the pass-fail status on the minimum competency test and the competent-incompetent classification of the persons tested.

The effectiveness of a minimum competency test resides ultimately in the degree to which it can distinguish competent from incompetent students, that is, the accuracy of competent-incompetent classification decisions. Decision validity evidence is usually expressed as probabilities of correct and incorrect classifications, sensitivity and specificity indices, and validity coefficients (for details, see Berk, 1976, 1984e, chap. 6). Essentially, the value or usefulness of a minimum competency test is contingent on the nature of this evidence. For example, if 93% of the students are correctly labeled competent and incompetent on the mathematics subtest, then that subtest may be judged effective in accomplishing what it was designed to do. However, if only 74% of the students are correctly classified with 18% false incompetency and 8% false competency errors on the writing subtest, it is less effective and, depending on the loss function adopted, the cutoff score may be lowered to reduce the 18% error rate.

Such evidence is also crucial in attempting to justify the selection of the performance standard using the criterion-groups and contrasting-groups approaches. Furthermore, without decision validity evidence related to the cutoff score, it seems pointless even to compute an index of decision consistency (see next section on "Reliability"). Certainly one can compute an index based on any performance standard. However, if it is not known whether the decisions based on the cutoff score will be accurate, then one possible interpretation of a high index of decision consistency might be that the *test can consistently classify students into the wrong groups*. Consistent decision making without accurate decision making has questionable value.

The groups of mastery and nonmastery students described in the preceding section on concurrent validity and in conjunction with the recommended standard setting procedure can be used in a decision validity study of the academic skills areas. Also, different occupational groups of competent and incompetent adults can supply the data for the life skills subtest. It is possible, in fact, to employ the same criterion groups for both the standard setting and criterion-related validity (concurrent, predictive, decision) analyses.

Estimating the Reliability of the Scores and Decisions

Reliability refers to the degree of consistency between two or more measurements of the same thing. It may be the individual scores or decisions based on those scores that are analyzed over repeated measurements using a single test or parallel test forms. This meaning of reliability should be viewed in the context of the following points. Reliability is

1. a necessary but not sufficient condition for validity;
2. inferred from the way in which the test scores are used and interpreted;
3. specific to a particular type of consistency;
4. determined ultimately by judgment;
5. expressed by degree.

There are numerous types of reliability that account for different sources of error in the test scores. Several summaries and critiques of reliability statistics recommended for criterion-referenced and minimum competency tests have been conducted by Berk (1980b, 1984d), Hambleton et al. (1978), Linn (1979a), Millman (1979), Shepard (1980b), and Traub and Rowley (1980). In-depth presentations of two major categories of reliability have also been given by Subkoviak (1984) and Brennan (1984). This review concentrates on three components of reliability that are particularly important for minimum competency tests: parallel forms reliability, interscorer consistency, and decision consistency.

Parallel Forms Reliability

The development of parallel forms of a minimum competency test is essential for one or more of the following reasons. First, in this era of test disclosure (e.g., La Valle Act in New York), the public and the students may wish to scrutinize the test items and the answer key. Second, the ever present problem of test security can be reduced when several test forms are used. And third, students who are given multiple opportunities to pass a minimum competency test should not receive the same test each time.

These circumstances suggest that two or more test forms should be generated. The parallel forms reliability must then be estimated, and, finally, the scores on the different forms must be equated. The procedures for equating will be discussed in a subsequent section of the chapter.

Parallel forms reliability is estimated using two separate but equivalent, parallel, or alternate forms of a test. The forms are constructed systematically from the same competency specifications so that, at least from a judgmental perspective initially, they both appear to measure the same material. This can be accomplished by drawing two random samples of items from the domain of items developed from the specifications or by building the two forms item by item according to content and difficulty level. The former method results in *randomly parallel forms*; the latter produces *classically parallel forms*. The item sampling approach is often preferable because the reliability coefficient derived from the classical approach does not take into account item sampling error.

The test forms are then administered to the same group of students in close succession with no intervening time. Frequently the items from the two forms are included in one test, where Form A items may be even-numbered and Form B items odd-numbered. This procedure is intended to minimize the effects of certain factors that could lower the degree of equivalence. For example, fatigue at the end of the test should theoretically influence performance equally on items from both forms when the items alternate (A, B, A, B, etc.); if Form A items were administered first and Form B items administered second, only Form B would be affected.

The two administrations produce two sets of scores, one from each form. These scores can then be correlated to determine the degree to which the items on each form measure the same construct, an academic skill or life skills. A correlation coefficient of .90 or above is required to adequately demonstrate equivalence.

In addition to the correlation coefficient, other statistics need to be computed in order to assess the equivalence of classically parallel forms. These are the mean, variance, and the item analysis results (i.e., difficulty, discrimination, and interitem correlation matrix) for each test form.

Interscorer Consistency

Most minimum competency tests currently in use typically employ an objective item format, such as multiple choice (Gorth & Perkins, 1979b). This characteristic facilitates either manual or computer scoring which cannot be influenced by individual judgment; that is, the scoring is totally objective, not subjective. In certain academic skills, for example, writing and speaking (e.g., Illinois, Massachusetts, Oregon), and in performance-based life skills, such as using a telephone in a simulated emergency situation, where the behaviors must be observed directly, objectivity is not easily achieved. The individuals who score an essay test or record specific behaviors may allow their own judgments, biases, and/or opinions to contaminate the results. This is possible whenever writing samples or essays are required or behavioral checklists or rating scales are used.

The problem is that if scores vary markedly from one scorer to another, how can one discern the true score. This fluctuation or inconsistency between scorers, judges, observers, or raters must be minimized in order to provide useful data. The most effective strategies for achieving interscorer consistency are to delineate very specific, operational criteria for scoring (or recording), and then to train the persons involved so that their tasks can be executed as objectively as possible.

One method to measure the degree of objectivity attained and, in essence, the effectiveness of those strategies is to estimate interscorer consistency. Over the past 30 years more than 20 different statistical indices have been recommended (see review by Berk, 1979a). Among the various indices, the correlation coefficient used to express the previous types of reliability can also be applied here. Two sets of scores/ratings by two independent scorers/observers are obtained on one group of students at the same point in time. The results are then correlated to estimate the scoring consistency. In this case, the index, referred to as an *interclass correlation,* assesses the amount of error in the scores due to the person(s) who did the scoring. No other source of error is considered.

The criterion for an adequate level of interscorer consistency may vary as a function of the skills or behaviors being measured, the particular scoring procedures followed, and the index used. Very often, as scorers/observers are being trained, several reliability checks are conducted, so that by the completion of training (and sometimes retraining), a near perfect level of consistency is attained. When coefficients are finally estimated, they usually fall in the .90s. For minimum competency writing tests and other performance tests, interclass correlations in that vicinity are required to assure dependable individual decisions.

Decision Consistency

The type of reliability that reflects the purpose and the characteristics of a minimum competency test as well as the decisions for which the scores are used is decision consistency. It deals with the consistency of competency-incompetency classification decisions based on the performance standard.

There are two indices of decision consistency: p_o, the percentage of students consistently classified as competent and incompetent across repeated measures with one test or classically parallel test forms, and κ the percentage of students consistently classified beyond that expected by chance. They are derived from the threshold loss function that assumes (a) a dichotomous, qualitative classification of students as competent and incompetent based on a threshold or cutoff score and (b) the losses associated with all false competency and false incompetency classification errors are equally serious regardless of their size.

The selection of p_o or κ is a function of the method for setting the cutoff score (relative or absolute) and the conclusions reached from an analysis of the disadvantages of each index (see Berk, 1984d). The p_o index should be used where an absolute standard is chosen and for minimum competency tests that contain short subtests and/or yield low score variance. The κ index may be the preferred index of agreement where relative cutoff scores are set according to the consequences of passing or failing a particular proportion of the population, as in the case of some minimum competency tests where the cutoff score is adjusted according to the political, economic, social, and/or instructional consequences of not graduating or promoting a certain proportion of the students in the school district. The problems associated with κ, however, render it less useful then p_o.

In regard to estimating p_o or κ for minimum competency tests, the Hambleton and Novick (1973) and Swaminathan, Hambleton, and Algina (1974) two-administration procedures are recommended using classically parallel test forms. These procedures make it possible to measure both *stability and equivalence*. That is, $\hat{p}_o$ and $\hat{\kappa}$ will estimate the stability of the competency-incompetency decisions over time *and* the equivalence of the scores on the two item samples (test forms). Alternatively, when only one test form is available, Huynh's (1976a) single-administration approach or Peng and Subkoviak's (1980) approximation can be employed.

Equating the Scores on Different Test Forms

When parallel forms of a minimum competency test or two different levels of the test (e.g., 9th grade and 12th grade) are developed, score equating is necessary to assure fair and valid decisions based on the individual scores from those forms. A parallel forms reliability coefficient provides evidence only of the degree of equivalence; even when this equivalence is perfect (1.0) and the forms are tau-equivalent, individual scores will differ on the two tests. For example, one form of a minimum competency test, Form B, may be easier than another form, Form A. If no adjustment in the scores were made to account for those differences in difficulty, a passing score, of say, 60, on each form would mean something different. It would be harder to attain that score on Form A. The student taking Form B would have an unfair advantage over the student who was administered Form A. For this student, the consequences of not equating the scores would be failing the test and not graduating. All scores must be equated across Forms A and B, especially the cutoff score and those scores close to the cutoff, in order to adjust for these differences and to establish their comparability (see, for example, Bernknopf, 1980).

Although the need for test score equating has existed for some time, the La Valle Act, effective January 1980, in New York, added a legal impetus. This law required test disclosure—providing students the opportunity to see the test questions used in obtaining their scores on admission tests. Once the questions were released, new test forms had to be generated. Equating the scores on these different forms became essential if the decisions about test takers were to be fair and valid (Berk, 1983).

Horizontal and Vertical Equating

There are two types of equating: horizontal and vertical. *Horizontal equating* involves equating test forms that are developed to measure the same content at the same level for the same population, as in the preceding example of parallel forms (A and B) of a minimum competency test. *Vertical equating* is the process of equating tests that differ in difficulty so that they are roughly "exchangeable," i.e., converting to a common scale the scores on forms of a test designed for populations at different grade levels (Slinde & Linn, 1977, p. 23). This equating is applicable to states where two or more levels of a minimum competency test are constructed. For example, a 9th grade preliminary (practice) or diagnostic version of the test may be administered prior to the 11th or 12th grade version used for graduation certification. (*Note*: This strategy is similar to the administration of the PSAT and SAT.) Equating scores at adjacent grade levels has been accomplished satisfactorily (see, for example, Slinde & Linn, 1979); equating tests that differ more drastically in difficulty, say two or three grade levels apart, is troublesome.

There are three major approaches frequently used to equate test scores: linear, equipercentile, and logistic or item response theory. The first two methods are traditional; they have been applied for more than three decades and are, by far, the most popular (Angoff, 1971; Flanagan, 1951). The logistic or latent trait models constitute a relatively recent innovation in the field (Holland & Rubin, 1982; Marco, 1981). One- , two- , and three-parameter models have been studied extensively, and variations of those models have also been examined (Phillips,

1983). The empirical research over the past 5 years that has compared the precision of these various models suggests, in general, that similar results are found across methods for tests of approximately equal difficulty (horizontal equating), but substantially different results occur for tests of unequal difficulty (vertical equating) (see Arter, 1982; Butera & Raffeld, 1979, Jaeger, 1981; Kolen, 1981; Kolen & Whitney, 1982; Linn, 1981).

The net effect of all of this research on test score equating is that it is now possible to translate the raw scores on parallel forms or different levels of a minimum competency test into one scale. The resulting scores are often called scaled scores, which are usually assumed to constitute an equal-interval scale. Although there are systematic equating errors associated with the scaled scores (Hoover, 1982), they are typically less serious than the unfair and invalid decisions that can result from not equating the scores on different forms of a minimum competency test.

CRUCIAL ISSUES IN MINIMUM COMPETENCY TESTING

Embedded throughout the preceding description of the technical specifications are the major issues confronting minimum competency test makers. Since most state departments of education have chosen to construct their own tests and the technical analyses are conducted using in-house expertise (the alternative is to contract the work to an external agency)[4] (Gorth & Perkins, 1979b), the settlement of some of the issues may be contingent more on the commitment of resources than on psychometric research. Practical constraints and available resources will probably dictate what can be done. Hopefully this will closely approximate what should be done.

According to the latest edition of the *Standards* (AERA/APA/NCME Joint Committee, in preparation) and the methodological recommendations given previously, minimum competency testing practices must meet certain "minimum" standards; that is, the tests should be psychometrically as well as legally defensible. The issues that appear to be most critical to the success of a minimum competency testing program along with suggestions for their settlement are listed below:

1. *Can the domain of minimum competencies be defined objectively*? The choice of what competencies should be tested involves the judgments of professional educators and the lay public. While basic academic skills in reading, mathematics, and writing have a concrete educational foundation in the school curricula, the selection of the most important skills for the purpose of testing in

[4]It is also possible to split the effort between in-house expertise and outside contractors.

high school is highly subjective. The definition of life or survival skills which lack such a foundation tends to be even more subjective. There is no objective method for defining the domain of competencies or any other domain. The choices at each step rest on value judgments. Acknowledging this subjectivity in the process means that the task is to obtain the consensus of all interested parties so that the definition is meaningful and credible. Imposing "objective" procedures on the process will not remove the subjectivity.

2. *Is there a "most effective" strategy for defining the domain*? For the specification of academic skills, the strategies listed in Table 5.5 represent tradeoffs between precision and practicability. Once an outline of the skills has been developed and reviewed, perhaps one of the objectives-based schemes such as amplified objectives, IOX test specifications, or mapping sentences (Berk, 1978) offers a reasonable compromise (Berk, 1980a). Since none of the strategies has been applied extensively to life skills and some of them have been tested only in reading or mathematics, the most adaptable objectives-based approaches again seem worthy of recommendation.

3. *Are standardized test administrations essential*? Standardized procedures for administering a minimum competency test must be documented in a test administration manual and then followed precisely by the person who administers the test. Strict adherence to administration instructions, time limits, test presentation, item response mode, and similar specifications is essential to ensure comparability of test scores and fairness for all students. In addition, certain efforts should be made to maintain test security and to eliminate opportunities for cheating. These efforts might include monitoring the testing process, simultaneous administration to all individuals taking the same test form, and requiring particular seating arrangements (e.g., with adequate space between seats). Irregularities in any of these administration procedures can render the test results invalid. The meaning of scaled scores on multiple test forms and the passing score on the test is contingent on the observation of standardized administration procedures. If some students are given more than the designated time to complete the test or there were "minor" variations in the test taking instructions, the interpretation of their scores must necessarily be different from the interpretation of all other scores. Their scores, in fact, should be judged invalid; those students experienced an unfair advantage over other students, and the scaled scores and the passing score can not be applied.

4. *Are performance tests necessary*? Paper-and-pencil multiple-choice tests have many advantages in the measurement of certain academic skills. However, they are inadequate tools to assess writing, listening and speaking, and several application level life or survival skills. Alternative item and test formats must be employed in order to measure those areas validly. State departments should consider essay formats (restricted and extended response), performance tests such as work samples, situational tests, in-baskets, and trainability tests (see Berk, in press), and behavioral checklists. Certainly, impracticability has been a

drawback of these techniques in large-scale assessments. Recently, however, their popularity has increased and some states have already incorporated performance-based methods in their minimum competency testing programs (e.g., Maryland, Nebraska, Nevada, South Carolina, Texas).

5. *Is their a defensible approach to setting a standard for minimum competence*? Given the judgmental limitations of all of the methods reviewed, there are three options: (1) use a judgmental method such as Angoff (1971), (2) use a judgmental-empirical method such as Livingston and Zieky's (1982) contrasting groups, or (3) use a combination of judgmental and judgmental-empirical methods. The combination approach which has been recommended by Hambleton (1980), Koffler (1980), Shepard (1984), and others has the advantage of capitalizing on the strengths of different methods and the disadvantage of reconciling conflicting results from those methods. A judgmental approach by itself, while politically appealing, is actually a systematic way to "objectify arbitrary input" on what the standard should be. In view of the state of the art, the most defensible course of action seems to be to use a data-based method. The contrasting groups approach has numerous advantages over the judgmental methods, plus it is relatively easy to implement. The primary difficulties with the approach relate to the selection of competent and incompetent persons. Such difficulties are not insurmountable. They are worth tackling, for it is the performance of those groups that gives meaning to the standard.

6. *Is instructional validity evidence necessary for a minimum competency test*? In the *Debra P.* case, the Fifth Circuit Court ruled that the state was required to demonstrate that the material on the test was *actually* taught in the classrooms. Although referred to as content validity in the decision, this evidence of instructional validity (McClung, 1979) must be obtained. The appellate decision offered no advice on how a state was to gather such proof. Popham (1983) has identified four data sources for measuring instructional validity. Unfortunately, at present there are major methodological problems in executing the data gathering procedures, although evidence can be obtained (see Fisher, 1983). If direct measurement is not possible, then the state has two options: (1) either incorporate the skills being tested into the curricular documents and instruction or (2) do not test those skills not being taught formally in the schools. In other words, life skills either should be taught or not tested. Testimony on the teaching of the academic skills should prove adequate (e.g., *Anderson* v. *Banks,* 1982).

7. *Can teaching the test improve instructional validity*? Teaching the specific items on the test or very similar items can destroy the value of the test as a representative sample from the domain of academic or life skills. Such a practice will also invalidate the test scores. The match between the test content and what is actually taught can be improved by teaching from the *objectives* that the test items measure. Teaching to the test or the test itself can only lead to invalidity.

8. *Can miminum competency tests be biased against females and minorities*? Any achievement test can be biased against a particular sex, racial, or ethnic

subpopulation of students as well as groups from different geographic regions within a state. Precautions should be taken during the construction of the test to eliminate stereotypic, culture-specific, region-specific, and offensive language and to assure fair representation in the work roles and life styles of all groups. Furthermore, statistical analyses of item and test bias (see Berk, 1982; Selkow, 1984) should be conducted to furnish evidence that the test scores can be used validly with different groups (Citron, 1983b).

9. *What types of validity evidence are most important for minimum competency tests*? Considering the traditional categories of validity evidence and issues 6 and 8, the most important type of evidence pertains to decision validity. It addresses directly the purpose of a minimum competency test and the use of the scores. Decision validity evidence indicates the degree to which a test can differentiate accurately between competent and incompetent students, and therefore, reveals whether the test is effective and useful. Such evidence can also be used to justify or defend the choice of the performance standard. Concurrent and predictive validity evidence should follow.

10. *What types of reliability evidence are most important for minimum competency tests*? Despite the continued reliance on Kuder-Richardson Formula 20 and alpha coefficients for minimum competency tests, a pool of reliability indices exists that relate to the specific design of the tests and the score uses. Perhaps most important is decision consistency evidence. Once an acceptable level of accuracy in competency-incompetency classification decisions has been attained (decision validity), the dependability of those decisions needs to be assessed. The recommended agreement indices (p_o or κ) provide evidence of the stability of the decisions and the equivalence of item samples based on classically parallel test forms. Single administration estimates are also available (Huynh, 1976a; Peng & Subkoviak, 1980). If parallel forms of the test are constructed or sampled, an equivalence coefficient should also be computed. Finally, if performance tests (or subtests) which require judgmental scoring or direct observation are used, estimates of interscorer reliability are essential.

11. *Do the scores on different forms of a minimum competency test have to be equated*? Score equating is necessary only when the different forms are used for the same decision. If parallel test forms are administered to different students the same year or in different years and passing either form is required to receive a high school diploma, then the scores must be equated onto a common scale so that adjustments in test difficulty can be made. The passing score and each score on the scale should have the same meaning regardless of which form is used. Equating is one method to assure fair and valid individual decisions irrespective of test form (assuming, of course, there are no other sources of unfairness or invalidity).

12. *Should handicapped students be required to pass a minimum competency test to receive a regular high school diploma*? According to a survey of state competency testing programs completed by the National Association of State

Directors of Special Education (1979), 19 states currently have some form of competency testing for the handicapped, 6 states require handicapped students to take the tests, and 7 states are either providing or are in the process of developing special testing procedures for the handicapped population (see also Wiederholt, Cronin, & Stubbs, 1980). Of special significance, however, is the fact that 31 states issue regular diplomas to handicapped students and 17 states leave that decision to the local school board's discretion. Few states issue special diplomas.

The relationship between minimum competency testing and the requirements of Public Law 94–142 (The Education for All Handicapped Children Act of 1975) suggests a set of separate issues that must be tackled (McCarthy, 1980). Four provisions of the law which are directly relevant to competency testing programs are nondiscriminatory testing, the Individualized Education Program (IEP), procedural and placement safeguards, and free appropriate public education. Much of the literature on the topic has addressed these provisions, especially the IEP (e.g., Amos, 1980; Baratz, 1978; Ewing & Smith, 1981; Gillespie & Lieberman, 1983; Lewis, 1979; Linde & Olsen, 1980; McClung & Pullen, 1978; Olsen, 1980; Rosewater, 1979, Ross & Weintraub, 1980; Safer, 1980; Serow & O'Brien, 1983; Smith & Jenkins, 1980).

The first problem that needs attention is the definition of "handicapped." At present, the U.S. Department of Education (1980) has identified nine categories of handicapping condition: speech impaired, learning disabled, mentally retarded, emotionally disturbed, deaf and hard of hearing, visually handicapped, multihandicapped, deaf and blind, and other health impaired. The classification of students into many of these categories is imprecise, for example, learning disabled (Berk, 1984e, chap. 1), and individuals can vary markedly in the severity of their condition.

Once this definitional issue has been settled and the benefits and costs of testing handicapped students have been weighed, it is not unreasonable to conclude that all students should be required to pass the minimum competency test to receive a regular diploma. As McCarthy (1980) observed:

> The use of a single standard for the awarding of the diploma does not imply that the preparation process for all children must be the same. The IEP is a means to an end and should be individualized, while the diploma is an end itself and can be based on universal criteria. (p. 172)

Certainly there are alternatives to this conclusion, such as awarding certificates of attendance and special diplomas (Grise, 1980; Ross & Weintraub, 1980). These alternatives have been upheld by several recent appellate court decisions (e.g., *Board of Education of Northport-East Northport* v. *Ambach,* 1982). Policy makers should examine carefully the alternatives and the anticipated impact on handicapped students before reaching their own conclusion.

THE FUTURE OF MINIMUM COMPETENCY TESTING

It is very risky to predict the success or even the direction of most politico-educational movements. (Actually the only danger is being wrong.) While the minimum competency testing movement was politically instigated, the momentum for change in the schools now rests with the professional educators. More than a decade has passed since a state legislature mandated the first minimum competency testing program. At present, nearly 40 states have mandated such programs, a number large enough to ratify an amendment to the U.S. Constitution. Any ideas proferred here regarding the success of these programs are merely conjectural at this time.

First, the public's dissatisfaction with the "rising tide of incompetents" or the "regression toward mediocrity" and the mounting evidence of increasing rates of illiteracy and incompetent high school graduates has demonstrated that "a serious and substantial educational problem faces the country today" (Lerner, 1981, p. 1062). The National Commission on Excellence in Education (1983) recently emphasized the scope of the problem. The minimum competency testing movement is the public's response to this problem, its best hope for at least a partial solution when no superior alternative is available.

Second, the success of minimum competency testing programs will probably hinge on the credibility and technical quality of the test and on the extent to which the program can be executed effectively. These goals will require the galvanized efforts of educators at all levels—a strong commitment to make the program work. The goals are not within the purview of legislators. The design of the testing program and, particularly, the setting of competency standards are the responsibilities of testing experts with the approval of the public.

The testing technology exists to develop minimum competency tests that are both psychometrically and legally defensible. The dozen issues discussed in the preceding section must be confronted and tackled if a program is to succeed. Despite the role of judgment and subjectivity in all of the procedures, from defining the domain of competencies to equating the scores on different test forms, there are sufficient precedents in other fields of competency testing to suggest that such procedures will survive legal scrutiny. These precedents take the form of specifications to guide competency testing practices in Section 430 of the 1978 Civil Service Reform Act, in the U.S. Equal Employment Opportunity Commission et al.'s (1978) *Uniform Guidelines on Employee Selection,* and in the *Principles for the Validation and Use of Personnel Selection Procedures* (APA, 1980), as well as in the *Standards for Educational and Psychological Tests* (APA/AERA/NCME Joint Committee, 1974). Furthermore, competency test applications in occupational licensing and certification and in the performance appraisal of employees have a history of litigation in the 1970s that has implications for minimum competency testing practices in education (e.g., *Al-*

bermarle Paper Company v. *Moody*, 1975; *Brito* v. *Zia Company*, 1973; *Dickerson* v. *U.S. Steel*, 1978; *Griggs* v. *Duke Power*, 1971; *Wade* v. *Mississippi Cooperative Extension Service*, 1974).

Third, a testing program is just the first step toward solving the incompetency problem. It furnishes only the means of certification or the mechanism for accountability. No test can improve competency levels; it just measures them. The test must be augmented with a competency-based education program to teach the competencies (Goldhammer & Weitzel, 1981; Spady, 1977). Descriptions of 13 exemplary programs throughout the country have been presented by McClure and Leigh (1981). They represent a variety of approaches that may concentrate upon classroom organization, curriculum development, teacher responsibility, learning packages, or integrated tasks (see Lasser & Olson, 1977; Schalock, 1976). As Nickse (1981) points out, however:

> Whatever versions ultimately predominate, and it seems certain that there will continue to be several, the competency-based approach to instruction will serve as a powerful management tool for formal and informal education both within and outside traditional institutions. (p. 223)

These trends in minimum competency testing and competency-based education during the past decade strongly indicate that public pressure for results and educator response to that pressure will continue and probably intensify in the 1990s. The state mandates for educational change demand immediate action and long-term planning, at least until the discontent over incompetence has abated and the meaning of the high school diploma has been restored.

ACKNOWLEDGMENT

I express my appreciation to Stanley Bernknopf, Ronald K. Hambleton, Stephen L. Koffler, and Robert L. Linn for their very useful suggestions on an earlier version of this chapter.

REFERENCES

AERA/APA/NCME Joint Committee. (in preparation). *Joint technical standards for educational and psychological testing*. Washington, DC: American Psychological Association.

Airasian, P. W., Madaus, G. F., & Pedulla, J. J. (1979). *Minimal competence testing*. Englewood Cliffs, NJ: Educational Technology Publications.

Airasian, P. W., Pedulla, J. J., & Madaus, G. F. (1978). *Policy issues in minimal competency testing and a comparison of implementation models*. Boston: Heuristics.

Albemarle Paper Company v. Moody, U.S. Supreme Court Nos. 74–389 and 74–428, 10 FEP Cases 1181, 1975.

American Friends Service Committee. (1978). *A citizen's introduction to minimum competency programs for students.* Columbia, SC: Southeastern Public Education Program.

Amos, K. M. (1980). Competency testing: Will the LD student be included? *Exceptional Children, 47,* 194–197.

Anderson, R. C. (1972). How to construct achievement tests to assess comprehension. *Review of Educational Research, 42,* 145–170.

Anderson v. Banks. 540 F. Supp. 761 (S.D. Ga. 1982).

Andrew, B. J., & Hecht, J. T. (1976). A preliminary investigation of two procedures for setting examination standards. *Educational and Psychological Measurement, 36,* 35–50.

Angoff, W. H. (1971). Scales, norms, and equivalent scores. In R. L. Thorndike (Ed.), *Educational measurement* (2nd ed., pp. 508–600). Washington, DC: American Council on Education.

Angoff, W. H. (1982). Use of difficulty and discrimination indices for detecting item bais. In R. A. Berk (Ed.), *Handbook of methods for detecting test bias* (pp. 96–116). Baltimore, MD: Johns Hopkins University Press.

APA/AERA/NCME Joint Committee. (1974). *Standards for educational and psychological tests* (rev. ed.). Washington DC: American Psychological Association.

APA (Division of Industrial and Organizational Psychology). (1980). *Principles for the validation and use of personnel selection procedures* (2nd ed.). Berkeley, CA: Author.

Armstead et al. v. Starkville Mississippi Municipal Separate School District., No. EC 70–51–5 (N. D. Miss., 1972).

Arter, J. A. (1982, March). *Out-of-level versus in-level testing: When should we recommend each?* Paper presented at the annual meeting of the American Educational Research Association, New York.

Baca, L., & Chinn, P. C. (1982). Coming to grips with cultural diversity. *Exceptional Education Quarterly, 2,* 33–45.

Baker, E. L. (1974). Beyond objectives: Domain-referenced tests for evaluation and instructional improvement. *Educational Technology, 14*(6), 10–16.

Baratz, J. (1978, July). *In setting minimum standards, have we abandoned concerns for equity and access?* Washington, DC: Educational Testing Service.

Beard, J. G. (1979). Minimum competency testing: A proponent's view. *Educational Horizons, 58,* 9–13.

Behuniak, P., Jr., Archambault, F. X., & Gable, R. K. (1982). Angoff and Nedelsky standard setting procedures: Implications for the validity of proficiency test score interpretation. *Educational and Psychological Measurement, 42,* 247–255.

Bergan, J. R., Cancelli, A. A., & Luiten, J. W. (1980). Mastery assessment with latent class and quasi-independence models representing homogeneous item domains. *Journal of Educational Statistics, 5,* 65–81.

Berk, R. A. (1976). Determination of optimal cutting scores in criterion-referenced measurement. *Journal of Experimental Education, 45,* 4–9.

Berk, R. A. (1978). The application of structural facet theory to achievement test construction. *Educational Research Quarterly, 3,* 62–72.

Berk, R. A. (1979a). Generalizability of behavioral observations: A clarification of interobserver agreement and interobserver reliability. *American Journal of Mental Deficiency, 83,* 460–472.

Berk, R. A. (1979b, April). *Technical issues in setting standards for minimum competency testing.* Invited address at the Regional Minimum Competency Testing Conference sponsored by the Basic Skills Assessment Consortium and Educational Testing Service, San Francisco.

Berk, R. A. (1980a). A comparison of six content domain specification strategies for criterion-referenced tests. *Educational Technology, 20*(9), 49–52.

Berk, R. A. (1980b). A consumers' guide to criterion-referenced test reliability. *Journal of Educational Measurement, 17,* 323–349; Erratum, 1981, *18,* 131.

Berk, R. A. (Ed.). (1980c). *Criterion-referenced measurement: The state of the art.* Baltimore, MD: Johns Hopkins University Press.

Berk, R. A. (1980d). A framework for methodological advances in criterion-referenced testing. *Applied Psychological Measurement, 4,* 563–573.

Berk, R. A. (Ed.) (1982). *Handbook of methods for detecting test bias.* Baltimore, MD: Johns Hopkins University Press.

Berk, R. A. (1983, April). *Advances in the field of measurement: Job analysis, passing scores, and test score equating.* Invited address at the Educational Testing Service Seminar on New Frontiers in Assessment for the Medical and Health Professions, Washington, DC.

Berk, R. A. (1984a). Conducting the item analysis. In R. A. Berk (Ed.), *A guide to criterion-referenced test construction* (pp. 97–143). Baltimore, MD: Johns Hopkins University Press.

Berk, R. A. (1984b). Criterion-referenced tests. In T. Husén & T. N. Postlethwaite (Eds.), *International encyclopedia of education: Research and studies.* Oxford, England: Pergamon Press.

Berk, R. A. (Ed.) (1984c). *A guide to criterion-referenced test construction.* Baltimore, MD: Johns Hopkins University Press.

Berk, R. A. (1984d). Selecting the index of reliability. In R. A. Berk (Ed.), *A guide to criterion-referenced test construction* (pp. 231–266). Baltimore, MD: Johns Hopkins University Press.

Berk, R. A. (1984e). *Screening and diagnosis of children with learning disabilities.* Springfield, IL: Charles C Thomas.

Berk, R. A. (1985, April). *A consumers' guide to setting performance standards on criterion-referenced tests.* Paper presented at the annual meeting of the National Council on Measurement in Education, Chicago.

Berk, R. A. (Ed.). (in press). *Performance assessment: Methods and applications.* Baltimore, MD: Johns Hopkins University Press.

Bernknopf, S. (1980, April). *Cut-scores and alternate forms: A new frontier or back to the trenches.* Symposium paper presented at the annual meeting of the National Council on Measurement in Education, Boston.

Bernknopf, S., Curry, A., & Bashaw, W. L. (1979, April). *A defensible model for determining a minimal cut-off score for criterion-referenced tests.* Paper presented at the annual meeting of the National Council on Measurement in Education, San Francisco.

Bersoff, D. N. (1979). Regarding psychologists testily: Legal regulation of psychological assessment in the public schools. *Maryland Law Review, 39,* 27–120.

Bersoff, D. N. (1982a). *Larry P.* and *PASE*: Judicial report cards on the validity of individual intelligence tests. In T. R. Kratochwill (Ed.), *Advances in school psychology* (Vol. II) (pp. 61–95). Hillsdale, NJ: Lawrence Erlbaum Associates.

Bersoff, D. N. (1982b). The legal regulation of school psychology. In C. R. Reynolds & T. B. Gutkin (Eds.), *The handbook of school psychology* (pp. 1043–1074). New York: Wiley.

Block, J. (1972). Student learning and the setting of mastery performance standards. *Educational Horizons, 50,* 183–190.

Board of Education of Northport-East Northport v. Ambach, 458 N.Y.S. 2d 680 (A.D. 1982).

Bormuth, J. R. (1970). *On the theory of achievement test items.* Chicago: University of Chicago Press.

Brennan, R. L. (1984). Estimating the dependability of the scores. In R. A. Berk (Ed.), *A guide to criterion-referenced test construction* (pp. 292–334). Baltimore, MD: Johns Hopkins University Press.

Brennan, R. D., & Lockwood, R. E. (1980). A comparison of the Nedelsky and Angoff cutting score procedures using generalizability theory. *Applied Psychological Measurement, 4,* 219–240.

Brito v. Zia Company, 478 F.2d. 1200 (1973).

Burrill, L. E. (1982). Comparative studies of item bias methods. In R. A. Berk (Ed.), *Handbook of methods for detecting test bias* (pp. 161–179). Baltimore, MD: Johns Hopkins University Press.

Butera, B. J., & Raffeld, P. C. (1979, April). *Investigation of vertical scaling for out-of-level testing*. Paper presented at the annual meeting of the American Educational Research Association, San Francisco.

Castro, J. G., & Jordan, J. E. (1977). Facet theory attitude research. *Educational Researcher, 6* (11), 7–11.

Chickering, A., & Claxton, C. (1981). What is competence? In R. Nickse & L. McClure (Eds.), *Comptency-based education: Beyond minimum competency testing* (pp. 5–41). New York: Teachers College Press.

Christie, S. G., & Casey, J. A. (1983). Heading off legal challenges to local minimum competency programs. *Educational Evaluation and Policy Analysis, 5,* 31–42.

Citron, C. H. (1982). Competency testing: Emerging principles. *Educational Measurement: Issues and Practice, 1*(4), 10–11.

Citron, C. H. (1983a). Courts provide insight on content validity requirements. *Educational Measurement: Issues and Practice, 2*(4), 6–7.

Citron, C. H. (1983b, March). *Legal rules for student competency testing* (ECS Issuegram No. 36). Denver, CO: Education Commission of the States.

Cleary, T. A. (1968). Test bias: Prediction of grades of Negro and white students in integrated colleges. *Journal of Educational Measurement, 5,* 115–124.

Cleary, T. A., Humphreys, L. G., Kendrick, S. A., & Wesman, A. (1975). Educational uses of tests with disadvantaged students. *American Psychologist, 30,* 15–41.

Cody, E. (1983, May 22). Diplomaless Florida "graduate" shocked. *The Washington Post,* p. A5.

Cohen, D., & Haney, W. (1980). Minimums, competency testing, and social policy. In R. M. Jaeger & C. K. Tittle (Eds.), *Minimum competency achievement testing: Motives, models, measures, and consequences* (pp. 5–22). Berkeley, CA: McCutchan.

College Entrance Examination Board. (1977). *On further examination: Report of the advisory panel on the SAT decline*. New York: Author.

Colton, D. A., & Hecht, J. T. (1981, April). *A preliminary report on a study of three techniques for setting minimum passing scores*. Paper presented at the annual meeting of the National Council on Measurement in Education, Los Angeles.

Conoley, J. C., & O'Neil, H. F., Jr. (1979). A primer for developing test items. In H. F. O'Neil, Jr. (Ed.), *Procedures for instructional systems development* (pp. 95–127). New York: Academic Press.

Cureton, E. E. (1951). Validity. In E. F. Lindquist (Ed.), *Educational measurement* (pp. 621–694). Washington, DC: American Council on Education.

Debra P. v. Turlington, 474 F. Supp. 244, 265 (M.D. Fla. 1979).

Debra P. v. Turlington, 644 F.2d 397, 404 (5th Cir. 1981).

Debra P. v. Turlington, 564 F. Supp. 177 (M.D. Fla. 1983), appeal pending.

Dickerson v. United States Steel Corp., 582 F.2d. 827, 17 FEP 1589 (3d Cir. 1978).

Ebel, R. L. (1979). *Essentials of educational measurement* (3rd ed.). Englewood Cliffs, NJ: Prentice-Hall.

Elford, G. (1977, May). *A review of policy issues related to competency testing for high school graduation*. Paper presented at the annual meeting of the New England Educational Research Association, Manchester, NH.

Emrick, J. A. (1971). An evaluation model for mastery testing. *Journal of Educational Measurement, 8,* 321–326.

Ewing, N. J., & Smith, J. E., Jr. (1981). Minimum competency testing and the handicapped. *Exceptional Children, 47,* 523–524.

Finn, P. J. (1975). A question writing algorithm. *Journal of Reading Behavior, 7,* 341–367.

Fisher, T. H. (1978). Florida's approach to competency testing. *Phi Delta Kappan, 59,* 599–602.

Fisher, T. H. (1983). Implementing an instructional validity study of the Florida high school graduation test. *Educational Measurement: Issues and Practice, 2*(4), 8–9.

Flanagan, J. C. (1951). Units, scores, and norms. In E. F. Lindquist (Ed.), *Educational measurement* (pp. 695–763). Washington, DC: American Council on Education.

Fremer, J. (1978). In response to Gene Glass. *Phi Delta Kappan, 59,* 605–606, 625.

Gale, L. E., & Pol, G. (1975). Competence: A definition and conceptual scheme. *Educational Technology, 15*(6), 19–25.

Gallup, G. H. (1978). *A decade of Gallup polls of attitudes toward education, 1969–1978.* Bloomington, IN: Phi Delta Kappan.

Gallup, G. H. (1979). The 11th annual Gallup poll of the public's attitudes toward the public schools. *Phi Delta Kappan, 61,* 33–46.

Gallup, G. H. (1980). The 12th annual Gallup poll of the public's attitudes toward the public schools. *Phi Delta Kappan, 62,* 33–46.

Gallup, G. H. (1981). The 13th annual Gallup poll of the public's attitudes toward the public schools. *Phi Delta Kappan, 63,* 33–47.

Gallup, G. H. (1982). The 14th annual Gallup poll of the public's attitudes toward the public schools. *Phi Delta Kappan, 64,* 37–50.

Gallup, G. H. (1983). The 15th annual Gallup poll of the public's attitudes toward the public schools. *Phi Delta Kappan, 65,* 33–47.

Gallup, G. H. (1984). The 16th annual Gallup poll of the public's attitudes toward the public schools. *Phi Delta Kappan, 66,* 23–38.

Georgia Association of Educators v. Nix, 407 F. Supp. 1102 (1976).

Gillespie, E. B., & Lieberman, L. M. (1983). Individualizing minimum competency testing for learning-disabled students. *Journal of Learning Disabilities, 16,* 565–566.

Glass, G. V. (1978a). Matthew Arnold and minimum competence. *Educational Forum, 42,* 139–144.

Glass, G. V. (1978b). Minimum competence and incompetencies in Florida. *Phi Delta Kappan, 59,* 602–605.

Glass, G. V. (1978c). Standards and criteria. *Journal of Educational Measurement, 15,* 237–261.

Goldhammer, K., & Weitzel, B. (1981). What is competency-based education? In R. Nickse & L. McClure (Eds.), *Competency-based education: Beyond minimum competency testing* (pp. 42–61). New York: Teachers College Press.

Gonzales, E. (1982). Issues in the assessment of minorities. In H. L. Swanson & B. L. Watson, *Educational and psychological assessment of exceptional children: Theories, strategies, and applications* (pp. 375–389). St. Louis: Mosby.

Gorth, W. P., & Perkins, M. R. (1979a, December). *A study of minimum competency testing programs* (Final program development resource document). Amherst, MA: National Evaluation Systems.

Gorth, W. P., & Perkins, M. R. (1979b). *A study of minimum competency testing programs* (Final typology report). Amherst, MA: National Evaluation Systems.

Greene, K., & Gay, R. (1980). *Occupational regulation in the United States.* Washington, DC: Employment and Training Administration, U.S. Department of Labor.

Griggs v. Duke Power Company, 401 U.S. (1971), 3 EPD 8137.

Grise, P. J. (1980). Florida's minimum competency testing program for handicapped students. *Exceptional Children, 47,* 186–191.

Guttman, L. (1970). Integration of test design and analysis. In *Proceedings of the 1969 Invitational Conference on Testing Problems.* Princeton, NJ: Educational Testing Service.

Halpin, G., Sigmon, G., & Halpin, G. (1983). Minimum competency standards set by three divergent groups of raters using three judgmental procedures: Implications for validity. *Educational and Psychological Measurement, 43,* 185–196.

Hambleton, R. K. (1980). Test score validity and standard-setting methods. In R. A. Berk (Ed.), *Criterion-referenced measurement: The state of the art* (pp. 80–123). Baltimore, MD: Johns Hopkins University Press.

Hambleton, R. K. (1982). Advances in criterion-referenced testing technology. In C. R. Reynolds & T. B. Gutkin (Eds.), *The handbook of school psychology* (pp. 351–379). New York: Wiley.

Hambleton, R. K. (1984). Validating the test scores. In R. A. Berk (Ed.), *A guide to criterion-referenced test construction* (pp. 199–230). Baltimore, MD: Johns Hopkins University Press.

Hambleton, R. K., & Eignor, D. R. (1978). Guidelines for evaluating criterion-referenced tests and test manuals. *Journal of Educational Measurement, 15,* 321–327.

Hambleton, R. K., & Eignor, D. R. (1980). Competency test development, validation, and standard setting. In R. M. Jaeger & C. K. Tittle (Eds.), *Minimum competency testing: Motives, models, measures, and consequences* (pp. 367–396). Berkeley, CA: McCutchan.

Hambleton, R. K., & Novick, M. R. (1973). Toward an integration of theory and method for criterion-referenced tests. *Journal of Educational Measurement, 10,* 159–170.

Hambleton, R. K., & Powell, S. (1983). A framework for viewing the process of standard setting. *Evaluation and the Health Professions, 6,* 3–24.

Hambleton, R. K., Swaminathan, H., Algina, J., & Coulson, D. B. (1978). Criterion-referenced testing and measurement: A review of technical issues and developments. *Review of Educational Research, 48,* 1–47.

Haney, W., & Madaus, G. F. (1978). Making sense of the competency testing movement. *Harvard Educational Review, 48,* 462–484.

Hardy, R. (1983, April). *Measuring instructional validity: A report of an instructional validity study for the Alabama High School Graduation Examination.* Paper presented at the annual meeting of the American Educational Research Association, Montreal.

Hively, W., Maxwell, G., Rabehl, G., Sension, E., & Lundin, S. (1973). *Domain-referenced curriculum evaluation: A technical handbook and a case study from the MINNEMAST Project* (CSE Monograph Series in Evaluation, No. 1). Los Angeles: Center for the Study of Evaluation, University of California.

Holland, P. W., & Rubin, D. B. (Eds.). (1982). *Test equating.* New York: Academic Press.

Hopkins, K. D., & Stanley, J. C. (1981). *Educational and psychological measurement and evaluation* (6th ed.). Englewood Cliffs, NJ: Prentice-Hall.

Hoover, H. D. (1982, March). *Some comments on vertical equating using item response theory.* Symposium paper presented at the American Educational Research Association, New York.

Huynh, H. (1976a). On the reliability of decisions in domain-referenced testing. *Journal of Educational Measurement, 13,* 253–264.

Huynh, H. (1976b). Statistical consideration of mastery scores. *Psychometrika, 41,* 65–78.

Huynh, H., & Saunders, J. C. (1979, April). *Bayesian and empirical Bayes approaches to setting passing scores on mastery tests.* Paper presented at the annual meeting of the National Council on Measurement in Education, San Francisco.

Ironson, G. H. (1982). Use of chi-square and latent trait approaches for detecting item bias. In R. A. Berk (Ed.), *Handbook of methods for detecting test bias* (pp. 117–160). Baltimore, MD: Johns Hopkins University Press.

Jackson, C. D. (1975). On the report of the Ad Hoc Committee on Educational Uses of Tests with Disadvantaged Students: Another psychological view from the Association of Black Psychologists. *American Psychologist, 30,* 86–90.

Jaeger, R. M. (1978). *A proposal for setting a standard on the North Carolina High School Competency Test.* Paper presented at the annual meeting of the North Carolina Association for Research in Education, Chapel Hill.

Jaeger, R. M. (1981). Some exploratory indices for selection of a test equating method. *Journal of Educational Measurement, 18,* 23–38.

Jaeger, R. M. (1982). The final hurdle: Minimum competency achievement testing. In G. R. Austin & H. Garber (Eds.), *The rise and fall of national test scores* (pp. 223–246). New York: Academic Press.

Jaeger, R. M. (in press). Certification of student competence. In R. L. Linn (Ed.), *Educational measurement* (3rd ed.). Washington, DC: American Council on Education and National Council on Measurement in Education.

Jensen, A. R. (1980). *Bias in mental testing*. New York: Free Press.

Jordan, J. E. (1978). Facet theory and the study of behavior. In S. Shye (Ed.), *Theory construction and data analysis in the behavioral sciences* (pp. 192–209). San Francisco, CA: Jossey-Bass.

Kleinke, D. J. (1980, April). *Applying the Angoff and Nedelsky techniques to the National Licensing Examination in Landscape Architecture*. Paper presented at the annual meeting of the National Council on Measurement in Education, Boston.

Klemp, G. O., Jr. (1979). Identifying, measuring, and integrating competence. In P. S. Pottinger & J. Goldsmith (Eds.), *New directions for experiential learning (No. 3): Defining and measuring competence* (pp. 41–52). San Francisco, CA: Jossey-Bass.

Knapp, T. R. (1977). The reliability of a dichotomous test item: A "correlationless" approach. *Journal of Educational Measurement, 14*, 237–252.

Koffler, S. L. (1980). A comparison of approaches for setting proficiency standards. *Journal of Educational Measurement, 17*, 167–178.

Kolen, M. J. (1981). Comparison of traditional and item response theory methods for equating tests. *Journal of Educational Measurement, 18*, 1–11.

Kolen, M. J., & Whitney, D. R. (1982). Comparison of four procedures for equating the Tests of General Educational Development. *Journal of Educational Measurement, 19*, 279–293.

Kriewall, T. E. (1972). *Aspects and applications of criterion-referenced tests* (IER Technical Paper No. 103). Downers Grove, IL: Institute for Educational Research.

Larry P. et al. v. Wilson Riles, Superintendent of Public Instruction for the State of California, et al., No. C–71–2270 (N. D. Cal.) (1979, Oct 11).

Larry P. et al. v. Wilson Riles, Superintendent of Public Instruction for the State of California, et al., Appeal No. 80–4027 (9th Cir) (1984, Jan 28).

Lasser, B. R., & Olson, A. L. (1977, April). *Strategies for implementation of competency based education programs*. Salem, OR: Oregon Competency Based Education Program, Northwest Regional Educational Laboratory. (ERIC Document Reproduction Service No. ED 147 950).

Lazarus, M. (1981). *Goodbye to excellence: A critical look at minimum competency testing*. Boulder, CO: Westview Press.

Leinhardt, G. (1983). Overlap: Testing whether it is taught. In G. F. Madaus (Ed.), *The courts, validity, and minimum competency testing* (pp. 153–170). Hingham, MA: Kluwer- Nijhoff.

Leinhardt, G., & Seewald, A. M. (1981). Overlap: What's tested, what's taught? *Journal of Educational Measurement, 18*, 85–96.

Leinhardt, G., Zigmond, N., & Cooley, W. W. (1981). Reading instruction and its effects. *American Educational Research Journal, 18*, 343–361.

Lerner, B. (1981). The minimum competence testing movement: Social, scientific, and legal implications. *American Psychologist, 36*, 1057–1066.

Lewis, D. (1979). Certifying functional literacy: Competency testing and implications for due process and equal educational opportunity. *Journal of Law and Education, 8*, 145–148.

Linde, J., & Olsen, K. R. (1980). *Minimum competency testing and handicapped students*. Lexington, KY: Mid-South Regional Resource Center.

Linn, R. L. (1979a). Issues of reliability in measurement for competency-based programs. In M. A. Bunda & J. R. Sanders (Eds.), *Practices and problems in competency-based measurement* (pp. 90–107). Washington, DC: National Council on Measurement in Education.

Linn, R. L. (1979b). Issues of validity in measurement for competency-based programs. In M. A. Bunda & J. R. Sanders (Eds.), *Practices and problems in competency-based measurement* (pp. 108–123). Washington, DC: National Council on measurement in Education.

Linn, R. L. (1980). Issues of validity for criterion-referenced measures. *Applied Psychological Measurement, 4*, 547–561.

Linn, R. L. (1981). Measuring pretest-posttest performance changes. In R. A. Berk (Ed.), *Educational evaluation methodology: The state of the art* (pp. 84–109). Baltimore, MD: Johns Hopkins University Press.

Linn, R. L., Madaus, G. F., & Pedulla, J. J. (1982). Minimum competency testing: Cautions on the state of the art. *American Journal of Educaton, 91*(1), 1–35.

Livingston, S. A. (1975). *A utility-based approach to the evaluation of pass/fail testing decision procedures* (Report No. COPA-75-01). Princeton, NJ: Center for Occupational and Professional Assessment, Educational Testing Service.

Livingston, S. A. (1980). Choosing minimum passing scores by stochastic approximation techniques. *Educational and Psychological Measurement, 40,* 859–873.

Livingston, S. A., & Zieky, M. J. (1982). *Passing scores: A manual for setting standards of performance on educational and occupational tests.* Princeton, NJ: Educational Testing Service.

Macready, G. B., & Dayton, C. M. (1977). The use of probabilistic models in the assessment of mastery. *Journal of Educational Statistics, 2,* 99–120.

Macready, G. B., & Dayton, C. M. (1980). The nature and use of state mastery models. *Applied Psychological Measurement, 4,* 493–516.

Madaus, G. F. (1981). NIE clarification hearing: The negative team's case. *Phi Delta Kappan, 63,* 92–94.

Madaus, G. F. (1983). Minimum competency testing for certification: The evolution and evaluation of test validity. In G. F. Madaus (Ed.), *The courts, validity, and minimum competency testing* (pp. 21–61). Hingham, MA: Kluwer-Nijhoff.

Madaus, G. P., Airasian, P. W., Hambleton, R. K., Consalvo, R. W., & Orlandi, L. R. (1982). Development and application of criteria for screening commercial, standardized tests. *Educational Evaluation and Policy Analysis, 4,* 401–415.

Marco, G. L. (1981). Equating tests in an era of test disclosure. In B. F. Green (Ed.), *New directions for testing and measurement (No. 11)—Issues in testing: Coaching, disclosure, and ethnic bias* (pp. 105–122). San Francisco, CA: Jossey-Bass.

McCarthy, M. M. (1980). Minimum competency testing and handicapped students. *Exceptional Children, 47,* 166–173.

McClung, M. S. (1978). Are competency testing programs Fair? Legal? *Phi Delta Kappan, 59,* 397–400.

McClung, M. S. (1979). Competency testing programs: Legal and educational issues. *Fordham Law Review, 47,* 651–712.

McClung, M. S., & Pullin, D. (1978). Competency testing and handicapped students, *Clearinghouse Review, 11,* 922–927.

McClure, L., & Leigh, J. (1981). A sampler of competency-based education at its best. In R. Nickse & L. McClure (Eds.), *Competency-based education: Beyond minimum competency testing* (pp. 81–147). New York: Teachers College Press.

Meskauskas, J. A. (1976). Evaluation models for criterion-referenced testing: Views regarding mastery and standard-setting. *Review of Educational Research, 46,* 133–158.

Miller, B. S. (Ed.). (1978). *Minimum competency testing: A report of four regional conferences.* St. Louis: CEMREL.

Millman, J. (1979). Reliability and validity of criterion-referenced test scores. In R. E. Traub (Ed.), *New directions for testing and measurement: (No. 4): Methodological developments* (pp. 75–92). San Francisco, CA: Jossey-Bass.

Millman, J. (1980). Computer-based item generation. In R. A. Berk (Ed.), *Criterion-referenced measurement: The state of the art* (pp. 32–43). Baltimore, MD: Johns Hopkins University Press.

National Assessment of Educational Progress. (1976). *Functional literacy: Basic reading performance.* Denver, CO: Author.

National Association of State Directors of Special Education. (1979). *Competency testing, special education, and the awarding of diplomas: A report of survey information.* Washington, DC: Author.

National Commission on Excellence in Education. (1983, April). *A nation at risk: The imperative for educational reform.* Washington, DC: U.S. Government Printing Office.

National Institute of Education. (1981). *Transcript of minimum competency testing: Clarification hearings, Washington, DC, July 8–10, 1981*. Washington, DC: Alderson Reporting Company.

National School Boards Association. (1978). *Minimum competency. A research report*. Denver, CO: Author.

Nedelsky, L. (1954). Absolute grading standards for objective tests. *Educational and Psychological Measurement, 14,* 3–19.

Nickse, R. (1981). Conclusion. In R. Nickse & L. McClure (Eds.), *Competency-based education: Beyond minimum competency testing* (pp. 220–223). New York: Teachers College Press.

Novick, M. R., & Lewis, C. (1974). Prescribing test length for criterion-referenced measurement. In C. W. Harris, M. C. Alkin, & W. J. Popham (Eds.), *Problems in criterion-referenced measurement* (CSE Monograph Series in Evaluation, No. 3, pp. 139–158). Los Angeles: Center for the Study of Evaluation, University of California.

Oakland, T. D. (1980). Nonbiased assessment of minority group children. *Exceptional Education Quarterly, 1,* 31–46.

Oakland, T. D., & Laosa, L. M. (1977). Professional, legislative, and judicial influences on psychoeducational assessment practices in schools. In T. D. Oakland (Ed.), *Psychological and educational assessment of minority children* (pp. 21–51). New York: Brunner/Mazel.

Oakland, T. D., & Matuszek, P. (1977). Using tests in nondiscriminatory assessment. In T. D. Oakland (Ed.), *Psychological and educational assessment of minority children* (pp. 52–69). New York: Brunner/Mazel.

Olsen, K. R. (1980). Minimum competency testing and the IEP process. *Exceptional Children, 47,* 176–183.

Osburn, H. G. (1968). Item sampling for achievement testing. *Educational and Psychological Measurement, 28,* 95–104.

Parents in Action on Special Education (PASE) et al. v. Joseph P. Hannon, General Superintendent of Schools in Chicago, et al., No. 74 C 3586 (N.D. Ill., July 7, 1980).

Peng, C-Y. J., & Subkoviak, M. J. (1980). A note on Huynh's normal approximation procedure for estimating criterion-referenced reliability. *Journal of Educational Measurement, 17,* 359–368.

Perkins, M. R. (1982). Minimum competency testing: What? Why? Why not? *Educational Measurement: Issues and Practice, 1*(4), 5–9, 26.

Phillips, S. W. (1983). Comparison of equipercentile and item response theory equating when the scaling test method is applied to a multilevel achievement battery. *Applied Psychological Measurement, 7,* 267–281.

Pine, S. M. (1977, March). Applications of item characteristic curve theory to the problem of test bias. In D. J. Weiss (Ed.), *Applications of computerized adaptive testing* (RR 77–1) (pp. 37–43). Minneapolis: Department of Psychology, Psychometric Methods Program, University of Minnesota.

Pipho, C. (1978). Minimum competency testing in 1978: A look at state standards. *Phi Delta Kappan, 59,* 585–588.

Pipho, C. (1983, March). *State activity: Minimum competency testing* (unpublished table). Denver, CO: Education Commission of the States.

Poggio, J. P., Glasnapp, D. R., & Eros, D. S. (1981, April). *An empirical investigation of the Angoff, Ebel, and Nedelsky standard-setting methods*. Paper presented at the annual meeting of the American Educational Research Association, Los Angeles.

Popham, W. J. (1974). Selecting objectives and generating test items for objectives-based tests. In C. W. Harris, M. C. Alkin, & W. J. Popham (Eds.), *Problems in criterion-referenced measurement* (CSE Monograph Series in Evaluation, No. 3, pp. 13–25). Los Angeles: Center for the Study of Evaluation, University of California.

Popham. W. J. (1978a). *Criterion-referenced measurement*. Englewood Cliffs, NJ: Prentice-Hall.

Popham, W. J. (1978b). *Setting performance standards.* Los Angeles: Instructional Objectives Exchange.

Popham, W. J. (1981a). The case for minimum competency testing. *Phi Delta Kappan, 63,* 89–91.

Popham, W. J. (1981b). *Modern educational measurement.* Englewood Cliffs, NJ: Prentice-Hall.

Popham, W. J. (1983, April). *Issues in determining adequacy-of-preparation.* Symposium paper presented at the annual meeting of the American Educational Research Association, Montreal.

Popham, W. J. (1984). Specifying the domain of content or behaviors. In R. A. Berk (Ed.), *A guide to criterion-referenced test construction* (pp. 29–48). Baltimore, MD: Johns Hopkins University Press.

Poynor, L. (1978, March). *Instructional dimensions study data management procedures as exemplified by curriculum analysis.* Paper presented at the annual meeting of the American Educational Research Association, Toronto.

Public Law 94–142. Education for All Handicapped Children Act, S.6, 94th Congress [Sec 613(a)(4)] 1st session, June 1975. Report No. 94–168.

Reschly, D. J. (1979). Nonbiased assessment. In G. D. Phye & D. J. Reschly (Eds.), *School psychology: Perspectives and issues* (pp. 215–253). New York: Academic Press.

Reynolds, C. R. (1982a). Methods for detecting construct and predictive bias. In R. A. Berk (Ed.), *Handbook of methods for detecting test bias* (pp. 199–277). Baltimore, MD: Johns Hopkins University Press.

Reynolds, C. R. (1982b). The problem of bias in psychological assessment. In C. R. Reynolds & T. B. Gutkin (Eds.), *The handbook of school psychology* (pp. 178–208). New York: Wiley.

Roid, G. H. (1984). Generating the test items. In R. A. Berk (Ed.), *A guide to criterion-referenced test construction* (pp. 49–77). Baltimore, MD: Johns Hopkins University Press.

Roid, G. H., & Haladyna, T. M. (1982). *A technology for test-item writing.* New York: Academic Press.

Rosewater, A. (1979). *Minimum competency testing programs and handicapped students: Perspectives on policy and practice.* Washington, DC: Institute for Educational Leadership.

Ross, J. W., & Weintraub, F. J. (1980). Policy approaches regarding the impact of graduation requirements on handicapped students. *Exceptional Children, 47,* 200–203.

Roudabush, G. E. (1974, April). *Models for a beginning theory of criterion-referenced tests.* Paper presented at the annual meeting of the National Council on Measurement in Education, Chicago.

Rowley, G. L. (1982). Historical antecedents of the standard-setting debate: An inside account of the minimal-beardedness controversy. *Journal of Educational Measurement, 19,* 87–95.

Safer, N. (1980). Implications of minimum competency standards and testing for handicapped students. *Exceptional Children, 46,* 288–290.

Samuda, R. S. (1975). *Psychological testing of American minorities: Issues and consequences.* New York: Harper & Row.

Sattler, J. M. (1982). *Assessment of children's intelligence and special abilities* (2nd ed.). Boston: Allyn & Bacon.

Saunders, J. C., Ryan, J. P., & Huynh, H. (1981). A comparison of two approaches to setting passing scores based on the Nedelsky procedure. *Applied Psychological Measurement, 5,* 209–217.

Scandura, J. M. (1973). *Structural learning I: Theory and research.* New York: Gordon & Breach.

Scandura, J. M. (1977). *Problem solving: A structural/process approach with instructional implications.* New York: Academic Press.

Schalock, H. D. (1976, October). *Alternative models of competency based education* (2nd ed.). Salem, OR: Oregon Competency Based Education Program, Northwest Regional Educational Laboratory. (ERIC Document Reproduction Service No. ED 147 951)

Scheuneman, J. D. (1979). A new method of assessing bias in test items. *Journal of Educational Measurement, 16,* 143–152.

Scheuneman, J. D. (1982). A posteriori analyses of biased items. In R. A. Berk (Ed.), *Handbook of methods for detecting test bias* (pp. 180–198). Baltimore, MD: Johns Hopkins University Press.

Schlesinger, I. M. (1978). On some properties of mapping sentences. In S. Shye (Ed.), *Theory construction and data analysis in the behavioral sciences* (pp. 181–191). San Francisco, CA: Jossey-Bass.

Schmeiser, C. B. (1982). Use of experimental design in statistical item bias studies. In R. A. Berk (Ed.), *Handbook of methods for detecting test bias* (pp. 64–95). Baltimore, MD: Johns Hopkins University Press.

Schmidt, W. H. (1983a). Content biases in achievement tests. *Journal of Educational Measurement, 20,* 165–178.

Schmidt, W. H. (1983b, April). *Methods of examining mismatch.* Symposium paper presented at the annual meeting of the National Council on Measurement in Education, Montreal.

Schmidt, W. H., Porter, A. C., Schwille, J. R., Floden, R. E., & Freeman, D. J. (1983). Validity as a variable: Can the same certification test be valid for all students? In G. F. Madaus (Ed.), *The courts, validity, and minimum competency testing* (pp. 133–151). Hingham, MA: Kluwer-Nijhoff.

Schoon, C. G., Gullion, C. M., & Ferrara, P. (1979). Bayesian statistics, credentialing examinations, and the determination of passing points. *Evaluation and the Health Professions, 2,* 181–201.

Selkow, P. (1984). *Assessing sex bias in testing: A review of the issues and evaluations of 74 psychological and educational tests.* Westport, CN: Greenwood Press.

Senior, J. R. (1976). *Toward the measurement of competencies in medicine* (Report of the Computer-Based Examination Project). Philadelphia, PA: National Board of Medical Examiners and American Board of Internal Medicine.

Serow, R. C., & O'Brien, K. (1983). Performance of handicapped students in a competency testing program. *Journal of Special Education, 17,* 149–155.

Shepard, L. A. (1980a). Standard setting issues and methods. *Applied Psychological Measurement, 4,* 447–467.

Shepard, L. A. (1980b). Technical issues in minimum competency testing. In D. C. Berliner (Ed.), *Review of research in education* (Vol. 8, pp. 30–82). Washington, DC: American Educational Research Association.

Shepard, L. A. (1982). Definitions of bias. In R. A. Berk (Ed.), *Handbook of methods for detecting test bias* (pp. 9–30). Baltimore, MD: Johns Hopkins University Press.

Shepard, L. A. (1983). Standards for placement and certification. In S. B. Anderson & J. S. Helmick (Eds.), *On educational testing* (pp. 61–90). San Francisco, CA: Jossey-Bass.

Shepard, L. A. (1984). Setting performance standards. In R. A. Berk (Ed.), *A guide to criterion-referenced test construction* (pp. 169–198). Baltimore, MD: Johns Hopkins University Press.

Shimberg, B. (1981). Testing for licensure and certification. *American Psychologist, 36,* 1136–1146.

Shimberg, B. (1982a). *Occupational licensing: A public perspective.* Princeton, NJ: Educational Testing Service.

Shimberg, B. (1982b). What is competence? How can it be assessed? In M. R. Stern (Ed.), *Power and conflict in continuing professional education* (pp. 17–37). Belmont, CA: Wadsworth.

Skakun, E. N., & Kling, S. (1980). Comparability of methods for setting standards. *Journal of Educational Measurement, 17,* 229–235.

Slinde, J. A., & Linn, R. L. (1977). Vertically equated tests: Fact or phantom? *Journal of Educational Measurement, 14,* 23–32.

Slinde, J. A., & Linn, R. L. (1979). A note on vertical equating via the Rasch model for groups of quite different ability and tests of quite different difficulty. *Journal of Educational Measurement, 16,* 159–165.

Smith, J., & Jenkins, D. (1980). Minimum competency testing and handicapped students. *Exceptional Children, 46,* 440–443.

Spady, W. G. (1977). Competency based education: A bandwagon in search of a definition. *Educational Researcher, 6*(1), 9–14.

Subkoviak, M. J. (1984). Estimating the reliability of mastery-nonmastery classifications. In R. A. Berk (Ed.), *A guide to criterion-referenced test construction* (pp. 267–291). Baltimore, MD: Johns Hopkins University Press.

Swaminathan, H., Hambleton, R. K., & Algina, J. (1974). Reliability of criterion-referenced tests: A decision-theoretic formulation. *Journal of Educational Measurement, 11,* 263–267.

Swaminathan, H., Hambleton, R. K., & Algina, J. (1975). A Bayesian decision-theoretic procedure for use with criterion-referenced tests. *Journal of Educational Measurement, 12,* 87–98.

Thurston, P., & House, E. R. (1981). The NIE adversary hearing on minimum competency testing. *Phi Delta Kappan, 63,* 87–89.

Tiemann, P. W., & Markle, S. M. (1983). *Analyzing instructional content: A guide to instruction and evaluation* (2nd ed.). Champaign, IL: Stipes.

Tittle, C. K. (1982). Use of judgmental methods in item bias studies. In R. A. Berk (Ed.), *Handbook of methods for detecting test bias* (pp. 31–63). Baltimore, MD: Johns Hopkins University Press.

Traub, R. E., & Rowley, G. L. (1980). Reliability of test scores and decisions. *Applied Psychological Measurement, 4,* 517–545.

United States v. State of North Carolina, 400 F. Supp. 343 (E.D.N.C. 1975).

United States v. State of North Carolina, 425 F. Supp. 789 (E.D.N.C. 1977).

United States v. State of South Carolina, 15 FEP Cases 1196 (D.C.S.C. 1977).

U.S. Department of Education. (1980). *To assure the free appropriate public education of all handicapped children* (Second Annual Report to Congress on Implementation of Public Law 94–142: The Education for All Handicapped Children Act). Washington, DC: Author.

U.S. Department of Health, Education and Welfare, Public Health Service. (1977, July). *Credentialing health manpower* (DHEW Publication No. (05) 77–50057). Washington, DC: Author.

U.S. Equal Employment Opportunity Commission, U.S. Civil Service Commission, U.S. Department of Labor, & U.S. Department of Justice. (1978). Uniform guidelines on employee selection procedures. *Federal Register, 43* (166), 38290–38309.

van der Linden, W. J. (1980). Decision models for use with criterion-referenced tests. *Applied Psychological Measurement, 4,* 469–492.

van der Linden, W. J. (1982). A latent trait method for determining intrajudge inconsistency in the Angoff and Nedelsky techniques of standard setting. *Journal of Educational Measurement, 19,* 295–308.

van der Linden, W. J., & Mellenbergh, G. J. (1977). Optimal cutting scores using a linear loss function. *Applied Psychological Measurement, 1,* 593–599.

Wade v. Mississippi Cooperative Extension Service, 372 F. Supp. 126 (1974), 7EPD 9186.

Wiederholt, J. L., Cronin, M. E., & Stubbs, V. (1980). Measurement of functional competencies and the handicapped: Constructs, assessments, and recommendations. *Exceptional Education Quarterly, 1,* 59–73.

Wilcox, R. R. (1977a). New methods for studying equivalence. In C. W. Harris, A. P. Pearlman, & R. R. Wilcox (Eds.), *Achievement test items—Methods of study* (CSE Monograph Series in Evaluation, No. 6, pp. 66–76). Los Angeles: Center for the Study of Evaluation, University of California.

Wilcox, R. R. (1977b). New methods for studying stability. In C. W. Harris, A. P. Pearlman, & R. R. Wilcox (Eds.), *Achievement test items—Methods of study* (CSE Monograph Series in Evaluation, No. 6, pp. 45–65). Los Angeles: Center for the Study of Evaluation, University of California.

Wilcox, R. R. (1979a). Comparing examinees to a control. *Psychometrika, 44,* 55–68.

Wilcox, R. R. (1979b). A lower bound to the probability of choosing the optimal passing score for a mastery test when there is an external criterion. *Psychometrika, 44,* 245–249.

Williams, R. L. (1970). Danger: Testing and dehumanizing Black children. *Clinical Child Psychology Newsletter, 9,* 5–6.

Williams, R. L. (1971). Abuses and misuses in testing black children. *Counseling Psychologist, 2,* 62–77.

Yalow, E. S., & Popham, W. J. (1983). Content validity at the crossroads. *Educational Researcher, 12*(8), 10–14, 21.

Ysseldyke, J. E. (1979). Issues in psychoeducational assessment. In G. D. Phye & D. J. Reschly (Eds.), *School psychology: Perspectives and issues* (pp. 87–121). New York: Academic Press.

6 The Future of Testing for Licensure and Certification Examinations

Michael T. Kane
The American College Testing Program
Iowa City, Iowa

Licensure and certification examinations constitute a major use of tests in the United States, and since licensure and certification provide obvious benefits to the persons licensed or certified and to the public, their use is not likely to decrease. Rather, the usage of such examinations to document competence is likely to continue to increase, although perhaps more slowly than it has in the recent past.

Critics of licensure have argued that licensure tends to benefit the licensed profession more than the public and that the benefits of licensure to the public do not always justify the costs (e.g., see Hogan, 1979; Williamson, 1976). However, the weight of criticism of licensure and certification tends to be that they do not provide sufficient protection rather than that protection is not needed. Furthermore, if licensure were eliminated in areas like the health professions, we would undoubtedly experience some increase in quackery, leading to demands for increased protection. Therefore, I expect that the criticisms of licensure and certification are more likely to change the social and legal context in which the various forms of credentialing operate than they are to decrease the extent of credentialing. In particular, the trend has been in the direction of greater public scrutiny of the activities of licensing and certifying bodies accompanied by demands for more public disclosure.

Both licensure and certification are credentials intended to document the possession of specialized knowledge and skills. Both forms of credentials confer on their holders certain privileges and responsibilities. They differ in the source of the credential and in the legal status of the credential.

Licensure is a state function and is usually administered by a state board with legal authority to regulate the practice of the profession. Although professional organizations have traditionally been involved in setting standards and nominating members of state licensure boards, ultimate authority rests with the state legislature. The laws vary from state to state and from profession to profession, but, in general, a license confers on its holder the right to use a title and to provide certain services that the licensure law makes illegal for nonlicensed persons to provide. It also subjects the licensed professional to regulation by the state licensing authority, often referred to as the "board." The license is interpreted as indicating that its holder has the basic knowledge and skills required for safe and effective practice.

Voluntary certification programs are administered by professional organizations and do not generally have a formal legal status. The professional organization uses certification to recognize training and experience beyond the basic requirements for licensure. To the extent that certification works well, it provides the public with a basis for identifying individuals who are especially well qualified to handle certain kinds of specialized problems. Like licensure, certification can provide substantial benefits to both the practitioner and the public.

It is worth noting that there is considerable variation, and potential confusion, in the terminology used to describe various credentials. Although teachers are certified in most states, the requirements for teacher certification are state-imposed and mandatory for practice. In most school contexts, teacher certification is an example of "licensure" rather than "certification" as these terms are used here.

Although certification does not generally have a formal legal status, it is pervasive enough in medicine as to have significant legal and professional implications. Hospitals may not permit a practitioner who is not certified to provide services usually provided by board-certified practitioners. Furthermore, physicians who engaged in specialized activities, like major surgery, without being certified would expose themselves to punitive malpractice judgments. Therefore, in terms of the restrictions that a lack of certification imposes, some kinds of certification are effectively very similar to licensure.

Since the requirements for voluntary certification in terms of education, experience, and examinations are also quite similar to those for licensure, it will not be necessary for most of the discussion that follows to draw a sharp distinction between these two forms of credentials. Where the differences between the two types of credentials have a significant impact on the issue under discussion, I will try to make this clear, but for simplicity I will emphasize licensure, the more pervasive of the two kinds of credentials.

CURRENT STATUS

Although the specific requirements for licensure vary considerably across the professions and trades that are licensed and across the jurisdictions awarding these licensures, the general pattern is fairly consistent. The requirements typically involve four components: educational requirements, an examination, evidence of good character, and an ongoing policing function. Requirements for relicensure, including continuing education and/or retesting, have become more common, but they are still not the rule (Lowenthal, 1981, provides a recent overview of issues in continuing education for professionals). The main concern of this paper is, of course, the examinations, but some remarks about the other components are in order because they establish the context in which licensure decisions are made and therefore help to place in perspective the issues associated with examinations.

Educational Requirements

The educational requirements for licensure generally involve successful completion of an approved educational program. The requirements can be quite extensive and usually are quite detailed, often specifying, for example, the length of the program, particular courses to be included, etc.

These educational requirements have a significant impact on the interpretation of licensure examination results in at least two ways. First, they have implications for the specification of the content domain to be covered by the examination. Content which is viewed as providing a useful background for practice but having only an indirect or secondary impact on performance could reasonably be omitted from the licensure examination on the basis that this content is thoroughly taught and tested in the educational program; for example, the research methodology of a discipline might be given relatively little emphasis on the examination, assuming that it is covered in the educational requirements. Furthermore, skills that are difficult to assess in a large-scale examination (e.g., performance skills like giving an injection or conducting an interview) are often omitted from the examination, based, at least in part, on the assumption that these skills are adequately documented by the educational program. These remarks suggest that the content of the licensure examination need not be the same as the content of the curriculum; indeed, it would be wasteful and counterproductive if one followed the other too closely. Nevertheless, we should. expect a high degree of overlap between the content of professional school curricula and the content of licensure and certification examinations, since both presumably emphasize knowledge and skills that are viewed as needed for effective practice.

This leads to the second implication of the educational requirements. The existence of rigorous educational requirements provides some assurance that the persons taking the licensure examination are generally well prepared. Therefore, if the system as a whole is working well, the failure rate on a licensure examination that has rigorous education prerequisites should typically be relatively low; if the failure rate were very high, it would be reasonable to suspect that either the examination procedures or the educational programs are not functioning properly.

Of course, this point raises the question of whether the examinations, particularly licensure examinations, are needed at all. If the examination is assessing knowledge and skills that have already been assessed in the educational program, what function does the examination serve? In many cases (e.g., college teachers), professionals are allowed to practice on the basis of educational credentials without having to take any specific examinations.

For certification examinations, the relationship between the expected failure rate on the examination and the extensiveness and degree of rigor of the educational requirements that must be met before the examination can be taken is less clear, because certification is intended to document levels of competence that are often much higher than that required for licensure. Given the high level of skills expected for certification, even relatively lengthy educational preparation may not be viewed as providing strong assurance that most candidates are in fact qualified.

Both licensure and certification examinations can be effective in doing two things. First, the examination provides an additional check on the preparation of individual candidates for licensure. Given the inevitable variability of educational programs, some candidates with deficiencies in some areas of preparation are likely to graduate; the examinations provide evidence of practitioner competence, based on assessment procedures that are the same for all candidates. In a sense, the interpretation given to the examinations is Bayesian in that the educational record constitutes prior information indicating that most candidates for licensure are qualified. This view is reflected in the fact that graduates of foreign professional schools, for which less documentation of program content and quality is available, are often required to take a somewhat more extensive battery of examinations than is required of graduates of approved schools in the United States.

Second, the examination provides a measure of the variability in educational programs and helps to encourage consistency of standards across the programs within the state. It provides an external check on the quality of educational programs. The examination also provides an incentive for programs with disproportionately high failure rates to take steps designed to improve their graduates' performance.

Evidence of Good Character and the Policing Function

The third and fourth types of requirements, evidence of "good character" and the ongoing policing function, are both designed to maintain ethical standards, although the policing function also covers questions of continuing competence. Evaluations of ethics, or "character," raise obvious problems (including possible invasion of privacy), and the effectiveness of the policing functions of licensing boards has been criticized widely. Nevertheless, the existence of these special mechanisms for maintaining ethical standards is significant in that they reflect the general perception, which I think is well justified, that a written examination does not provide an effective mechanism for evaluating ethics and related characteristics, like conscientiousness.

The importance of ethical considerations in determining the quality of professional practice is illustrated by a study of laboratory practice done at the Center for Disease Control in Atlanta (referenced by Williamson, 1976). For half of a set of blood samples, the laboratories knew their performance was being evaluated, and for the other half, the samples were simply sent in by a local physician with a patient's name on them. According to Williamson (1976), "A 4% deficienty rate occurred when the lab was aware it was being tested, whereas a 50% deficiency rate was found when the lab was not aware it was being assessed" (p. 24). The point is that the typical performance of both organizations and individuals falls short of what they would be capable of at their best, and the difference between typical performance and optimal performance is determined by the degree of care and effort that is devoted to an activity. Thus, conscientiousness and the larger issues of professional ethics are likely to be a major determinant of the quality of practice. Unfortunately, methods for assessing an individual's current level of ethics are rather weak, and our ability to predict future behavior is even weaker.

Probably the best available indication of a candidate's ethics is provided by the record of the candidate's performance in the required educational program. Because candidates have strong incentives to present themselves in the best possible light on a licensure or certification examination, candidates are likely to perform at levels close to their optimal levels of performance during the examination. Since the faculty in the education program have the opportunity to observe the candidates' performance in a variety of situations over a long period of time, they have a good opportunity to detect dishonesty, laziness, carelessness, etc., and this current indicator of ethics is probably the best predictor of future behavior. However, a policing function of some kind is needed to limit the negative consequences caused by practitioners who subsequently get into trouble (due, for example, to physical or mental illness, personal problems, financial difficulties, etc.), because we cannot predict such future developments with any accuracy.

It is worth noting that some licensure examinations also include items, or a separate test, on the ethical code for the profession. However, since such items cover knowledge and understanding of the rules of ethics rather than inclination to observe the rules, the requirement should be considered as part of the examination process rather than as part of the "good character" component.

Examinations

Having sketched some aspects of the other components that commonly occur in licensure procedures, we can turn to the central concern of this paper, the examinations themselves. In terms of format, most licensure examinations are written multiple-choice tests, although some also involve other forms of written test (e.g., see Hubbard, 1971), and some have a performance component (e.g., see Kastrinos & Livingston, 1979; Reed, 1978).

The content specifications for the examination, which provide an operational definition of the domain of knowledge and skills covered by the examination, are typically developed by members of the profession being licensed, with assistance from testing specialists. The items are also written and reviewed by members of the profession, who may or may not be the same persons responsible for the content specifications. In most cases, the items are reviewed and edited by testing specialists, and as part of this technical review, item analysis procedures are routinely employed.

Determination of Passing Score. After the examination has been prepared, a passing score is determined. There is wide variation in how this is done, but two general approaches can be identified: those based on the distribution of scores for some "norms" group, and those based on professional judgment. The norms-based methods, which are the more traditional, typically set the passing score at something like one or two standard deviations below the mean score of the norms group. An obvious disadvantage of the norms-based approach is that the performance of each candidate is judged relative to the performance of other candidates, those in the norms group, rather than being judged against the requirements of practice. Given the purpose of licensure and certification examinations, such relative standards do not seem to be appropriate.

The judgment-based standard setting procedures in common use are variants of those proposed by Nedelsky (1954) and Angoff (1971). In these procedures, experts review each item and determine a minimal pass level, or MPL, defined in terms of the probability that a minimally competent candidate would answer the item correctly. Presumably these estimates reflect the experts' judgments about the importance of the content being tested and the difficulty of the item. (A method proposed by Ebel, 1972, explicitly incorporates judgments about importance and difficulty but is not as widely used.) The MPLs are then summed over items to obtain the passing score for the test. These methods have the advantage of being based on expert judgment and therefore of having a rational relationship to practice, but they have a number of problems of stability. Although they are intended to serve the same purpose, the methods tend to give different results (e.g., see Andrew & Hecht, 1976; Brennan & Lockwood, 1980; Shepard, 1980), and the consistency among raters using a given method is not especially high. Furthermore, neither the norms-based nor the judgment-based standard setting methods generate passing scores with an obvious interpretation in terms of practice requirements. This last issue is discussed in some detail later in this paper.

Public Disclosure. A trend of the recent past, which is likely to continue in the future, is greater public disclosure of the characteristics of the examinations as well as the details of licensure procedures in general. This trend has involved such developments as the appointment of public members to licensure boards and sunset legislation mandating periodic legislative review of the work of licensure boards, as well as public disclosure of test items and test forms. Licensure tests have also been examined intensively by various state agencies (e.g., Werner, 1981). Certification examinations have been less subject to outside scrutiny but have also tended to move toward greater public disclosure. This trend should lead to more research on licensure and certification examinations and more thorough documentation of their characteristics, and therefore should facilitate informed debate.

As a final note on the current status of licensing and certification examinations, it is fair to say that the procedures used to develop the tests generally involve the traditional approach to developing standardized achievement tests; in some cases, they are classic examples of this methodology.

CRITICAL ISSUES

The central issue for licensure and certification examinations is validity, that is, the evidence for the interpretation of the results of the examination. Simply put, the question is: What can we justifiably infer about candidates for licensure or certification on the basis of their scores on the examination?

Validity and Utility

Validity is a fundamental concern and, as such, is related to a number of other issues, including the more general concern for the utility of specific forms of regulation embodied in certification and licensure. Presumably, the aim of such credentials is to protect the public, and the effectiveness of examinations in abetting this goal is based on two basic assumptions. First, it is assumed that the public needs protection, and that this need is sufficiently great that society should bear the considerable expense imposed by licensure and certification procedures. In medicine, where practitioners act relatively independently, where the public is generally not in a good position to judge the competence of practitioners, and where the consequences of incompetence can be severe, the protection provided by licensure and certification seems to be most justified. However, it should be noted that even in the case of medicine, the argument has been made that licensure serves the interests of the profession more than it serves the interests of the public (e.g., see Gross, 1978). In other fields (e.g., cosmetology), the need for relatively expensive forms of protection, like licensure, is less clear, but in any case, the public through the political process must decide how much protection/regulation it wants to buy and how much of this protection is achieved most effectively by the regulation of individuals. In some cases, it is clearly more efficient for the state to regulate organizations as is done via safety regulations in industry.

Second, given that protection from incompetent practitioners is needed, it is assumed that licensure or certification, and in particular the examinations required for these credentials, afford the desired protection. This assumption, and therefore the utility of an examination as part of the overall process, depends on the interpretation given to the results of the examination and on the evidence for the proposed interpretation. In particular, when evaluating a licensure or certification examination, the case must be made that those who pass the examination are more likely to be safe and effective practitioners than those who fail the examination. This case will rest on the evidence for the validity of the examination and, as a related issue, on the justification for the procedures used to establish the passing score.

Trade-off Between Utility and Validity. In a sense, there is a trade-off between the utility, or import, of the type of interpretation assigned to examination results and the ease with which the interpretation can be validated (see Kane, 1982b). If the interpretation given to the examination involves strong inferences, validation will be relatively difficult, but if validation were achieved, the examination would make a large contribution to the utility of the resulting decisions. More limited interpretations generally have less utility but are also easier to validate. For example, if a test consisting of questions about the ethical code for a profession were interpreted as a measure of knowledge of the ethical code, it would be relatively easy to validate and would have substantial utility for licensure decisions; under this interpretation, knowledge of the ethical code as reflected in performance on the examination is viewed as a necessary but not sufficient condition for observance of the code. If the test were given a stronger interpretation as a predictor of how ethical the candidate would be in practice, it could have great utility if validated but would be difficult, if not impossible, to validate adequately; under this interpretation, knowledge of the ethical code is viewed as a sufficient condition for observance of the code.

Bias. Closely related to the issue of validity is the issue of bias. To the extent that candidates who have acquired the skills needed for practice fail an examination because of irrelevant factors such as race, sex, or the existence of a handicap that would not interfere with effective practice, the examination would not be valid. However, to the extent that the examination scores reflect candidates' degree of preparedness for safe practice, they would not be considered biased even if they had adverse impact in the sense that the failure rate is higher in some groups than in others. The distinction between adverse impact and bias is embedded in the Uniform Guidelines for Employee Selection Procedures (Equal Employment Opportunity Commission, Civil Service Commission, Department of Labor, and Department of Justice, 1978) used by federal agencies in enforcing civil rights legislation, and reflects the recognition that differential educational experiences can lead to differential achievement. Since different failure rates for various groups, i.e., adverse impact, may result either from differential levels of preparation for the group or from bias in the examination, the Uniform Guidelines require evidence for validity in cases where significant adverse impact occurs.

Because they were developed to aid in the enforcement of federal civil rights legislation, the Uniform Guidelines do not require evidence for validity unless there is adverse impact against groups specifically protected by federal legislation. Furthermore, because of the special role of state government in our federal system, the Uniform Guidelines may not apply to licensure examinations. However, since the justification for the use of licensure and certification examinations depends on their interpretation, evidence for the validity of the proposed interpretation is needed to justify the use of such examinations, even if adverse impact is not found. Where adverse impact is found, the need for careful evaluation of the validity of the examinations is especially important.

Validating Licensure Examinations

Given that validity consists of the evidence supporting the proposed interpretation of examination scores, and that a candidate's score on a licensure examination is interpreted as indicating the candidate's readiness to practice safely and effectively, the required evidence for validity should establish a relationship between scores on the examination and readiness for practice. The issue, then, is the nature of this relationship and the evidence needed to establish that the intended relationship exists.

Since licensure laws are written by state legislatures and administered by state boards, the presumed relationship between scores on the examination and readiness for practice is determined by the legislature and by the state boards and therefore varies from state to state and from profession to profession. Similarly, for certification examinations, the presumed relationship between examination scores and performance in practice depends on the interpretation proposed by the certifying agency. The remarks that follow apply to the general goal of promoting "safe and effective" practice and would apply in general terms to most licensure and certification programs; these remarks represent a more fully developed discussion of suggestions made in Kane (1982a).

Validity consists of an argument for an interpretation of examination scores, and the evidence included in such an argument may take many forms. In most discussions of validity, the types of evidence are discussed under three headings: content validity, criterion validity, and construct validity. Content validity evidence supports the interpretation of test scores in terms of some domain of content and indicates that test scores reflect the degree of mastery of the content domain.

Evidence for criterion validity supports the interpretation of test scores as predictors of some criterion of interest. In the case of licensure and certification examinations, the criterion may be a measure of future performance (e.g., ratings of performance in practice), or it may be a score on some assessment of performance given at about the same time as the examination (e.g., a performance examination simulating some aspects of practice situations might be used to examine the validity of a multiple-choice examination). These two subclasses of criterion validity are called predictive validity and concurrent validity, respectively.

Construct validity supports the interpretation of test scores in terms of certain assumptions about what is being measured and indicates that the test scores reflect an attribute defined by the assumptions. The methods of construct validity that depart dramatically from the more traditional methods of content validity and criterion validity are most clearly applicable where the attribute being measured is implicitly defined by a theory. In such cases the assumptions used to generate validity evidence would be drawn from the theory (i.e., see Cronbach & Meehl, 1955). However, construct validity can also be viewed as subsuming content validity and criterion validity. In criterion validity, the assumption being investigated is that readiness for practice as measured by the examination is related, usually linearly, to subsequent performance in practice. In content validity, the assumption being tested is that the test measures knowledge of a domain that is important for performance in practice.

Because validity is associated with the interpretation of measurements (Cronbach, 1971), evidence that supports the intended interpretation of test scores supports claims for validity, and evidence that disagrees with the intended interpretation tends to refute claims for validity. As noted earlier, there are two common interpretations of the scores on licensure and certification examinations. First, they can be interpreted as providing predictions of an examinee's future professional performance. Second, they can be interpreted as providing evidence of an examinee's present competence on specific abilities that are needed for practice. The interpretation of licensure examinations as predictors of future professional performance suggests the use of predictive validity in evaluating licensure examinations. The interpretation in terms of abilities that are needed in practice suggests the use of content validity.

I have argued (Kane, 1982a) that content validity, considered broadly, would provide a more effective approach to investigating the validity of licensure examinations than can be provided by criterion validity. Manning (1978) has made a similar argument in discussing the legal aspects of validation for employment testing. However, before summarizing the reasons for this position, it is worth emphasizing that, to some extent, all three types of validity evidence are likely to occur in validating any test interpretation; the issue is one of emphasis rather than a choice between clearly separate and distinct approaches.

Criterion Validity—The Interpretation of Licensure Examination Scores as Predictors of Future Performance

The interpretation of licensure examination scores as predictors of future performance in practice is appealing because it implies a high degree of utility for the licensure process. To the extent that this interpretation does provide the justification for a licensure examination, arguments for validity would be based on empirical evidence indicating how well examination scores predict future performance, that is, on predictive validity (Hogan, 1979; Menges, 1975; Pottinger, 1979). Hecht (1979) has stated this position clearly:

> It would appear to me that predictive criterion-related validation studies would be the type most closely fitting the expressed purpose of licensure exams, that of assuring minimal competency on the job for the protection of the public. Interest is with the criterion not yet obtainable at the time of testing. (p. 21)

Similarly, Andrew (1976) has emphasized criterion validity as the ultimate aim in validating certification examinations:

> The challenge that faces us now should encourage us to get on with the business of establishing content validation for our examinations, and to turn our attention even more vigorously to the establishment of criterion-related validity for our certifying examinations. In doing so we must focus our attention on the development of techniques to assess criterion measurements of performance. (p. 46)

As illustrated by these quotations, predictive validity is often presented as the best approach for validating licensure and certification examinations, but this preference for predictive validity is not reflected in practice.

The Criterion Problem. The usefulness of predictive validity for licensure and certification examinations is limited greatly by the fact that criteria of proven validity are not available for licensure examinations. The development and validation of a criterion measure of professional performance presents fundamental conceptual problems as well as great practical difficulties, in part because practice requires a high level of professional judgment for effective performance. The distinction between good practice and poor practice is not clear-cut in most cases (e.g., see Strupp, Hadley, & Gomes-Schwartz, 1977), and the development of general measures of the quality of practice that are reliable, valid, and complete is probably not possible for most professions. Assumptions about the validity of the criterion are likely to be questionable at best, and to the extent that the validity of the criterion is questionable, any conclusions drawn from a predictive validity study would be questionable.

The seriousness of the criterion problem is illustrated by the experience of the National Board of Medical Examiners, as reported by Hubbard (1971). After 3 years of attempting to develop a reliable bedside evaluation using real clients, they found that when one observer rated a candidate in one situation and another observer rated the same candidate in a different situation, the interrater agreement was at the chance level. In another study, Hoffman (1977) found that an oral examination based on a physician's interaction with a client had low reliability because of variability in the assessments of performance from one situation to another. Where such results occur, one must conclude that the ratings are, to a large extent, measuring characteristics of the raters, the situations, or other contextual factors rather than the competence of the candidate.

Technical Problems. In addition to the criterion problem, there are two technical issues that limit the application of criterion validity to licensure and certification examinations. First, licensure is not intended to indicate readiness for a specific task or job, but rather for a wide range of activities in a variety of settings. A criterion validity study showing that a test predicts performance in one setting does not necessarily demonstrate that the test also predicts performance in other settings, and it is not clear whether evidence for criterion validity can be generalized from one setting to another (Cronbach, 1980a; Hunter, 1980). For a licensure examination, therefore, the logic of criterion validity could require not one validity study but a large number of validity studies—one for each of the settings in which those who are licensed might practice. For certification examinations, the range of practice situations is more restricted but is still quite broad.

A second technical problem is that the data needed to evaluate the predictive validity of a licensure examination are not generally available, because those who do not pass the examination are not allowed to practice. A licensure examination is not designed to predict varying degrees of expertise, but simply to distinguish those candidates who are prepared for practice from those who are not. The crucial question for a study of the predictive validity of a licensure examination is whether those who pass the examination are more likely to be safe and effective in practice that those who fail, and this question is not answered by a correlation coefficient based only on passing candidates. A more appropriate index of the predictive validity of a licensing examination would be a measure of the agreement between the pass/fail dichotomy on the licensure examination and a competent/incompetent dichotomy in subsequent practice; however, an index, like coefficient kappa (Cohen, 1960), that would address this issue cannot be estimated without having criterion scores for those who fail the licensure examination as well as for those who pass. Attempts to collect such data might be considered unethical (and probably illegal) in many professions.

This second technical problem does not apply with equal force to certification examinations, because, as this term is used here, certification is not a mandatory requirement for practice in a specialized area. For example, physicians can treat children without being board-certified in pediatrics. Williamson (1976) discusses a number of studies that are relevant to the predictive validity of certification examinations. However, as noted earlier, the practice of a physician who is not certified in an area where certification is common is likely to be somewhat restricted by hospital policies, difficulties in getting malpractice insurance, etc. Also, individuals who choose to specialize in an area of practice will typically meet at least some of the requirements for certification in terms of education and experience even if they are not certified. Therefore, the differences in the scope of practice between certified and noncertified specialists and the overlap in credentials will make decisive studies of predictive validity difficult to implement even in the case of certification examinations.

A related issue that is not as serious as the two technical difficulties just described involves the determination of how strong the relationship between examination scores and the criterion measure must be in order to establish a reasonable case for criterion validity. In some cases, even a weak relationship (e.g., a relatively low correlation) might be sufficient to justify the use of a licensure or certification examination, since even a small increase in the average level of performance in a profession could yield major benefits for society. Furthermore, there are good reasons to expect that the relationship between scores on a licensure or certification examination and subsequent performance in practice would not be particularly strong. As indicated by Gonnella, Goran, Williamson and Cotsonas (1970), successful performance on an examination does not provide a guarantee that the examinee's current level of performance in practice would be satisfactory.

Inferences to future performance are even more problematic since there are a number of factors (e.g., serious illness) that could have a major impact on the quality of future performance but cannot be predicted in advance. The interest in mandatory continuing education is based on the realization that practitioners vary in how well they maintain or enhance their skills after they enter practice. The requirement that small correlations be estimated with precision, combined with the intrinsic difficulties in conducting criterion validity studies for licensure and certification examinations, makes it unlikely that such a study would yield dependable results.

The Interpretation of Licensure Examination Scores as Measures of Critical Abilities

The severe problems associated with predictive validity can be avoided by interpreting the test scores in terms of a domain of knowledge and skills required for practice. The knowledge and skills included in the domain are assumed to be "critical" in that they are necessary, although not sufficient, for effective performance in practice. Abilities are considered *critical* to the extent that their absence would be a serious limitation in the practice of the profession. The critical abilities for a profession typically include cognitive abilities involving knowledge and the ability to apply knowledge, as well as psychomotor skills involving the ability to apply various skills for clients.

In interpreting licensure and certification examinations in terms of critical abilities, the connection between test performance and performance in practice involves two steps. First, the test scores are interpreted as indicating overall level of proficiency in a domain of critical abilities, and second, some level of proficiency in the domain is viewed as necessary for effective performance in practice.

Abilities as Necessary but Not Sufficient Requirements. The fact that skills that are necessary for effective performance do not generally guarantee effective performance is illustrated by the study on the treatment of urinary tract infections mentioned earlier (Gonnella et al., 1970). In this study, the performance of patient care teams in detecting and treating urinary tract infections was evaluated by a review of clinic charts, and the team members were given a 50-item multiple-choice examination and a simulated clinical problem dealing with urinary tract infection. The authors concluded that:

> In the comparison of knowledge and performance major discrepancies were found in our study. It is disturbing to learn that on an examination the students and physicians indicate that a history of catheterization, nephrolithiasis, past treatment of urinary tract infection, hypertension, and diabetes mellitus are critical data but in an actual treatment situation either fail to ask these questions or fail to follow through once the information has been obtained. (p. 2043)

The possession of critical knowledge and skills does not guarantee that the knowledge or skills will be used effectively. The clinic situation, in which the physician deals with multiple patients, interacts with many other professional staff, and must wait for lab results for hours or days, is quite different from the examination situation, in which the facts are presented in an orderly fashion and there are no distractions. However, it is safe to assume that persons who do not possess the required knowledge and skills will not be likely to make use of them. Thus, the critical abilities are necessary but not sufficient requirements for effective practice.

Critical Abilities and the Department of Learning. What kind of abilities should be considered critical abilities for a profession? The *American College Dictionary* defines a profession as a "vocation requiring knowledge of some department of learning or science." Presumably many of the critical abilities will be included in the department of learning or science associated with the profession. The abilities may be quite general (e.g., communication skills) or quite specific (e.g., the ability to carry out a particular procedure). Including a particular ability in a licensure or certification examination would be justified by evidence connecting the ability to client outcomes, and typically this evidence would be drawn from the department of learning for the profession. The inclusion of some abilities is based on empirical evidence (e.g., ability to carry out medical procedures that are based on clinical trials). In other cases, abilities are justified by logical analysis and by procedural rules (e.g., in law). Specification of test content in terms of critical abilities does not require an exhaustive listing of the abilities required for practice, but each ability should be clearly related to practice. Where certification follows basic licensure, the critical abilities for certification include all those required for licensure and, in addition, include specialized knowledge and skills in the area of certification.

Structure of Validity Arguments. The structure of validity arguments involving the critical ability approach is quite simple, including two premises and a conclusion. The first premise states that, because the critical abilities are necessary for effective performance, individuals who lack the critical abilities to a substantial degree will not be able to perform adequately in practice. The second premise states that individuals who have low scores on the examination lack the critical abilities to a substantial degree. The conclusion which follows from these two premises states that individuals who have low scores on the examination will not be able to perform adequately in practice.

Although the structure of the argument is simple, the development of such arguments in specific cases is not simple because it requires substantial evidence for the two premises. The second premise involves issues usually considered under the label of "content validity" (i.e., relationship between test and domain) and issues of standard setting (i.e., what does it mean to say that individuals lack the critical abilities "to a substantial degree"?).

The first premise assumes a relationship between the critical abilities and performance in practice. If it were necessary to start from scratch, justification for the relationship between critical abilities and performance in practice could be an enormous undertaking; a large-scale study might be required to establish a relationship between a particular intervention (e.g., polio vaccination) and the quality of professional practice defined in terms of client outcomes (e.g., incidence of polio). Fortunately, it is not necessary to start from scratch. The department of learning for a profession often includes a large body of data on the relationship between abilities and outcomes. In fact, much of the research effort included in the relevant department of learning can be interpreted as an attempt to identify critical elements in the practice of the profession. To the extent that this research has been replicated and subjected to careful review without being refuted, we have a reasonable basis for confidence in the results.

It is undoubtedly the case that the department of learning for every profession is incomplete, and in some respects incorrect, but for most professions it does represent a substantial body of knowledge about the critical requirements for practice. Therefore, the department of learning establishes a connection between various critical abilities and the quality of practice, and provides the justification for demanding some level of mastery of the critical abilities as a requirement for licensure or certification. As a result, a validation strategy based on critical abilities can concentrate on the second premise, showing that the examination results can be interpreted as indices of the level of proficiency in the required critical abilities.

Of course, the critical abilities approach to validation has its problems and, like criterion validation, is no panacea. The departments of learning are often large and are seldom organized in a way that is appropriate for test development. Therefore, expert judgment is involved in organizing the department of learning for test development purposes (i.e., defining a table of specifications for the test). This effort requires evaluation of the relative importance of various parts of the domain, and such judgments are always fallible. Empirical studies of patterns of practice can help to evaluate the relative importance of different abilities, and therefore provide a useful check on these judgments.

Combining Validation Strategies

The critical abilities approach incorporates aspects of content validity, criterion validity, and construct validity. The evidence supporting the interpretation of test scores in terms of a domain of critical abilities would incorporate many elements of content validity. Several of the issues that arise in this context are discussed in the next section, labeled *Changes Needed,* and in comments on empirical job analyses that appear later in the paper.

The evidence relating critical abilities to client outcomes can be interpreted as providing indirect criterion validation of the licensure examination. A predictive validity study seeks to determine the relationship between performance on a test and some criterion of future performance for each individual, while the critical abilities approach depends on the relationship between an ability and client outcomes averaged over large numbers of professionals and clients (i.e., in clinical trials) or on rational analysis (as in law and some aspects of teaching). Such studies are likely to provide the most accurate analysis available of the importance of various abilities for professional practice.

In a sense, the difference between the predictive validity approach and the critical abilities approach is that the predictive validity approach is almost purely empirical, while the critical abilities approach depends heavily on both the theoretical and empirical content of the department of learning associated with the profession. Studies of predictive validity draw on the "department of learning" in defining the criterion but usually take the examination as a given and proceed to evaluate the empirical relationship between examination scores and criterion scores.

The critical abilities approach makes more extensive use of the theory and the accumulated body of empirical findings in the department of learning, which it uses to define an appropriate content domain. The domain definition is subject to challenge, and empirical job analyses can be employed to investigate some possible challenges. The examination designed to measure mastery of the domain is also subject to challenges of various kinds, and the discussion in the next section will elaborate on the nature of some of the possible challenges and the steps that can be taken to evaluate such challenges.

In its emphasis on the department of learning and the empirical testing of assumptions based on this body of knowledge, the critical abilities approach requires arguments/analyses that are more complicated than those typically employed in studies of criterion validity and content validity. This more general form of validity evidence can be viewed as an example of construct validity, where the construct at issue, professional competence, is defined in terms of the network of theoretical and empirical relationships incorporated in the department of learning.

Testing Standards and Guidelines

In part because of their increasing visibility, licensure and certification examinations have been discussed explicitly in several recent documents containing standards or guidelines for test preparation and use. The most prominent of such documents is the *Joint Technical Standards for Educational and Psychological Testing,* published in draft form in February of 1984 by the American Educational Research Association, American Psychological Association, and National Council on Measurement in Education.

Joint Technical Standards. Chapter 13 of the draft standards (AERA, APA, & NCME, 1984) is devoted to standards for licensure and certification examinations. The introduction to chapter 13 acknowledges the difficulties in conducting sound predictive validity studies for licensure and certification and suggests that:

> The difficulty in conducting criterion-related validation studies does not, however, lessen the importance of validity, which remains a central concern. The test user should develop the evidential basis to support the particular use. For licensure and certification, however, primary reliance must usually be placed on content evidence supplemented by evidence of the appropriateness of the construct being measured. (p. 13–2)

This suggestion, combined with the first standard in chapter 13, quoted below, reflects the basic rationale for a validation strategy based on critical abilities.

> Standard 13.1. The content domain to be covered by the test should be clearly defined and explained in terms of the importance of the content for competent

> performance in the occupation. A rationale should be provided to support a claim that the knowledge or skills being assessed are required for competent performance in the occupation and are consistent with the purpose for which the licensing or certification program was instituted. (p. 13–2)

The comment following Standard 13.1 emphasizes the importance of job analyses and, in particular, of relating the knowledge and skill covered by the examination to the requirements of practice:

> The fact that successful practitioners possess certain knowledge or skills is relevant but not persuasive. Such information needs to be coupled with an analysis of the purpose of the licensing or certification program and the reasons that the knowledge or skill is required for competent performance in the occupation. (p. 13–3)

As suggested by this comment, the purposes of licensure and certification are sufficiently different from those in employment testing and sufficiently important as to merit the development of job analysis procedures that are specifically tailored to the purposes of licensure and certification. As discussed later in this chapter, I would expect these specialized job analysis procedures to incorporate the results of previous research (e.g., clinical trials) and logical analysis of the demands of practice in addition to the kinds of empirical job descriptions usually derived from job analyses in employment settings.

The other standards in chapter 13 of the draft Standards emphasize disclosure policies and issues, like reliability and reading level, which are related to how well the examination measures the knowledges and skills included in the content domain for the examination. In general, the approach taken here is consistent with the fourth draft of the Standards, which is expected to be similar to the final form of the Standards.

Uniform Guidelines. A validation strategy based on critical abilities is also consistent with the *Uniform Guidelines for Employee Selection Procedures* (EEOC et al., 1978), which are used by the federal agencies in enforcing civil rights legislation. There is some question about whether the Uniform Guidelines apply to state licensure examinations. As stated in question 7 of the *Adoption of Questions and Answers to Clarify and Provide a Common Interpretation of the Uniform Guidelines on Employee Selection Procedures* (Equal Employment Opportunity Commission, Office of Personnel Management, Department of Justice, Department of Labor, & Department of the Treasury, 1979):

> 7. Q. Do the Guidelines apply to the licensing and certification functions of state and local governments?
>
> A. The Guidelines apply to such functions to the extent that they are covered by Federal law. Section 2B. The courts are divided on the issue of such coverage. The

> Government has taken the position that at least some kinds of licensing and certification which deny some persons access to employment opportunity may be enjoined in an action brought pursuant to Section 707 of the Civil Rights Act of 1964 as amended. (p. 11997)

However, even if these guidelines are not legally binding, they are likely to be employed in legal reviews of testing procedures, and they have been made part of state law in California (Werner, 1981).

The Uniform Guidelines, which were developed primarily for employment selection, emphasize criterion validity but allow for procedures measuring specific abilities if it can be shown that:

> (a) the selection procedure measures and is a representative sample of that knowledge, skill, or ability; and (b) that knowledge, skill, or ability is used in and is a necessary prerequisite to performance of critical or important work behavior(s). (p. 38302)

Therefore, the Uniform Guidelines explicitly allow for selection tests based on critical abilities, and as argued here, this approach is especially appropriate for licensure examinations.

NCHCA Guidelines. The National Commission for Health Certifying Agencies (1981) has published guidelines for credentialing examinations suggesting that certifying agencies should progress from content to predictive (or criterion-related) to construct validity.

This approach is laudable in setting ambitious goals for certifying agencies but may have some potentially negative consequences. In particular, by encouraging certifying agencies to take predictive validity and construct validity as goals, the NCHCA guidelines may draw attention away from the basic issue of content relevance. Since I am not optimistic about the value for licensure and certification examinations of predictive validity and versions of construct validity that require the adoption of strong theoretical assumptions, I think that this would generally be a bad trade-off if it occurred. The critical abilities approach to validation incorporates aspects of content, predictive, and construct validity, and aims to develop a validation strategy specifically designed for licensure and certification examinations.

CHANGES NEEDED

As is probably clear by now, the basic theme of this discussion is that the validation of licensure and certification examinations should be tailored to the purpose of these examinations and should be consistent with the intended interpretations of the examinations. Given this assumption, it was argued in the last section that criterion validity is inappropriate for licensure and certification examinations for both practical and conceptual reasons, but that a strategy based on critical abilities is both feasible and consistent with the intended interpretations of such examinations.

A validation strategy based on critical abilities incorporates many elements of content validity. The standard method for establishing the content validity of tests is to have experienced practitioners decide which abilities need to be evaluated. These content decisions may be based in part on empirical studies of patterns in the conduct of practice in order to ensure that the content of the examination reflects the actual day-to-day practice of the profession. Both expert judgment based on the department of learning and empirical job analysis may play large roles in studies of content validity but do not supply all of the evidence needed to establish the validity of licensure examinations as measures of critical abilities. In addition, there are a number of issues, in particular, the relationship between the critical abilities and performance in practice, that an argument for the validity of a licensure or certification examination should address.

Abilities Should Be Clearly Related to Client Outcomes

The abilities measured by the examination should be "critical" in the sense that they have a significant influence on client outcomes, and any ability required for licensure should be explicitly linked to client outcomes. The linkage may be based on clinical research, on logical analysis, or on a combination of the two, but it should be explicit.

For many professions the linkage between critical abilities and client outcomes has a large empirical component. The requirement that pharmacists be able to dispense drugs correctly is based on clinical research relating dosage to the effectiveness and safety of the drugs. The expectation that physicians know the symptoms and typical course of development of various diseases is based on empirical research showing that the detection and subsequent treatment of the diseases has positive effects. In some professions, the linkage between various abilities and client outcomes is based mainly on logical analysis. A strong logical case can be made for the linkage between knowledge of the law and effectiveness in such professions as accounting or law. Similarly, the relationship between knowledge of academic content and the ability to teach that content is based more on logical analysis than on empirical studies. Generally, the critical abilities will be determined by a combination of empirical data and logical analysis that constitutes the department of learning for the profession.

Where evidence linking an ability to client outcomes is less straightforward than it is for these simple examples, decisions about criticality become more complicated. If there are several approaches to some issue of professional practice and the evidence does not consistently favor one approach, it would still be reasonable to require that candidates for licensure know enough about the various approaches to recognize their potential benefits and limitations. Given that none of the approaches is clearly superior to all others, it is necessary to allow for the use of professional judgment in selecting a particular approach for each client, but it is also appropriate to require that practitioners be familiar with the available options. Given the purpose of licensure, it is especially important that practitioners be aware of any dangers inherent in various interventions.

Knowledge of research results and theory provides a basis for informed clinical judgment. Since the situations encountered in the practice of most professions tend to be highly variable, the most effective approach to each situation cannot be standardized, and the practitioner is often called upon to employ professional judgment. For many areas of practice, mastery of a domain of knowledge that is relevant to a broad range of situations may be required to inform the practitioners' decisions about how to handle specific situations. That is, an approach to validity based on critical abilities should not be viewed as an attempt to reduce highly developed, complex content domains to a set of discrete pieces of knowledge linked to specific practice situations. The linkage of theory to client outcomes may be more general and less direct than it is for specific skills, but the linkage should be clearly established.

Abilities Should Be Weighted According to Their Importance for Practice

Since some critical abilities will be more important than others, the weight given to various content areas in a licensure examination should reflect the importance of the content areas for practice. The importance of an ability depends on how often it is needed in practice and on how much difference it is likely to make in terms of client outcomes.

The frequency of occurrence of a situation in professional practice is obviously one factor in determining how important it is that a practitioner be able to deal with the situation. For example, it is clearly appropriate that examinations for medical licensure in the United States devote considerable attention to heart disease, diabetes, cancer, and flu because they have a high rate of incidence. The content of examinations used to certify practitioners for specialized practice would naturally give a heavy emphasis to the conditions included in the specialty area even if these conditions are not encountered often in general practice. But even here, the more common conditions in the specialty would generally be given more emphasis than rare conditions.

Job analyses usually place heavy emphasis on frequency data (Williamson, 1979). There are several empirical methods for determining the frequencies of occurrence of various situations in practice. The most obvious method is to ask practitioners how they spend their working hours (e.g., see Williamson, 1979). A more direct approach involves observing the professional's activities over an extended period (Miller, 1963). The direct observation approach is less subject to the kinds of bias that often occur in self-reported data but is more expensive to implement, and it is therefore likely to involve a smaller, and perhaps less representative, sample of practitioners. Both of these approaches provide data on the kinds of demands placed on practitioners and on the time devoted to different kinds of activities, and are therefore clearly relevant to the issue of content validity.

In addition to formal job analyses, there are often existing sources of information about the demands encountered in professional practice. For example, the statistics routinely collected by local, state, and federal government provide a wealth of information about the incidence of various health problems (e.g., a morbidity and mortality weekly report is published by the Center for Disease Control in Atlanta); these data indicate the kind of patients that are likely to be encountered in the practice of the health professions, and therefore provide data relevant to the frequency with which various situations will be encountered in practice.

A major difficulty with data on how professionals spend their time is that the activities included in such data will vary in their importance relative to the purpose of licensure, protection of the public. Williamson (1979) reports that 32% of a physician's time at work is spent on activities other than patient care. Even if attention is restricted to the abilities required in providing professional services to clients, frequency data do not indicate how serious the lack of an ability would be in a particular situation.

The second component in evaluating the relative importance of different abilities is the gravity of the possible consequences of the situations that require the ability. Although common colds occur more frequently than concussions, the consequences that would result from improper treatment of a concussion suggest that a licensing examination for physicians should give more attention to the concussions than the frequency of this condition might indicate. This is not to say that the treatment of colds should be ignored, but rather that the weight given to various abilities should be a function of both frequency and seriousness. Rakel (1979) has stated this point succinctly:

> The temptation to achieve content validity in examinations by matching test items to the frequency of problems encountered in practice could also be counterproductive. There is a justifiable need to test more heavily on problems that have a high morbidity and fall into the ''uncommon but harmful if missed'' category. Because of their serious nature, they deserve greater representation in an examination than practice surveys indicate. (p. 93)

Given that the purpose of licensure is to protect the public, the ''harmful if missed'' category should be emphasized in licensure examinations. Licensure examinations should emphasize the abilities required by situations involving the ''uncommon but harmful if missed'' category, as indicated by the department of learning, even though these abilities are likely to have relatively low frequency of occurrence in practice.

Empirical job analyses are useful in providing data on relative importance as well as frequency. In addition to estimating the frequencies with which various situations arise, respondents are usually asked to indicate the criticality of the actions taken in each type of situation. The critical incident technique (Flanagan, 1954) specifically addresses the perceived importance of an activity as well as its frequency. However, this technique, which is focused on critical incidents, does not provide a clear-cut definition of professional competence. A serious limitation in empirical job analyses is that they focus on what is currently going on in practice, but they do not provide a thorough analysis of what practice would need to be like to best serve the public interest.

In order to address this larger issue of the public interest, the results of job analyses need to be interpreted in terms of the department of learning for the profession. The department of learning will generally provide the best guidance on how the various situations that arise in practice should be handled. Taking an example that is close to home, it seems unarguable that examinations used for teacher certification should reflect the best current thinking on how tests and other assessment instruments should be used in making educational decisions, and should not rely solely on surveys of current practice. In general, empirical job analyses are particularly useful in providing information about the kinds of situations that will be encountered in practice, while the department of learning for the profession is a more reliable source of information about how these situations should be handled. Therefore, in weighting various critical abilities, both empirical job analysis and the department of learning have major roles to play.

Extraneous Factors Should not Unduly Affect Examination Scores

The interpretation of test scores as measures of critical abilities assumes that differences in scores are due to differences among candidates in their attainment of the critical abilities. Cronbach (1980b) points out the need "to establish that an achievement test contains no irrelevant difficulty, if we are to say that it measures command of certain subject matter" (p. 106). To ensure that tests of the critical abilities are measuring what they claim to measure, plausible alternative hypotheses should be investigated.

Some potential competing hypotheses are examined under the heading of reliability or generalizability (Brennan, 1983; Cronbach, Gleser, Nanda, & Rajaratnam, 1972). Measures of stability address the competing hypothesis that observed scores are a function of the occasion on which the measurement is made. Measures of interrater reliability address the hypothesis that scores are largely determined by the observer rather than by the candidate's performance.

Measures of internal homogeneity (e.g., KR-20, coefficient alpha) or parallel-forms reliability address the competing hypothesis that the choice of a particular set of test items strongly influences the outcome.

A potentially serious problem for many types of measurement procedures is the presence of response sets, or tendencies of some persons to respond in a stereotypical way. Affective traits, which are defined in terms of typical performance rather than the best possible performance, are especially subject to response sets. A pattern of "correct" answers to questions about ethical issues may reflect a response set favoring socially desirable responses rather than a commitment to ethical behavior. In research on affective traits, this problem has sometimes been handled by camouflaging what is being measured, but the use of this approach in a licensing examination would raise serious practical and legal problems (Levine, 1980). As noted earlier, the fact that candidates for licensure and certification have a vested interest in performing well makes it especially difficult to evaluate affective traits like conscientiousness or "good character."

Standard test development procedures (e.g., see Ebel, 1972) are designed to minimize the chances that candidates who have the abilities being tested will get an item wrong or that candidates who do not have the abilities being tested will get the item right. In particular, many of the rules for developing objective tests are designed to minimize the influence of response sets. Poorly constructed tests are likely to give an unfair advantage to candidates who are skillful at taking tests (Sarnacki, 1979).

It is also important to ensure that the language used in the test does not constitute an artificial barrier to performance. Except for the use of technical vocabulary, the reading level of the examination should be kept sufficiently low so that anyone with the abilities required for practice will be able to read it. Similarly, the instructions for the examination should be as clear and simple as possible (especially when the instructions are unusual, as they often are for simulations).

It is important to avoid any extraneous factors that could cause minority or women candidates who had developed the critical abilities being tested to get items wrong. Although important content should not be omitted simply to eliminate differences between subgroups, the wording of items should be reviewed to avoid any source of bias (Schmeiser, 1982) that would be likely to interfere with the performance of subgroups within the population.

Licensure and Certification Examinations Should Cover as Wide a Range of the "Critical" Abilities as Is Feasible

Since licensure laws typically qualify the professional to practice in a broad range of settings and to deal with the full spectrum of problems that arise in these settings, the results of the licensure examination are likely to be interpreted as indicating candidates' command of a correspondingly broad range of abilities.

Although it is usually not possible to test the relevant domain exhaustively, the examination content should provide reasonable coverage of the domain as a whole. The scope of the content domain for certification examinations is likely to be more specialized than it would be for a licensure examination, but it is still important to sample the domain adequately. The test specifications typically used to develop licensure and certification examinations are designed to ensure a representative sampling of content. Of course, if the interpretation given to the results of the examination is consistent with a narrowly defined domain, the resulting examination could be highly valid as a measure of knowledge of that domain, but it would probably not serve the purpose of licensure examinations, the protection of the public, very well.

The content of the licensing examination must also be consistent with the scope of practice specified in the legislation authorizing licensure for the profession. Although the laws governing the scope of professional practice may be stated in general terms that leave considerable latitude for interpretation, the legal definition of professional practice still limits the content for a licensure examination to the extent that it limits practice. It would be inappropriate for a licensure examination to require demonstration of a skill that is legally prohibited in practice.

The Cognitive Level of the Items Should be Appropriate

Although the definition quoted earlier refers to "knowledge of some department of learning or science," it is clear that to be safe and effective in practice, the professional must also be able to use this knowledge to solve problems. The professional must be able to apply elements from the department of learning or science to the situations that arise in practice.

If the questions in a licensure or certification examination require the application of knowledge to specific situations, the performance required of the candidate taking the examination is closer to that required in practice than would be the case if the examination involved simple recall of facts. To the extent that the performance required on the licensure examination is similar to the performance required in practice, inferences drawn from test performance to readiness for practice are more direct and therefore easier to justify.

The Level of Proficiency Required by the Examination Should not Be Higher Than That Required for Practice

For a licensure or certification examination that measures a selected set of critical abilities rather than all of the characteristics required for good practice, it is important that the standards on the examination not be set unreasonably high. Although some level of mastery of a critical ability may be necessary for practice, it is not always true that higher levels of the ability will lead to improved performance. Thus, for example, some skill in arithmetic is necessary in many professions, but mastery of higher mathematics would probably not improve performance significantly in most professions.

In general, licensure examinations emphasize knowledge and the ability to apply knowledge because these skills can be measured accurately with written tests. Since cognitive skills are important for professional practice, this is not inappropriate, but the standards for these abilities should not be higher than the level of competence required for practice. If the standards for the cognitive abilities are artificially high, the licensing examination is likely to exclude many candidates who would make good practitioners.

Validity Should Be a Public Function

As noted earlier, to validate an interpretation for an examination is to produce convincing evidence that the interpretation is justified. Since licensure examinations serve a public function, the evidence for validity should be public. That is, the types of evidence suggested in this section should be available for review by the public that licensure procedures, including the licensure examination, are designed to protect. Although the argument for public disclosure is not as strong for certification examinations, a reasonable level of disclosure would also be desirable for these examinations since their effectiveness depends to a large extent on public confidence in the certification process.

Where feasible, the release of sample copies of the examination would serve a useful function in informing discussion and debate about licensure and certification. The periodic release of retired forms of the examinations would not generally have a significant impact on the quality of the examinations and would provide an opportunity for external review of examination content, format, and quality. Complete disclosure of all examinations is probably not justified in most cases because of the costs involved and because the additional benefits of complete disclosure compared to partial disclosure would be marginal.

Since the evidence for the validity of licensure examinations is generally available to interested outside reviewers and since sample copies, or at least sample items from the examinations, are available, I don't see the disclosure issue as a major problem. However, one area in which additional information is probably needed is standard setting. This is especially true because a major criticism of licensure examinations is that the passing scores have been used to restrict entry to the professions in order to protect the interests of the professions (Friedman & Friedman, 1980).

RECENT DEVELOPMENT AND NEW AREAS OF EMPHASIS

There are at least three areas in which improvements in the methodology applied to licensing and certification examinations are needed and, I believe, possible. Two of these have already been touched upon—standard setting and domain specifications. The third area involves the possibility of expanding the scope of critical abilities that are included in the examinations through computerized simulations of practice situations.

Standard Setting

There are two basic problems with current judgmental methods for standard setting. First, the different judgmental methods for setting standards tend to yield different passing scores, and there is no good basis for choosing among them. In addition, different groups of raters using the same method yield different results. Second, the judgmental standard setting methods do not provide a clear basis for interpreting the resulting passing score; rather, the reference populations that provide the basis for norms-based interpretations are simply replaced by a new reference population of raters. Recent developments in judgment-based standard setting (e.g., see Jaeger, 1982) would involve more thorough surveys of the opinions of different types of raters and could therefore probably improve the stability of the results across replications of the procedure, but would not help with the second problem.

Interpretability of Standards. The judgment-based standard setting procedures do not yield an interpretation of what the standards mean in terms of what passing candidates can do, because the results are not explicitly tied to item content. In the Angoff procedure, for example, expert judges are asked to consider the expected level of performance on each item (the probability of answering the item correctly) of hypothetical "minimally competent candidates." The judges are instructed to assign a minimum passing level (MPL) to each item in terms of the probability that a minimally competent candidate could answer that item correctly. Since the cutoff score for the examination is simply the sum of the MPLs for the individual items, it will depend on the sample of items and on the sample of raters.

Unless a behavioral interpretation of the test scores is available, the results of the Angoff procedure do not indicate the kind of behavior that distinguishes passing candidates from failing candidates. Although individual raters undoubtedly use some performance criteria in setting the MPL for each item, (e.g., their individual experiences with persons they considered to be minimally competent), the judgment-based standard setting procedures do not provide a mechanism for making these performance criteria explicit. Therefore, the interpretation of the resulting passing score depends on the criteria for selecting judges, and the burden of interpretation falls on the new reference population, the population of raters.

This concern about the interpretability of test scores in terms of explicit behavioral criteria is not new. Ebel (1962) suggested two methods for obtaining what he called "content standard test scores" over 20 years ago. One of his two methods is similar to the approach suggested below. Nitko (1980) has described a number of ways in which test scores can be referenced to specific content, some of which could be applied to licensure and certification examinations.

Of course, if some copies of the examination are made public along with the corresponding passing score, a reviewer could infer a behavioral interpretation by evaluating the content and difficulty of the examination and comparing the perceived difficulty to the passing score. The reviewer might even use one of the judgmental standard setting procedures to obtain an independent estimate of the difficulty of the examination.

Improving the Interpretability of Standards. An alternative approach that would make the interpretive information more readily available would be to provide data about the differences in performance of passing candidates and failing candidates on a representative sample of items. For example, on a written examination interpretive data of this kind might indicate the proportion of passing candidates and the proportion of failing candidates who correctly answered a particular item. If the topic is important, the question addresses a significant aspect of the topic, and passing candidates answer it correctly more often than failing candidates, such data would indicate that the examination is performing as intended. If the results were reversed and failing candidates did as well or better on the item than passing candidates, the data would suggest that the examination is not working as it should. Therefore, in addition to generating concrete referents for the distinction between passing and failing candidates, this kind of analysis provides a check on the overall validity of the examination process (e.g., see Council of State Boards of Nursing, 1979, pp. 123–127.)

A somewhat more sophisticated approach would be to provide graphs of the proportion of candidates answering an item correctly as a function of total score on the examination. Such graphs would provide detailed information about the implications of total test score for performance on that particular item and would therefore say something about the consequences of setting the passing score at a particular level. Graphs of this kind for a representative sample of items would provide a basis for the interpretation of the test scores in terms of candidate performance on the skills tested by the items.

A Check on Validity. In addition to its impact on interpretability, such approaches could lead to improvements in the setting of standards by providing a check on the internal consistency of the ratings. The minimal pass level (MPL) for an item represents the probability that a "minimally competent examinee" would be able to answer the item correctly. The passing score for the test is the sum of the MPLs for the items in the test. According to the assumptions underlying this procedure, candidates with scores at or just above the passing score can be considered minimally competent. By examining the proportion of these candidates who answer a given item correctly, we obtain an estimate of the probability that a "minimally competent candidate," as defined by the Angoff procedure for the test as a whole, can answer the item correctly. To the extent that this probability differs from the original MPL for the item, there is some inconsistency in the results.

Perfect consistency on judgments about MPLs for different items is not to be expected, and experience with the approach will be needed to determine what constitutes adequate agreement. However, major inconsistencies would suggest a possible problem; for example, if an item is important enough that the raters think that minimally competent candidates should be sure of answering it correctly (i.e., the MPL is 1.0), but candidates with relatively low probabilities of answering the item correctly are passing the examination, the passing score may be too low. At the very least, such comparisons will inform the raters of the fallibility of the standard setting procedure. It also gives the raters the opportunity to reexamine the overall passing score in light of its implications for particular items. This approach would be a natural extension of the Angoff procedure, which is based on raters' judgments of the probability that a minimally competent candidate would get an item correct, but it could be used for any judgmental standard setting procedure.

Because this approach has not been tried out yet, I would not recommend it for immediate application. However, I do think that it would be a fruitful topic for further research.

Definition of Content Domains

The task of defining an appropriate content domain for licensure and certification examinations is extremely important, but the methodology for accomplishing this task is not highly developed. However, in part because I have already discussed it to some extent, and in part because I do not have a very definite program for improvement to propose, I will restrict myself to a few general remarks on this topic.

First, we need to face the fact that the definition of the content domain, like the setting of standards, involves judgments and is therefore subjective to some extent. Attempts to substitute data for judgment in defining the content domain may succeed in diffusing responsibility for the judgments, but it doesn't necessarily improve the domain definition or the examinations developed to reflect the content domain.

I raise this issue as an important focus for study in part because of the emphasis that has sometimes been given to empirical job analyses as a necessary component of content validity (e.g., see Equal Employment Opportunity Commission et al., 1978). Data on how practitioners spend their working time are clearly relevant to the definition of content domains for licensure and certification examinations because they indicate the frequency with which various situations occur in practice. Most job analyses generally collect data on practitioners' ratings of the importance as well as the frequency of various activities, and therefore provide information about practioners' perceptions of what aspects of current practice are most important. Therefore, empirical studies of patterns of practice provide valuable guidance in specifying the range of situations encountered in practice and can be supplemented by the extensive data available, for some professions, at least, on the incidence and severity of various kinds of problems with which practitioners are expected to deal. Such studies indicate what is going on in practice, but unless we adopt the view that "all is for the best in this best of all possible worlds," they do not indicate what should be going on. Given that various activities are given different levels of emphasis in practice, the appropriateness of this distribution of emphasis is still open to question, and the resolution of such issues involves complex judgments.

The Role of the Department of Learning. Given specific situations, decisions about the knowledge and skills needed to deal effectively with three situations can be based on the relevant department of learning. The profession is defined in terms of the department of learning, which provides a body of theoretical and empirical knowledge of the causes, likely courses of development, and appropriate interventions for the situations encountered in professional practice. Such information provides a basis for identifying knowledge and skill required to deal effectively with situations resulting from the job analysis. For example, assuming that one determined, by an empirical job analysis or logical analysis, that a significant part of CPA practice involved the preparation of tax returns, it would probably be better in terms of validity to base a CPA examination on what the federal and state tax codes say can and should be done in determining tax liability than on surveys of what is done; the empirical job analyses would indicate which parts of the tax codes deserve most emphasis, but the items would be based on the tax code.

The most serious limitation in the use of the department of learning is that it is often unwieldy because it is extensive and is not organized in a way that is convenient for test development. The development of a test plan from the department of learning and a job analysis is usually accomplished by content specialists drawn from the profession. However, a readily available and organized source of information that can facilitate the translation of the department of learning into a domain definition for the examination is the textbooks used in professional schools. For reasons outlined earlier, licensure and certification examinations should not simply follow the curricula of professional schools, but the content of these curricula presumably reflect the combined judgments of faculty about what practitioners need to know. If professional school faculty are totally misguided about the demands of practice, society has a much more serious problem than the misspecification of the content domain for licensure and certification examinations. Therefore, textbooks may provide a useful source of information in defining the content domain for licensure examinations.

Empirical Check on Content Domain. A potentially useful, albeit expensive, procedure for empirically evaluating the procedures used to specify content domains was discussed by Cronbach (1971). Applying this general approach to licensure and certification examinations, two versions of the content domain and resulting examinations could be developed independently, based on the common objective of evaluating professional competence, using similar but preferably not the same procedures. The detailed specifications of the content domain and corresponding examinations would be developed by different groups of content specialists and test development experts, and the scores on the two examinations for a sample of candidates would be compared. If the scores on the two examinations were in good agreement, we would have evidence that the choice of experts, the details of the content domain specifications, and the procedures used for examination development do not have an undue influence on the outcome.

Impact of Public Policy on Content Domains. In using information from a job analysis and department of learning to develop licensure examinations, it is important to keep in mind that licensure is a public function controlled by law. Although professional practitioners and content specialists necessarily have much to say about requirements for licensure, including examination content, the public and, more specifically, the public's representatives in state legislatures also have a major interest in such requirements. A change in state law requiring that certain topics be taught in the high school science curriculum (e.g., content relevant to alcohol and drug abuse) would clearly have implications for teacher certification; such requirements, which are motivated by a desire to address social problems, would not necessarily be reflected in current job analyses or in content experts' judgments of what constitutes the core of their academic discipline. The point of this example is not that such specific requirements are commonly included in licensure laws; they are not. However, licensure laws do provide the legal basis for licensure, and although such laws are stated broadly, they incorporate a general view of requirements and restrictions in the practice of the profession being licensed. If the licensure procedures are to follow legislative intent, the content domain for the examination should be consistent with this general view of professional practice.

Obviously these remarks do not constitute a model for content domain specification. At best, they reflect some issues that could be considered in developing such a model.

Expanding the Scope of the Content Domain

My last suggestion of an area for future investigation may be too obvious to mention, but I will do so anyway. The suggestion is that it would be desirable to expand the scope of what is assessed on licensure and certification examinations to give more emphasis to realistic applications of professional judgment and less emphasis to factual knowledge. Current technological developments may offer good opportunities to do so more efficiently and more effectively than has been possible in the past. In particular, computerized simulations of practice situations may provide an effective means for evaluating skills that are not easily assessed in printed examinations (see McGuire, Solomon, & Bashook, 1976).

Simulations generally begin with a description of a client and the circumstances under which the client is first encountered, followed by a series of questions about what actions the examinee would take in order to assist the client. After the examinee has chosen an action, feedback on the results of the action is provided. As the test progresses, the situation is developed by providing additional information in the questions and in the feedback that is provided to the examinee after each response. The aim is to make the descriptions of the situations as realistic as possible and to require that professional judgment be used in deciding what to do. An advantage of simulations is that they make it possible to observe ''performance'' for a large number of simulated clients within a reasonable period of time.

The technology available for use with simulations includes relatively inexpensive microcomputers that are capable of presenting simulated situations and monitoring candidates' performance as they attempt to deal with the problems presented. It also includes videodisc equipment which can present high resolution photographs, as well as video segments involving sound and motion. Therefore, this technology may make it possible to present standardized but highly realistic simulations on an individual basis.

Licensure and certification examinations tend to be quite long, and in many cases good estimates of candidates' mastery of the content domain could be obtained with fewer items. The efficiency of many of the examinations could probably be further enhanced by a judicious allocation of items to subcategories in the domain (see Jarjoura & Brennan, 1982). Therefore, by diverting some of the resources currently devoted to developing multiple-choice items, computerized simulations might be used to expand the range of skills included in the content domain without significantly lengthening the examinations.

Limitations of Simulations for Assessment. In the short term, the practical difficulties of having enough terminals for candidates, maintaining security, and developing software will limit the applicability of this approach. There are also some conceptual problems associated with the use of simulations in licensure and certification examinations that need to be addressed. For example, in branching simulations, a candidate who chooses an option that introduces complications would be asked to deal with these complications and therefore would be required to demonstrate skills that other candidates might not be required to demonstrate. This raises issues of comparability that would not arise in a multiple-choice examination or in a linear simulation, where everyone answers the same questions. Given the importance of both fairness and the appearance of fairness in licensure and certification examinations, this lack of comparability may be viewed as a problem.

A related issue is the adequacy of sampling of content. Because realistic simulations take a significant amount of time to work through, the examination is likely to involve a relatively small number of separate simulations, thus making it difficult to assess a broad domain of content. This problem might be adequately resolved by using multiple-choice items to achieve broad coverage of the domain, while emphasizing the assessment of problem-solving strategies and professional judgment in the simulations.

A potentially serious problem in using simulations as a major component of licensure and certification examinations is their reliability. To the extent that the simulations are highly realistic and involve a high degree of professional judgment, assessments of candidate performance is likely to be variable across simulations. The result would be low reliability reflecting the fact that each candidate's score would be determined to a substantial degree by the choice of simulations employed rather than by the candidate's overall competence. Since the separate responses called for by each simulation are not independent (e.g., a candidate who gets off on the wrong track might find it very difficult to achieve a satisfactory performance on the simulation as a whole), the simulation examination would consist of a relatively small number of "items" (i.e., simulations). Therefore, the improvement in reliability achieved by averaging across a large number of independent items that is possible in multiple-choice examinations would not operate for simulation-based examinations.

In spite of these potential difficulties, I think that simulations offer considerable promise for extending the range of abilities that can be reliably assessed in licensure and certification examinations. It may be necessary to make some compromises between realism and standardization of the content covered, but further research on the properties of simulations should lead to improvements in their effectiveness as assessment instruments.

SUMMARY

There are two basic themes that have guided my discussion of licensure and certification examinations. The first of these themes is that the approach taken to validating the examinations should reflect the intended purpose and interpretation of the examinations. The second theme is that the results of research on the validity of the examinations and related issues, such as standard setting, should provide a basis for public discussion of these issues and cannot be expected to provide simple answers to complex questions.

Validation Strategy

Matching the validation approach to the intended interpretation of the examinations requires an explicit description of the intended interpretation. I have argued that the interpretation of the examination scores in terms of abilities that are necessary for safe and effective practice is appropriate, given the purpose of licensure, the protection of the public.

Given that an interpretation in terms of critical abilities is adopted, the argument for validity involves two components. The first component requires evidence that the abilities to be covered by the examination are critical for practice. Empirical job analyses are particularly useful in identifying situations that commonly arise in practice and in providing some data about the consequences of proper and improper handling of such situations. The department of learning provides information on how these situations should be handled to obtain optimal results. In combining these two sources of data, professional judgment plays a large role.

The second component requires evidence that the test scores reflect competence in the domain of critical abilities defined in the first step. Much of the evidence relevant to this issue is derived from the critical examination of the properties of the test in the light of plausible counterhypotheses about what the test measures.

As noted earlier, the critical abilities approach involves aspects of content validity, criterion validity, and construct validity. Each of these three standard validation strategies focuses on a specific type of interpretation, and the interpretation of licensure and certification examinations in terms of critical abilities involves aspects of all three of these standard interpretations. The methodology of content validity is central to defining the content domain and providing evidence that the examination samples the domain adequately. The emphasis on establishing the link between the critical abilities included in the domain and practice involves indirect criterion validation. The examination of plausible counterhypotheses involves issues usually considered under construct validity.

Informing Public Discussion

The second major theme of this discussion is that licensure and certification are public functions, subject to public scrutiny. The appropriate scope of practice for licensed professionals is a matter of public policy, which is encoded, at least in general terms, in licensure laws. The scope of practice reflected in voluntary certification procedures is determined by the professional judgment of the certifying agency but is also shaped by public expectations.

Therefore, research on the validity of licensure and certification examinations is more akin to policy research than it is to basic scientific research. The situations that practitioners should deal with, the types of interventions that they should employ, and the standard of skill to be expected in the implementation of these interventions are not questions that can be answered by empirical data; they are policy issues.

This suggests that the results of validity studies that attempt simply to determine whether the test is valid will be less helpful ultimately than studies that provide information useful in making policy decisions and, more fundamentally, in informing debate about policy issues. Questions particularly relevant to policy decisions include:

1. What types of situations occur most frequently in practice and/or have the most serious potential consequences?
2. What abilities, including knowledge and skills, are needed to deal with these situations effectively?
3. How well are these abilities taught and assessed in the educational programs preparing practitioners?
4. How well does the examination assess competence in the critical abilities, and what sources of variance other than differences in competence (e.g., reading level, response sets, ethnic bias) influence examination scores?
5. What are the implications of setting standards at different levels?

Although research on the validity of licensure and certification examinations will not resolve such issues, by addressing questions like those listed above, it can promote informed decision making.

REFERENCES

American Psychological Association, American Educational Research Association, & National Council on Measurement in Education. (1984). *Joint technical standards for educational and psychological testing* (4th draft). Washington, DC: American Psychological Association.

Andrew, B. J. (1976). Validation by task analysis. In *Conference on extending the validity of certification* (pp. 43–46). Chicago: American Board of Medical Specialties.

Andrew, B. J., & Hecht, J. T. (1976). A preliminary investigation of procedures for setting examination standards. *Educational and Psychological Measurement, 36,* 45–50.

Angoff, W. H. (1971). Scales, norms, and equivalent scores. In R. L. Thorndike (Ed.), *Educational Measurement* (pp. 508–600). Washington, DC: American Council on Education.

Brennan, R. L. (1983). *Elements of generalizability theory.* Iowa City, IA: American College Testing Program.

Brennan, R. L., & Lockwood, R. E. (1980). A comparison of the Nedelsky and Angoff cutting score procedures using generalizability theory. *Applied Psychological Measurement, 4,* 219–240.

Cohen, J. (1960). A coefficient of agreement for nominal scales. *Educational and Psychological Measurement, 20,* 37–46.

Council of State Boards of Nursing. (1979). *Examining the Validity of the State Board Test Pool Examination for Registered Nurse Licensure.* Kansas City: American Nurses' Association.

Cronbach, L. J. (1971). Test validation. In R. L. Thorndike (Ed.), *Educational Measurement* (2nd ed., pp. 443–507). Washington, DC: American Council on Education.

Cronbach, L. J. (1980a). Selection theory for a political world. *Public Personnel Management, 9,* 37–50.

Cronbach, L. J. (1980b). Validity on parole: How can we go straight? *New Directions for Testing and Measurement, 5,* 99–108.

Cronbach, L. J., Gleser, G. C., Nanda, H., & Rajaratnam, N. (1972). *The dependability of behavioral measurement.* New York: Wiley.

Cronbach, L. J., & Meehl, P. E. (1955). Construct validity in psychological tests. *Psychological Bulletin, 52,* 281–302.

Ebel, R. L. (1962). Content standard test scores. *Educational and Psychological Measurement, 22,* 15–25.

Ebel, R. L. (1972). *Essentials of educational measurement.* Englewood Cliffs, NJ: Prentice Hall.

Equal Employment Opportunity Commission, Civil Service Commission, Department of Labor, & Department of Justice. (1978). Adoption by four agencies of Uniform Guidelines on Employee Selection Procedures. *Federal Register, 43,* 38290–38315.

Equal Employment Opportunity Commission, Office of Personnel Management, Department of Justice, Department of Labor, & Department of the Treasury. (1979). Adoption of questions and answers to clarify and provide a common interpretation of the Uniform Guidelines on Employee Selection Procedures. *Federal Register, 44,* 11996–12009.

Flanagan, J. C. (1954). The critical incident technique. *Psychological Bulletin, 51,* 327–357.

Friedman, M., & Friedman, R. (1980). *Free to choose: A personal statement* (pp. 239–240). New York: Harcourt Brace Jovanovich.

Gonnella, J. S., Goran, M. J., Williamson, J. W., & Cotsonas, N. J. (1970). Evaluation of patient care. *Journal of the American Medical Association, 214,* 2040–2043.

Gross, S. J. (1978). The myth of professional licensuring. *American Psychologist, 33,* 1009–1016.

Hecht, K. A. (1979). Current status and methodological problems of validating professional licensing and certification exams. In M. A. Bunda & J. R. Sanders (Eds.), *Practices and problems in competency-based education* (pp. 16–27). Washington, DC: National Council on Measurement in Education.

Hoffman, P. (1977). Continued competence assurance: Some research and measurement considerations. In *Proceedings of a national conference for evaluating competence in the health profession* (pp. 25–32). New York: Professional Examination Service.

Hogan, D. (1979). *The regulation of psychotherapists: Vol. 1. A study in the philosophy and practice of professional regulation.* Cambridge, MA: Ballinger.

Hubbard, J. P. (1971). *Measuring medical education: The tests and procedures of the National Board of Medical Examiners.* Philadelphia: Lea & Febiger.

Hunter, J. E. (1980). Construct validity and validity generalization. In *Construct validity in psychological measurement* (pp. 119–125). Princeton, NJ: U.S. Office of Personnel Management & Educational Testing Service.

Jaeger, R. M. (1982). An iterative structured judgment process for establishing standards on competency tests: Theory and application. *Educational Evaluation and Policy Analysis, 4,* 461–475.

Jarjoura, D., & Brennan, R. L. (1982). A variance components model for measurement procedures associated with a table of specifications. *Applied Psychological Measurement, 6,* 161–171.

Kane, M. T. (1982a). The validity of licensure examinations. *American Psychologist, 37,* 911–918.

Kane, M. T. (1982b). A sampling model for validity. *Applied Psychological Measurement, 6,* 125–160.

Kastrinos, W., & Livingston, S. A. (1979). *The development of a proficiency examination for dental auxiliaries.* Princeton, NJ: Educational Testing Service.

Levine, M. (1980). A potential unanticipated consequence of legislation to protect human subjects. *American Psychologist, 35,* 583–584.

Lowenthal, W. (1981). Continuing education for professionals. *Journal of Higher Education, 52,* 519–538.

Manning, W. H. (1978). Test validation and EEOC requirements: Where we stand. *Personnel,* 70–77.

McGuire, C., Solomon, L., & Bashook, P. (1976). *Construction and use of written simulations.* New York: Psychological Corporation.

Menges, R. J. (1975). Assessing readiness for professional practice. *Review of Educational Research, 45,* 172–203.

Miller, R. B. (1963). Task description and analysis. In R. M. Gagne (Ed.), *Psychological principles in system development.* New York: Holt, Rinehart & Winston.

National Commission for Health Certifying Agencies. (1981). *Guidelines for membership criterion report for validity.* Washington, DC: Author.

Nedelsky, L. (1954). Absolute grading standards for objective tests. *Educational and Psychological Measurement, 14,* 3–19.

Nitko, A. J. (1980). Distinguishing the many varieties of criterion-referenced tests. *Review of Educational Research, 50,* 461–485.

Pottinger, P. S. (1979). Competence testing as a basis for licensing: Problems and prospects. In M. A. Bunda & J. R. Sanders (Eds)., *Practices and problems in competency-based education* (pp. 28–46). Washington, DC: National Council on Measurement in Education.

Rakel, R. E. (1979). Defining competence in specialty practice: The need for relevance. In *Definitions of competence in specialties of medicine, conference proceedings* (pp. 89–96). Chicago: American Board of Medical Specialties.

Reed, G. F. (1978). Experience with the patient simulation oral examination procedure American Board of Otolaryngology. In *Conference on Research in Evaluation Procedures* (pp. 14–20). Chicago: American Board of Medical Specialties.

Sarnacki, R. E. (1979). An examination of test wiseness in the cognitive test domain. *Review of Educational Research, 49,* 252–279.

Schmeiser, C. B. (1982). Use of experimental design in statistical item bias studies. In R. A. Berk (Ed.), *Handbook of methods for detecting test bias* (pp. 64–95). Baltimore: The Johns Hopkins University Press.

Shepard, L. (1980). Standard setting issues and methods. *Applied Psychological Measurement, 4,* 447–467.

Strupp, H. H., Hadley, S. W., & Gomes-Schwartz, B. (1977). *Psychotherapy for better or worse, the problem of negative effects.* New York: Aronson.

Werner, E. (1981). *Review of the examination for professional practice in psychology (EPPP).* Sacramento: California Department of Consumer Affairs.

Williamson, J. W. (1976). Validation by performance measures. In *Conference on extending the validity of certification* (pp. 21–42). Chicago: American Board of Medical Specialties.

Williamson, J. W. (1979). Improving content validity of certification procedures by defining competence in specialty practice: Directions, resources, and getting started. In *Definitions of competence in specialties of medicine, conference proceedings* (pp. 61–86). Chicago: American Board of Medical Specialties.

III CLINICAL, COUNSELING, AND ORGANIZATIONAL DIRECTIONS

7 The Future of Clinical Assessment

Jay Ziskin
Professor Emeritus, California State University, Los Angeles

At the outset, this writer's frame of reference should be made clear. It is the perspective of the author of a book that contains more than five hundred references from the scientific and professional literature which raise doubts about the expertise of clinicians in their role as diagnosticians in general, and also, specifically, within the forensic arena. The perspective further is that of one who provides a consultation service to lawyers pointing out to them the weakness and shortcomings of psychiatric and psychological reports and testimony both in terms of the inherent problems and of any specific deficiencies of omission or commission in the particular case. Thus, clinical assessments are often seen in their most public form and under conditions in which weaknesses and deficiencies are most vulnerable to exposure.

Clinical assessment has been defined as ''the process by which clinicians gain understanding of the patient necessary for making informed decisions'' (Korchin, 1976, p. 124). Korchin and Schuldberg (1981) elaborate,

> Clincal diagnosis, in the restricted sense, may be included, but more usually the intent is description and prediction towards the ends of planning, executing, and evaluating therapeutic interventions and predicting future behavior. Any of numerous techniques can be used, singularly or in combination, depending on the orientation of the clinician and the specific questions for which answers are sought. Thus, interviews with the client or with others; observation in natural or contrived situations; or the use of tests of different functions, varying in breath, objectivity, psychometric refinement, and inference might all be included. The immediate goal may be the relatively precise measurement of a particular psychological function or the construction of a 'working image or model of the person' (Sundberg & Tyler, 1962), (p. 1147).

Generally, clinical assessment is distinguished from other types of psychological assessment such as educational assessment or personnel assessment by its focus on determination of the presence or absence of psychopathology or deviance—that is, problems or discomforts the individual is having with himself/herself or problems or discomforts the individual is causing to society or other people. Clinical assessment concerns itself not only with the nature of the psychopathology but also with its extent, the implications of its nature and extent for the individual's prospective functioning, the potential for altering such functioning and the means to accomplish such alteration.

THE CURRENT STATUS OF CLINCIAL ASSESSMENT

The current status of psychological assessment generally, and psychological testing particularly, could be described as paradoxical. The demise of psychological testing has been announced often enough over the past two decades. Yet there has been no funeral. There is no corpse because life has not yet left the body and indeed, there are those who assert that not only is the patient not dying, but that in fact this patient is on the way to recovery (Korchin & Schuldberg, 1981). Rorer and Widiger (1983) state,

> It is no secret that personality assessment has been in big trouble as it has come under attack from both expert and lay critics. Assessment takes up a decreasing proportion of the professional practitioners time, occupies a place of decreasing importance in the university graduate curriculum and has been legally outlawed in many selection situations. Many have reacted by jumping what they believe to be a sinking ship, others have come to the defense of the establishment, and have argued that with a few refinements we can continue with business as usual. . . . Clearly (to these reviewers), it is not a time for business as usual nor is it time to abandon ship. Rather, it is a time to question our basic assumptions. (p. 433)

Several articles (Davids, 1973; Leavitt, 1973; Lewandowski & Saccuzzo, 1976; Petzelt & Craddick, 1978) have indicated that many graduate programs in clinical psychology have de-emphasized the teaching of psychological testing. Nevertheless, it seems that employers of clinical psychologists by an overwhelming majority continue to consider the capacity to perform psychological testing as one of the major requirements for employment at their facility. The courts, usually perceived as bastions of conservative hard headedness when it comes to the admission of evidence, no longer debate the admissibility of conclusions based on psychological testing, despite a mountain of evidence (Ziskin, 1981) which suggests that there is too much doubt about these procedures for them to meet admissibility requirements that continue to keep other types of evidence such as conclusions based on polygraph examinations out of the court room. Judging by developments in the field of psychology and law such as the establishment of Division 41 of the American Psychological Association, a division on psychology and law, and the establishment of the American Board of Forensic Psychology and my own experience, it seems more likely that the use of clinical assessment in the legal situation is expanding. This conflicted or sick, but not dying, status of the field arises from a number of factors.

In part clinical assessment owes its continued existence to a serious need for its product. With a multiplicity of treatment methods and particularly with a proliferation of psychotropic medications, it has become more necessary than ever to evaluate the nature of the patients problems in order to utilize the most appropriate kind of treatment and/or medication. Similarly in the forensic situation, the questions of state of mind or psychological capacities or propensities are of extreme importance so that any modality which has some degree of respectability and purports to provide answers is eagerly welcomed. Further, assessment is a field in which new methodologies and/or revisions of old methodologies occur with great frequency such that there is always the hope that today's new method will provide "the answer." The vast size of the graveyard of yesterday's great hopes does not appear to dampen enthusiasm.

Several factors impede the progress of clinical assessment toward a healthy state.

1. Lack of an Adequate Knowledge Base

Perhaps the most important of all, is the absence of a large, relevant, adequately validated body of knowledge on which to base clinical assessment. The mental health field consists of a conglomeration of unvalidated theories about human behavior, psychological conditions and so on. Havens (1981) states,

> Psychiatry as an agreed on body of knowledge hardly exists, instead we have a variety of psychiatrys. Psycho-analytic psychiatry, biological and behavioristic psychiatry, social and interpersonal psychiatry, existential analysis—the list can be made even longer. (p. 1279)

All of these theories possess some supporting data and all have a body of followers and indeed, many of them have substantial bodies of followers. But the existence of so many different theories (and they *are* different) defines the problem because in order to know what it is that one ought to be assessing one has to have a basis in knowledge of what variables are relevant in human psychology and how those variables interact. Thus, it could be argued for example, that many of the present ills of clinical assessment spring from the fact that during the growth period of clinical assessment, the most popular theory of human behavior was the psychoanalytic theory and thus much of the early development of testing revolved around attempts to assess psychoanalytic variables. Given the skepticism concerning psychoanalytic theory that has emerged in the past few decades, it may be no wonder that the enterprise failed.

2. Lack of an Adequate Classification System

A parallel problem has been the lack of an adequate classification system for diagnosing or classifying mental disorders. While it can be, and is, argued by some clinicians that formal diagnoses are not necessary for treatment purposes, the fact is that most treatment modalities, procedures and medications are based on formal diagnoses. The dismal history of the official diagnostic classification system, the diagnostic and statistical manuals published by the American Psychiatric Association ought to be well known to everyone in the field. DSM-I was published in 1952 and found to be quite poor and was replaced by DSM-II in 1968. It soon became apparent that DSM-II was not adequate to the task and work was shortly thereafter commenced culminating in the production of DSM-III in 1980, representing a radical departure from the previous manuals. While one can readily acknowledge that DSM-III is a considerable improvement over its predecessors, as it does specify with some clarity what the criteria are for the various mental disorders, recognizes psycho-social factors, and provides preliminary reliability research data, there is little reason to hope that DSM-III will prove to be an adequate classification system.

Regarding DSM-III, Eysenck, Wakefield, and Friedman (1983) provide an extensive reivew and state the conclusion that "This new scheme is based on foundations so insecure, so lacking in scientific support, and so contrary to well-established facts that its use can only be justified in terms of social need" (p. 167). They warn psychologists of the weaknesses of a scheme based on democratic voting rather than scientific research and they assert that the reliabilities are unacceptably low and there is a lack of indication of validity. In 1982 at the Annual Meeting of the American Psychiatric Association, a debate was held under the title "Do the advantages of DSM-III outweigh the disadvantages?" The debaters were Dr. Gerald L. Klerman, Dr. Robert L. Spitzer who was the chairman of the committee which developed DSM-III, Dr. Robert Michels and Dr. George E. Vaillant, all psychiatrists of some eminence. This debate was taped and can be obtained from the American Psychiatric Association. One outstanding characteristic of the debate was the absence of a strong assertion by speakers on either side indicating that DSM-III really is a "good" diagnostic system. Even more impressive was the unanimity with which each of the participants referred to the coming of DSM-IV.

The literature contains numerous negative assertions concerning DSM-III (Ziskin, 1981). One must consider the effects of alterations in the definitions of various psychopathological entities every decade or so on the validity, meaningfulness and applicability of previous research. It seems quite likely that a research population described as "schizophrenic" in the 1960s would be a different population from a clinical population described as schizophrenic in the 1980s. Keisling (1981) found that when a group of patients admitted to St. Elizabeths hospital in 1979 and 1980 were re-diagnosed on the basis of DSM-III criteria, the proportions of those diagnosed with schizophrenia and those diagnosed manic-depressive were altered radically. Therefore, in terms of application, one cannot apply the previous reasearch to patients presently diagnosed with the same label. At the very least, the clinician must entertain doubts regarding the application of research performed on a group of patients designated as having a certain disorder in the past to patients designated as having that disorder in the present when it is known that there is a strong possibility that the two patient populations might be different.

Probably the only safe course for clinicians is to disregard any research done prior to the publication of DSM-III for conditions where descriptions have been changed. Unfortunately, this would leave them in the position of trying to deal with entities for which virtually no research information would be available for several years. Further, by the time the research is completed and published, DSM-IV will be out and the process will have to start all over again.

3. Situation and Examiner Effects

Situation and the examiner effects are problems that have plagued clinical evaluation as evidenced in the literature over the past several decades continuing up to the present. See, for example, (Anastasi, 1982; Arkes, 1981; Bartol, 1983; Treece, 1982). "Situation effects" refers to the contamination of data obtained in a clinical examination by temporary events which surround the time of the examination. For example, if the subject has had a fight with his wife that morning, or if he has been scared out of his wits by a close call on the freeway or if she is involved in a lawsuit out of which she hopes to obtain a great deal of money for a relatively minor injury and is very anxious about the outcome, all such events can have an effect on the individual's psychological state at the time of the examination and cause him to produce data which could easily be seen by the clinician as an enduring characteristic. This has been more broadly referred to in the state vs. trait controversy which involves the problem of trying to tease out of the data that which represents relatively permanent characteristics of the individual versus that which is a resultant of some temporary condition. "Examiner effects" refers to the influence of the examiner and examiner subject interaction, not only on the data that is produced but on the data that is attended to and recorded and the interpretation of the data as well. Decades of studies (Ziskin, 1981, Chapter 6) have shown that the data produced and recorded and the interpretation of the data are influenced by such factors as the theoretical orientation of the examiner, personal characteristics of the examiner such as age, sex, race and socio-economic status, training and experience, personality characteristics and appearance as well as social or political values and attitudes and the expectations of the examiner. The effect of situational and examiner variables in reducing the reliability and validity of clinical evaluations should be obvious. It is difficult to avoid despair concerning an evaluation process in which the outcome is partially determined by time, place and purpose and which one of many examiners conducted the examination.

4. Multicultural Issues

Research, particularly in the last decade and a half, has raised questions concerning the assessment of members of ethnic minority groups and has left those questions unresolved. Thus, a plethora of studies (Ziskin, 1981, Chapter 9) indicate that it may be inadvisable to assess members of ethnic minority groups on the basis of white majority normative data. These studies indicate that there may be significantly different response patterns for members of ethnic minority groups particularly on tests of personality or psychopathology or that even where the response patterns may be similar that the behavioral correlates of the responses may be different for ethnic minority members (nearly 50 such studies are reported in Ziskin 1981). Some research suggests that these ethnic differences disappear when education level and socio-economic level are controlled ((Bertelson, Marks, & May 1982; Davis, 1975; Davis, Beck, & Ryan 1973; Davis & Jones 1974). On the other hand several researchers have found that statistically significant differences do exist in test data even when education level and/or socio-economic status are controlled (Brown, 1974; Cross, Barclay, & Berger, 1978; Holland, 1979; Lowe & Hildman, 1972). These issues were discussed in several papers at the first multiethnic conference on assessments held in Tampa, Florida in March of 1982 (Raymond D. Fowler, University of Alabama, Chair). The fact that such a conference was held suggests that some of these issues have not been resolved. Similarly the problem of assessment of the members of ethnic minority groups was mentioned by several presenters at the 1983 program of the Society for Personality Assessment held in San Diego, California. It is clear that this is a problem that has not gone away and it remains to be seen whether renorming of many tests with special norms for members of ethnic minority groups (Gynther, Lachar, & Dahlstrom, 1978) or some other solution would be the answer.

5. Ineffectiveness of Experience

A matter of considerable concern is the apparent inability of clinicians to improve their diagnostic reliability and validity as a result of experience. More than fifty publications (Ziskin, 1981) mostly within the last decade and a half indicate that experienced clinicians are no more reliable or accurate than are inexperienced clinicians and indeed a few studies indicate they are no more accurate than nonclinicians. These findings raise a serious question as to whether there is indeed a teachable and learnable skill of clinical assessment. The fact that experience does not sharpen such skills seems to suggest that the answer is negative and thus, raises a question of whether time is being wasted in graduate education and, indeed, raises the question as to whether the enterprise should even be continued. Of course, it may be possible that some clinicians are very good at assessment on some basis other than training and experience.

6. Illusory Correlation

Illusory correlations create another serious problem for assessment (Chapman & Chapman, 1967). As used originally by Chapman and Chapman, illusory correlation describes a process wherein the clinician *thinks* that she observes a relation between an item of behavior and some psychological variable when, in fact, no such relationship can be shown to exist. The Chapman's use the example of the hypothesis from the Draw-A-Person test that drawing of large eyes or emphasizing eyes is associated with paranoia or paranoid tendencies. While the hypothesis is logical, the research literature fails to substantiate it. Therefore, the sign is invalid, the relationship is ''illusory.'' Large amounts of what clinicians are taught in their training consists of such illusory correlations and then in the course of their practice, clinicians reinforce such false beliefs in each other by repeating what they have been taught or what they think they have observed and to be sure from time to time someone who draws large eyes does turn out to be paranoid so that there is always a certain amount of confirmation. These myths are perpetuated and become principles of assessment. There is an urgent need for clinical assessment to shed its mythologies.

7. Base Rates

A similar problem exists with regard to ignoring population base rates. What this means is that in many instances, behavior that is more likely than not within the realm of normal is seized upon and twisted and distorted to make it into a symptom of psychopathology (Rosenhan, 1973; Ziskin, 1981). I worked as a consultant on a case in which a very wealthy man who had been going around the country making substantial investments to the point where his family was worried that he might dissipate the fortune (all of which he had made himself), and managed to get him to return home in the middle of the night on the ruse that his wife was very sick and needed his presence. He caught several connecting flights and then drove for another 2 or 3 hours in the early hours of the morning to be greeted on the porch of his home by his perfectly healthy wife and family and four husky deputy sheriffs who told him that he needed to go to the hospital. He disagreed very strongly with their recommendations whereupon they placed him in restraints and took him off to the mental hospital. The psychiatrist's report starts with this sentence, ''On admission Mr. X was hostile and belligerent'' presumably as an indication of the psychopathology later diagnosed as a manic state. Who would not be hostile and belligerent under those circumstances? Certainly it is more normal to be angry under those circumstances than to be placid and accepting of having been fooled and forcibly hospitalized. Statements of this kind abound in clinical reports almost as though the clinician is determined to see everything through pathology colored glasses and rarely recognizes normal behavior when she sees it. This is carried to the point of absurdity in countless reports I have read in which the clinician states that the individual's scale score of 9 on the WAIS shows that he is ''below average'' on that particular dimension.

8. Art vs. Science

There is controversy as to whether clinical assessment is and/or should be an art or a science with the most common opinion holding that it is a mixture of both. One can readily point to an analogy in medicine which is often described as an art based on science. I suspect that in clinical psychological assessment as well as in medicine, art begins where science leaves off. By that, I mean that generally speaking the professional would prefer to be able to generate conclusions based on hard scientific data, but where such data is lacking, art or intuition or whatever the clinician wishes to call it must necessarily be employed, although there is a seldom used alternative called ''I do not know.'' Neuropsychological assessment may provide a useful example in this regard. In the late 1950s we were trained to assess brain damage with a combination of blunt instruments such as the Rorschach and the MMPI along with a Bender-Gestalt and the scatter patterns on a WAIS along with some behavioral observations, perhaps, in conjunction with or as a result of an EEG. Our knowledge of brain functioning was limited. We were forced to rely on certain signs that most of the time left us with equivocal conclusions and consequently with unsatisfactory validity (Goldstein & Deysach, 1973). Advances since then in computerized axial tomography and the development of neuropsychological batteries such as the Halstead-Reitan and the Luria-Nebraska allow not only conclusions of brain damage or disease at commonly reported rates of accuracy between 80% and 90% but enable the clinician to assess fairly well the locus and functional significance of the damage. It is, of course, true that some of the old tests are included within these neuropsychological batteries and that the behavioral observations of the clinician still play some role. The heart of the procedures however, seem to fall to a much greater extent within the area of science with its formulas and quantification. Given the relatively short life of these methods, it is not at all unreasonable to anticipate that in the near future they will be producing results with even higher rates of accuracy.

9. Assessment of "Whole" or "Part."

Whether and when to attempt assessment of some specific attribute or to assess the ''whole'' person is another issue that must be resolved. In the new era of accountability and reduced availability of funds for health care, there will be less tolerance for diagnostic procedures that are expensive, inefficient or unproven. Sometimes clinical assessment may not require anything more than the question "Tell me about your problems."

At least the foreseeable future is likely to see a continuation of this era of accountability and restriction of funds for both health research and health care. This suggests that clinical assessment is going to have to prove its value just as recent years have been increasing government pressure on psychotherapy to prove its value.

10. Derivation of Clinical Conclusions

Another issue that plagues the field is the general inability of clinicians to explicate the bases for their conclusions and the processes that led them to the conclusions. Possibly this problem is more noticeable in the courtroom setting where penetrating questions can be asked of the clinician as to how conclusions were derived in contrast to the clinic setting where, except for occasional case conferences, the clinician is not called upon to justify or explain the source of his conclusions. My experience has been that in the courtroom setting, it has been virtually impossible to get satisfactory information from the clinician in response to the question "What is the source of that conclusion?" The response almost invariably is more or less of the type "Not from any one thing, but from all of the data taken together." Persistent questioning by the cross-examining lawyer seldom clarifies the basis of conclusions, producing only the impression that the clinician does not know how the conclusions were derived.

11. Problems in Computer Interpretation

Adair (1978), reviewing automated or computerized MMPI interpretive services, generally notes that the question of validity in personality measurement continues to be a problem and that validity studies must be continued as a constant check on the accuracy of computer generated personality reports. He notes that some validational studies by various services have been done and "showed some promise." Butcher (1978) notes that most of the computerized interpretations are not pure actuarial systems but stem from programmed clinical decision rules utilizing clinical lore as the basic data in many cases. Butcher states:

> At this stage *computerized narratives using psychological test based information is little more than an art (or craft) disguised as a science.* For the most part, the narrative reports are clinical hunches (often many steps removed from data) which are automatically cranked out by an electronic beast that will, without conscience, weave a devastating and sometimes contradictory tale about an individual's personality and problems. The computer is a generally willing and efficient servant that will readily combine and give back stores of information from its vast memory. It cares not at all whether the information stored is from astrology charts, MMPI code

> books, Rorschach indices, or Somatatype descriptors . . . the "artisan" nature of this endeavor has been demonstrated, the "clinical" astuteness is often compelling, but the "science" is often neglected or of tertiary consideration. (p. 942)

Butcher additionally asserts that these clinical "hunches" are given more credibility than actually is deserved because of the aura of scientific mystique associated with computer outputs. He states:

> The computer approach to personality assessment has been "oversold" and users place more stock in the "scientific truth" than is actually deserved. (p. 943)

He states further:

> Once an MMPI interpretation program is written to print out sets of statements to given T score elevations, etc., a computer system can, in a matter of minutes, process thousands of cases producing an amount of halftruths and misstatements of staggering proportions. (p. 943)

He states further:

> By far the most haunting problem and serious shortcoming of the automated MMPI assessment approach remains that of system validation. Demonstrating the validity of computer-generated narratives (like that of demonstrating clinical interpretations generally) is a formidable task. (p. 944)

Butcher points out that several attempts at validating narrative reports have been published but the criteria employed, frequently consumer ratings of acceptability or judged accuracy, are inadequate to provide a demonstration of validity.

THE FUTURE

My attempt to predict the future course of clinical assessment basically will follow two paths. One path springs from what seems to be reasonable extensions of trends that can already be discerned. The other path springs from my imagination, including an out of character optimisim with regard to what can be accomplished in clinical assessment, if not now, at least in the foreseeable future.

1. Use of Computers

Despite the stringent warnings of Butcher and others given earlier, there can be no doubt that computers are going to make an enourmous impact on assessment (Jackson & Paunonen 1980). The "Actuarial vs. Clinical" controversy triggered by Meehl's famous publications on that topic portends this development. My summarization of the writing in this particular area of actuarial vs. clinical assessment suggests that the demonstrated superiority of the actuarial method has not yet made its impact on assessment because of the limited instances in which the actuarial formulas were available. However, it seems clear that this problem is being remedied to some extent by existing computer programs and one cannot really doubt that the capacities of the computer will facilitate the development of many more formulas. The speed with which the computer can analyze data and its capacities for storing and processing large amounts of information that are totally beyond the capacities of the individual clinician insure this development. One need only glance at the long running advertisements in the *APA Monitor* offering computerized interpretation services to recognize that this approach has already established its economic viability. The task that remains to be accomplished is that of systematic validation. Most services report high degrees of customer satisfaction and/or concurrence as evidence of validity. However, "customer satisfaction" cannot be a substitute for published validation studies.

Even more challenging than validation studies (and perhaps likely to increase validity) is the possibility of combining data from different sources into automated interpretive programs. At present the interpretive systems are mostly associated with the MMPI, although some indications of automated Rorschach interpretations have appeared. It does not seem utopian however, to imagine an automated assessment program which includes within it, for example, not only MMPI data and MMPI interpretive statements, but incorporates Rorschach data and interpretive statements as well as demographic and interview data. The size of the project may seem staggering but probably no more so than the idea of being able to feed data into a machine which was cabable of responding with more than twenty thousand different interpretive statements might have seemed 30 years ago. One day an operator may transmit into the computer an MMPI profile code along with the ratios from the Exner Comprehensive Rorschach System along with other similar data from these two tests and perhaps others plus demographic data plus quantified interview data. Such quantified interview data can be provided by means of structured interview procedures combined with the use of rating scales which will enable the interviewer to translate the data and even the behavioral observations into quantified form. Thus for example, the clinician rather than writing a full report saying that the patient showed "flat affect" will be able to punch in a number on a statement concerning range of affect. In this manner, it may become possible for the computer to actually do what clinicians say and think they actually do when they utilize information from a number of diagnostic sources in reaching their conclusions. Thus for example, when the clinicians say they use a battery of tests it has been my experience that they use relatively little information from the battery. The computer however, will be able to use vastly greater amounts of such information.

2. Classification Systems

There will almost certainly be a new classification system. Judging from the debate held by the American Psychiatric Association in 1981, soon there will be a DSM-IV that is likely to differ in substantial respects from DSM-III. Clinicians upset over particular omissions are already lobbying for their inclusion in DSM-IV. Similarly, certain classifications, such as "schizo-affective," and "borderline" have come under considerable fire with their validities being questioned. No doubt some of the current categories will be deleted. However, these are only specific content changes and one does not readily forsee an abandonment of the particular model. In contrast, several years ago, the American Psychological Association established a committee to look into the development of a more behavioral description type of classification system. This committee appeared to conclude that while such a system would be valuable, the cost of developing it was out of reach at the time, the late 1970s and early 1980s. However, sooner or later the issue of whether to continue with the DSM type of diagnostic system or to shift to some other approach will have to be faced. Already statements in the literature favoring a dimensional over a categorical approach have appeared.

3. Greater Focus on Assets

Much greater attention to strengths or assets should be expected. Clinical assessment has for much too long been almost totally absorbed in reciting psychopathology so that the reading of a clinical report is almost always like listening to a symphony of defects, deficiencies, problems, stresses and so on. Yet the modern drift of treatment approaches tends to focus on problem solving with many therapies attempting to build on the strengths the patient already has or to augment those strenghts to help the individual to function better. In light of this the assessment will have to take into account not only psychological strengths but also assets such as good looks, high intelligence, a wealthy family, a helpful spouse or whatever other assets the patient may bring to the situation.

4. Situational Variables

Similarly psychosocial stresses present in the patient's life space will also require assessment. A step in this direction was taken in DSM-III, but the calibration appears to be crude and without any particular scientific foundation. Nonetheless, the idea is a good one and needs a more careful calibration as to the rating of stresses inherent in various kinds of situations plus some rating of the stress for the particular individual in a given situation. Along these lines it seems most likely that a state vs. trait controversy will dissolve into recognition that while there may be enduring traits or characteristics that an individual has, these traits will become operative or will operate in a different manner according to various situations or contexts. It will be important for a therapist to have any significant situational information.

5. Projective Tests

Projective tests which have long been highly controversial within psychology face a future of change or disappearance. Those devices which are amenable to change to make them more scientific are likely to survive. Those that are not amenable or can be modified only with great difficulty are likely to disappear. Thus for example, the Draw-A-Person (DAP) technique which has a long history, would appear to me to have a very short future. Over its lengthy history numerous attempts have been made to validate its propositions. The net sum of all of this effort is most discouraging (Adler, 1970; Swenson, 1957). Very few of the existing DAP hypotheses have been validated. Economy and ease of administration do not compensate sufficiently for the failure of the test to provide conclusions about an individual in which one who reads the literature could have confidence. On the other hand, the Rorschach, long the object of vehement and derisive attacks by those alleged to be "hardnosed" unsympathetic scientists may very well be on the verge of a re-birth with John Exner and his colleagues as the attending physicians. Exner (1974, 1978) has pulled together diverse approaches and methodologies of the Rorschach into what appears on the way to becoming a unitary and standardized procedure which may bring it within the purview of science. Indeed, Exner and his associates are treating the Rorschach in a scientific manner, performing research to determine the reliabilities of various components and the validities of various interpretive principles. Clearly, they have met with impressive early successes. It is premature to conclude that the Exner Rorschach has now been fully validated, but at least it is being subjected to validation procedures and in time it is likely that we will know what it can do and what it cannot do.

6. Recognition of Limits

This leads me to another prediction for the future, one which I make with some hesitation. That is, I think the future will see those who do clinical assessment shrinking the territory somewhat and abandoning, at least in an applied sense, those areas for which the field is simply not ready. The two examples which come most readily to my mind are from the forensic area. One is the attempt to assess "dangerousness." It is clear from an overwhelming body of literature that neither the knowledge nor the methdology exists to do this with a respectable degree of accuracy. Already the American Psychiatric Association and the American Psychological Association have acknowledged the limitations of the field in this regard. Similarly, attempts to assess a defendant in a criminal case with regard to his state of mind at the time he committed the crime, usually weeks or months before he was seen by the clinician, will be recognized as beyond the clinician's present capability.

7. Standardized Interviews

The future is likely to see more use of structured or standardized interview methods. These methods offer some promise of increasing reliability by virtue of the fact that they will require that each clinician be gathering the same kind of data. They will also be an aid to the clinician in helping to insure that no relevant areas are overlooked. Additionally, it is likely that more frequent interviews will be conducted prior to drawing firm conclusions about an individual. Seeing the individual more than one time provides the clinician with opportunities to observe the possible operation of situation effects which may be present on one occasion but not on another. I do not expect the clinician to give up the flexibility to adapt the examination to the individual patient's needs and psychological state. I am however suggesting that the examination not be considered complete until all of the information required of a standardized interview has been obtained.

8. Demographic Variables

The issue of race and perhaps other demographic variables such as age and socioeconomic status will have to be dealt with in the future of assessment. One possibility is that a series of definitive studies may dispose of the problem as a pseudo-problem, thus, eliminating these variables from further consideration. The alternative is to begin the development of separate norms for various assessment devices such as that initiated by Gynther et al. (1978). If further research supports this approach by indicating that indeed there are racial, age, socioeconomic, or geographic differences that make a difference, then assessment for members of any of these groups is in for a period of considerable uncertainty while such norms are being developed. Once again, the computer may come to the rescue by expediting the research necessary to establish such norms.

9. Relation to Treatment

It seems likely that research establishing relationships between treatment modality effectiveness and some sort of typology whether it be psychiatric diagnosis, behavioral description, or some other classificatory scheme will have to be done if the purpose of clinical assessment as a treatment guide is to be accomplished. That is, even if some excellent classification system were to spring forth tomorrow, in terms of being a reliable and accurate descriptor of many deviant or nondeviant characteristics of an individual this would still not resolve the question of how to treat that individual, although it obviously would be a critical first step. The second step simply does not seem achievable without the first unless it should turn out that the type of treatment makes no difference or that there is one treatment, perhaps some as yet unknown, wonderful pill which treats all psychological disorders effectively.

10. Some Other Views of the Future

Anastasi (1982) describes current trends involving application of item response theory, Bayesian approaches to validity generalization, growing emphasis on construct validation, progress in analysis of trait, state, and situational variables, and recognition of the need for psychometrically sound assessment techniques in behavior modification programs.

Korchin and Schuldberg (1981) suggests trends that indicate "the development of more focused techniques of psychometric purity," more reliance on lower level interpretations rather than sweeping generalizations; more concern with situational and environmental factors; more attention given to the individual's own views of his character or problems rather than relying as heavily on external measures; and greater acceptance that there is an inevitable role for clinical judgment in collecting, integrating and interpreting assessment data, although they suggest that more disciplined thinking will be required.

On the other hand Rorer and Widiger (1983) clearly disagree with Korchin and Schuldberg's position that with a few refinements, assessment can continue with its business as usual. Their view is that "psychology is burdened with an outmoded philosophy, and a distorted view of science, to both of which it adheres with messianic fervor." The essence of their position, as I understand it, is that psychology has adopted a philosophy of science, namely that of emulating physics when that philosophy of science may never have existed and pretty clearly, according to these authors, no longer exists in physics. They assert that psychology adopted logical empiricism at a time when philosophy abandoned it. I am not going to attempt to deal with the entirety of their article. I can do no more than suggest that anyone interested in the future of clinical assessment ought to read the article, because it encompasses a radical change at the very core of assessment—the nature of the science (or nonscience, or different science) of psychology. According to these authors these changes are so radical that they would require logical empiricism be replaced by more contemporary philosophical positions on methodology; "analysis of variance, null hypothesis significance testing, and classical test theory would be replaced by taxometric methods, Bayesian statistics, analysis of covariance structures (including causal modeling), generalizeability theory, decision theory, and other methods appropriate for construct validation; sole reliance on the experimental method would be replaced by an emphasis on using methods appropriate for the study of personality structure, in particular those of clinical psychologies: and theoretical and integrated papers would be encouraged in place of the fragmented laboratory studies of unrelated personality traits that have added so little to our knowledge."

They do conclude their article with a statement "Finally and most difficult of all, we would become comfortable with the idea that there is no test that can separate science from non-science, and consequently that science is distinguished from religion precisely by the fact that it does not require acceptance of certain beliefs as an act of faith."

I take some comfort from the last statement if I correctly interpret it to mean that clinical assessment will continue to require validation of some kind. My simpleminded understanding of an applied science requires no less. It does not have to be logical empiricism or even physics but it does have to work. If it can be demonstrated (not just taken on faith) that it works in the field, then its existence is justified.

REFERENCES

Adair, F. L. (1978). In O. K. Buros (Ed.), *Eighth mental measurements yearbook*. Highland Park, NJ: Gryphon Press.

Adler, P. T. (1970). Evaluation of figure drawing techniques. *Journal of Consulting and Clinical Psychology, 35,* 52–57.

Anastasi, A. (1982). *Psychological testing*. New York: Macmillan.

Arkes, H. R. (1981). Impediments to accurate clinical judgement and possible ways to minimize their impact. *Journal of Consulting and Clinical Psychology, 49,* 323–330.

Bartol, C. R. (1983). *Psychology and American law*. Belmont, CA.: Wadsworth Publishing Company. (36)

Bertelson, E. D., Marks, P. A., & May, G. D. (1982). MMPI and race: A controlled study. *Journal of Consulting and Clinical Psychology, 50,* 316–318.

Brown, N. W. (1974). An investigation of personality characteristics of negros attending a predominantly white university and negros attending a black college. *Dissertation Abstract International, 34,* 3980A.

Butcher, J. M. (1978). In O. K. Buros (Ed.), *Eighth mental measurements yearbook*. Highland Park, NJ: Gryphon Press.

Chapman, L. J., & Chapman, G. P. (1967). Genesis of popular but erroneous psychodiagnostic observations. *Journal of Abnormal Psychology, 72,* 193–204.

Cross, D. R., Barclay, A., & Berger, G. K. (1978). Differential effects of ethnic membership, sex and occupation on a California psychological inventory. *Journal of Personality Assessment, 42,* 597–603.

Davids, A. (1973). Projective Testing: Some issues facing academicians and practioners. *Professional Psychology, 4,* 445–453.

Davis, W. E. (1975). Race and the differential "power" of the MMPI. *Journal of Personality Assessment, 39,* 139–140.

Davis, W. E., Beck, S. J., & Ryan, T. A. (1973). Race related and educationally related MMPI profile differences among hospitalized schizophrenics. *Journal of Clinical Psychology, 29,* 478–479.

Davis, W. E., & Jones, M. H. (1974). Negro vs. Caucasian psychological tests performance revisited. *Journal of Consulting and Clinical Psychology, 42,* 675–679.

Exner, J. (1974). *The Rorschach: A comprehensive system*. New York: Wiley.

Exner, J. (1978). *The Rorschach: A comprehensive system. Volume 2: Current research and advanced interpretations.* New York: Wiley.
Eysenck, H. J., Wakefield, J. A., Jr., & Friedman, A. F. (1983). Diagnosis and clinical assessment: The DSM-III. *Annual Review of Psychology, 34,* 167–193.
Goldstein, S. G., & Deysach, R. E. (1973). Effect of experience and amount of information on identification of cerebral impairment. *Journal of Consulting and Clinical Psychology, 41,* 30–34.
Gynther, M. D., Lachar, R. D., & Dahlstrom, W. G. (1978). Are special norms for minorities needed development of an MMPI scale for blacks. *Journal of Consulting and Clinical Psychology, 46,* 1403–1408.
Havens, L. L. (1981). Twentieth century psychiatry: A view from the sea. *The American Journal of Psychiatry, 138,* 1279–1287.
Holland, T. R. (1979). Ethnic group differences on MMPI profile patterns and factorial structure among adult offenders. *Journal of Personality Assessment, 43,* 72–77.
Jackson, D. N., & Paunonen, S. V. (1980). Personality Structure and Assessment. *Annual Review of Psychology, 31,* 503–551.
Keisling, R. (1981). Under-diagnosis of manic depressive illness in a hospital unit. *American Journal of Psychiatry, 135,* 672–673.
Korchin, S. J. (1976). *Modern Clincal Psychology.* New York: Basic Books.
Korchin, S. J., & Schuldberg, D. (1981). The future of clinical assessment. *American Psycholosigt, 36,* 1147–1158.
Leavitt, E. E. (1973). Internship versus campus, 1964 and 1971. *Professional Psychology, 4,* 129–132.
Lewandowski, D. G., & Saccuzzo, D. P. (1976). The decline of psychological testing. *Professional Psychology, 7,* 177–184.
Lowe, J. D., Jr., & Hildman, L. K. (1972). EPI scores as a function of race. *British Journal of Social and Clinical Psychology, 11,* 191–192.
Petzelt, J. T., & Craddick, R. (1978). Present meaning of assessment in psychology. *Professional Psychology,* 587–591.
Rorer, L. G., & Widiger, T. A. (1983). Personality structure and assessment. *Annual Review of Psychology, 34,* 431–463.
Rosenhan, D. L. (1973). On being sane in insane places. *Science, 179,* 250–258.
Swenson, C. H. (1957). Empirical validation of human figure drawings. *Psychological Bulletin, 4,* 431–468.
Treece, C. (1982). DSM-III as a research tool. *American Journal of Psychiatry, 139,* 577–583.
Ziskin, J. (1981). *Coping with psychiatric and psychological testimony.* Third Edition, Marina del Rey, CA.: Law and Psychology Press. (Supplement 1983)

8 Perspectives on the Future of Neuropsychological Assessment

Raymond S. Dean
Ball State University and Indiana University School of Medicine

PERSPECTIVES ON THE FUTURE OF NEUROPSYCHOLOGICAL ASSESSMENT

The relationship between human behavior and the functioning of the brain has intrigued scholars for centuries. Although the antecedents are clear, our present knowledge of the relationship between observable behavior and the physiology of the central nervous system owes more to the research vigor of the past 40 years than any other time in history. Of continuing interest to psychologists and other neuroscientists has been evidence of the correspondence between the functioning of the brain and cognitive, sensorimotor and affective dimensions of behavior. With such knowledge as a foundation has come the development and continuing validation of psychological methods which allow inferences concerning individuals' cortical functioning. Thus, what has become known as neuropsychological assessment attempts to relate behavior culled under standardized conditions to the functional efficiency of the brain. Although we are far from consensus on the role neuropsychological assessment should play in routine clinical examinations, numerous authors have argued in favor of a neuropsychological perspective in our understanding of psychiatric as well as neurological disorders (e.g., Dean, 1982a, Golden, 1978; Hartlage & Hartlage, 1977; Reitan, 1955, 1976).

The intent of this chapter is to examine the present status and approaches to neuropsychological assessment in light of a half century of research in the area. Following a review of the critical issues and psychometric adequacy of assessment procedures, this paper pursues developments within the neurosciences which may shape the direction of neuropsychological assessment during the next two decades.

Current Status of Neuropsychological Assessment

Historical Elements

As in most areas of measurement, neuropsychological assessment grew out of a need in an applied area. In the case of neuropsychology, the most salient influence has been the desire on the part of the medical community to more fully describe the behavioral effects of brain damage (Reitan, 1966). The administration of experimental and standardized psychological measures to patients with documented structural brain lesions gave rise to a data base which allowed investigation of the sensitivity of these measures to brain damage (Halstead & Settlage, 1943; Hunt, 1943; Reitan, 1955). In the post World War II years, these data were expanded with the relatively large number of patients with documented brain lesions resulting from head wounds (Boll, 1974; Luria, 1963; Reitan, 1966). Such events when combined with the growing empirical emphasis beginning in the decade just prior to World War II nurtured a quantitative approach which continues to characterize neuropsychological assessment in North America (Dean, 1982a; Luria & Majovski, 1977). Moreover, theoretical notions concerning brain function mattered less than the utility of assessment procedures in predicting and localizing cortical damage.

Neuropsychological assessment has often been considered an adjunct to the neurological examination. Basically a noninvasive technique, neuropsychological assessment was often seen as a viable alternative to physical diagnostic procedures which often held a mortality probability in themselves (e.g., angiogram) (Reitan, 1955). Although the emphasis on diagnostic prediction continues, increasing importance has been placed on outlining the extent of behavioral impairment, as well as defining the adaptive behavior remaining following brain damage. For, although definitive knowledge concerning the anatomical integrity of the brain may be available, rarely is the neurologist or neurosurgeon in a position to predict the behavioral expression of a given lesion in the patient's premorbid environment (Dean, in press).

The history of brain-behavior relations may be followed back some 2,500 years (Gibson, 1962). This is as clear as the fact that only since the late 19th century has public evidence begun to accumulate (Broca, 1861; Hebb, 1949; Jackson, 1874). Since this time, a good deal of scientific attention has been focused on the localization of specific functions to areas of the cerebrum (see Boring, 1942; Jackson, 1974; Nielsen, 1953; Wernicke, 1874). The early promise that observable behaviors could be localized to specific structures of the cerebral cortex (e.g., Broca, 1861) has not been realized (e.g., Reitan, 1976). Moreover, the luxury of retrospect allows us to criticize these rather naive attempts to document one-to-one correspondence between behavior and microstructures of the brain. Over the past 100 years, it has become clear that individual differences within cortical structures, variability in the magnitude of localized lesions, and the functional interaction within the brain threaten the heuristic value of pursuing a specific localization approach to the study of brain-behavior correlates (e.g., Boll, 1974; Reitan, 1955). Conversely, although the notion of static localization has been rejected by most neuroscientists (Dean, 1982a; Parsons, 1970; Reitan, 1974; Sperry, Gazzaniga, & Bogen, 1969), the validity of "organicity" as a single nosological category is also of questionable value (Reitan, 1955). "Organicity" as a unitary behavioral constellation grew from the early assumption that the effects of damage to the brain were similar regardless of location (Reitan, 1955). From this early point of view, the severity of cortical involvement and the patient's premorbid personality were the factors which accented the behavioral manifestations of cortical damage (Goldstein, 1942; Tymchuk, 1974). Operating under this predilection, numerous attempts were made to isolate a classical syndrome of behavioral and psychogenic signs which would be indicative of "organicity" (see Meier, 1963, for a review). In the main, such attempts have been unsuccessful; and, in fact, damage to grossly different microanatomical structures has been shown to manifest itself in a predictably distinct fashion on psychometric measures (e.g., Halstead, 1947; Hartlage & Hartlage, 1977; Parsons, 1970; Reed & Reitan, 1963; Reitan, 1955). It also has become quite clear that considerations such as the patient's age at onset, the acuteness of damage, and the length of the interval between onset and neuropsychological assessment are of such importance in test results that the search for a unitary organic syndrome is at best a quixotic pursuit (see Meier, 1963). It seems extraordinary that even with the wealth of knowledge that has been collected favoring complex organization of functions in the brain, the search for a "single measure of organicity" continues (e.g., Bender-Gestalt) (Tymchuk, 1974).

Due more to pressures existing in the applied setting than theoretical visions of neuropsychological assessment, the field has emphasized diagnostic aspects. Historically, the emphasis in assessment has been on offering data relevant to the diagnosis of neurologically based conditions (Boll, 1974; Diller & Gordon, 1981; Reitan, 1974). As mentioned previously, other than information useful in predicting the locus and extent of cortical involvement, neuropsychological assessment offers an objective baseline of neurologically related behavior functions (e.g., Lezak, 1976). This aspect emphasizes both impaired functions and adaptive behavior remaining after onset and, hence, provides a benchmark by which a patient's condition can be followed. This approach has been shown to be of considerable utility in following the course of progressive condition and outlining the extent of recovery (Dean, in press). As opposed to a diagnostic emphasis, this perspective also holds considerable implication for treatment planning and offers a synthesis of neuropsychological, educational, and behavioral information useful in establishing realistic treatment goals given the patient's level of functioning (Golden, 1978; Goldstein, 1979). Thus, neuropsychological assessment findings have the potential for providing rehabilitation specialists (physical medicine, occupational therapy, etc.) with an overview of the patient's physical, cognitive, and emotional rehabilitative needs (see Diller & Gordon, 1981).

Approaches to Neuropsychological Assessment

Quantitative Approach. Consistent with the North American empirical tradition in measurement, neuropsychological assessment has evolved with a distinct quantitative emphasis (Boll, 1974; Dean, 1982a; Luria, 1966; Reitan, 1976). Although exceptions exist, specific tests have been chosen and batteries assembled on the basis of estimates of reliability and the ability to distinguish groups with documented brain lesions (Golden, Hammeke, & Purisch, 1978; Klove, 1974; Reitan, 1955; Smith, 1975). Thus, specific tests have evolved more because of their psychometric adequacy than any global theoretical point of view of the functioning of the brain. Although this approach has a number of attributes that recommend it, a few seem rather unique to neuropsychology and the setting in which assessment is applied. Working within the medical setting, it became important early on to demonstrate the sensitivity of neuropsychological assessment procedures to differential diagnoses made by medical specialists (Reitan, 1955). For example, a method of validation employed extensively has been that of the ''clinically blind'' technique. More familiar in medicine than psychometrics, this validation procedure involves the administration and interpretation of test results without knowledge of the patient's history or admitting diagnosis (e.g., Reitan, 1955). Neuropsychological inferences and diagnoses resulting from test data are then evaluated against unambiguous neurological findings available for each patient. The objectives of this procedure are twofold. First, an estimate of the utility of the assessment procedure, and, second, an attempt to gain credibility within the medical setting (Crockett, Clark, & Klonoff, 1981; Parsons & Prigatano, 1978; Reitan, 1966).

The roots of neuropsychology lie in behavioral neurology and both clinical and experimental psychology. This fact is reflected in the measures presently utilized in neuropsychological assessment (Boring, 1942; Dean, 1978a; Reitan, 1976). This conclusion is quite clearly portrayed in presently available batteries (Golden, Hammeke, & Purisch, 1978; Reitan, 1969; Smith, 1975) and specialized tests (Dean, 1978b; Jortner, 1965) of neuropsychological functioning. Although the comprehensiveness varies, most batteries attempt to assess language, intellectual, sensorimotor, and personality factors that have been shown to relate to cerebral functioning (Boll, 1981; Reitan, 1974; Reitan & Davison, 1974). Table 8.1 offers an outline of the neuropsychological functions most consistently examined in the course of assessment. The assessment of individual functions while excluding the influence of others is often a tautological pursuit (Dean, Schwartz, & Smith, 1981). This is truc because organic impairment is rarely isolated to one specific function (e.g., Luria, 1966), and measuring many

TABLE 8.1
Neuropsychological Functions Considered During Assessment

I. COGNITIVE FUNCTIONS

- A. General Ability
- B. Verbal Functions
  - 1. Language
    - a. Receptive
    - b. Expressive
    - c. Knowledge Base
  - 2. Abstract Reasoning
    - a. Concept Formation
    - b. Symbolic Manipulation
  - 3. Memory/Learning
    - a. Registration
    - b. Immediate (Short-term) Memory
    - c. Long Term (Intermediate Memory
    - d. Remote
    - e. Acquisition Rate
  - 4. Integrative Functions
    - a. Visual-Verbal
    - b. Auditory-Verbal
    - c. Motor-Verbal
  - 5. Numerical Ability
    - a. Receptive
    - b. Expressive
    - c. Knowledge Base
- C. Nonverbal Functions
  - 1. Perceptual Organization
    - a. Receptive
    - b. Expressive
  - 2. Abstract Reasoning
    - a. Concept Formation
    - b. Spatial Manipulation
  - 3. Memory
    - a. Registration
    - b. Immediate (Short-term) Memory
    - c. Long Term (Intermediate) Memory
    - d. Remote
    - e. Acquisition Rate
  - 4. Integrative Functions
    - a. Visual-Motor
    - b. Auditory-Motor
    - c. Tactile-Motor
  - 5. Construction

II. PERCEPTION

- A. Visual
  - 1. Acuity
  - 2. Fields of Vision
  - 3. Ocular Dominance
- B. Auditory
  - 1. Acuity
  - 2. Discrimination
    - a. Verbal
    - b. Nonverbal
  - 3. Lateralization of Ability

continued...

continued...

- C. Tactile-Kinesthetic
  1. Acuity
  2. Discrimination
  3. Lateralization of Ability

III. MOTOR FUNCTIONS

- A. Strength
- B. Speed of Performance
- C. Perceptual Motor Speed
- D. Coordination
- E. Dexterity
- F. Lateral Preference

IV. EMOTIONAL/CONTROL FUNCTIONS

- A. Personality
- B. Emotional Functioning
- C. Attention
- D. Concentration
- E. Frustration Tolerance
- F. Interpersonal/Social Skills
- G. Flexibility

---

higher-order functions (e.g., verbal abstract reasoning) is predicated on the assumption of normal contributory functions (e.g., receptive and expressive language) (Dean, 1983a). Therefore, poor performance on a measure of verbal abstract reasoning is a necessary but not a sufficient condition to infer impairment of this function. These factors when combined with routine psychometric considerations make the assessment of functions outside of a wide-band approach offered by a battery a tenuous procedure (Reitan, 1966). Clearly, the interpretation of measures of neuropsychological functioning for individual cases must be made within the context of patterns of strengths and weaknesses (Baron, 1978; Reitan, 1974) rather than the use of the "level of performance" for a single test. Similarly, a number of inferential techniques have evolved which hold predictive utility in inferring brain dysfunction and outlining neuropsychological abilities (Reitan, 1974). Pathogenic signs, long employed in clinical medicine, involve the comparisons of an individual's performance with behavior constellations known to be consistent with specific neurological conditions. Such behavioral signs rely upon symptoms that would rarely be displayed in patients without neuropathology and therefore increase the probability of cortical impairment. Another method relies on knowledge of hemispheric lateralization of functions and allows left—right localization of functions normally served by different hemispheres of the brain (Boll, 1972a). This knowledge in conjunction with the fact that sensory and motor functions on one side of the body are served by the contralateral hemisphere provides additional inferential support for a neuropsychological hypothesis (Reitan, 1974). Finally, pattern analysis provides what many consider as an integration of the level of performance, pathognomic sign, and lateralization of function approaches (Reitan,

1976). Here, the patient's assessed strengths and deficits are compared with known test patterns indicative of neuropathology (Smith, 1975). This methodology seeks to identify configurations of individual test scores that are characteristic of specific types of diffuse and localized cortical dysfunction. While problems exist in such a pattern approach (Dean, in press), numerous researchers have stressed the diagnostic, localization, and descriptive power of this method which also serves to reduce the subjectivity inherent in clinically blind interpretations (Anthony, Heaton, & Lehman, 1980; Dean, 1978b; Reed & Reitan, 1963; Russell, Neuringer, & Goldstein, 1970; Selz & Reitan, 1979).

Pattern analysis in conjunction with actuarial norms has generally shown it to be superior to more subjective methods of differential diagnosis (Kiernan & Matthews, 1976; Klove, 1974; Knights, 1973; Russell et al., 1970; Spitzer & Endicott, 1974; Wedding, 1983). Moreover, as a number of researchers have shown in other areas of applied assessment, a mechanical/statistical approach to neuropsychology assessment results has been found superior to more clinically based attempts (Dean, 1978b; Finkelstein, 1976; Russell et al., 1970).

In sum, the emphasis of the quantitative approach to neuropsychological assessment has been on the development of standardized test batteries which allow the actuarial identification of aberrant cortical functioning from a neostructural point of view (Klove, 1974; Reitan, 1955). Interpretation of measures is facilitated by normative standards and information gleaned from groups with documented neurological conditions (Parsons & Prigantano, 1978; Satz, Fennell, & Reilly, 1970). In keeping with such an orientation, neuropsychological functions are viewed as individual difference variables. Assessment allows the description of the patient on norm-based psychological measures shown to relate to neurological functioning when groups of normals are compared with known neuropathological groups. Such data have been shown to be of considerable utility in making inferences concerning cortical integrity and the behavioral consequences of known lesions.

The Halstead—Reitan Batteries. The Halstead—Reitan Neuropsychological Test Battery is the best known and most widely used comprehensive measure of neuropsychological functioning in North America (Dean, 1983a). This battery represents a melding of experimental and clinical approaches to psychology and behavioral neurology (Reitan, 1974). The vast majority of tests within the battery were originally developed for distinct, experimental, and clinical purposes and were adapted, standardized, and interpretations expanded such that inferences could be gleaned concerning cortical functioning.

The battery is comprised of a number of psychological measures originally developed by Halstead (1947) on the basis of extensive experiments with patients having documented brain lesions. From a battery of experimental procedures, Halstead selected 13 which were seen to be sensitive to cerebral lesions and appeared to be complementary in terms of the range of functions assessed. This number was later shortened to the 10 measures which best discriminated aberrant cortical functioning (Halstead, 1947).

Ralph Reitan is primarily responsible for the present configuration and validation of measures which have come to be known as the Halstead—Reitan Neuropsychological Test Battery (Reitan, 1969). Reitan's (1955) early validation study led to the discontinuation of three measures (Time Sense Test, Critical Flicker Frequency, and Deviation) originally proposed by Halstead (1947). Although many of the individual tests were adapted and extended from other measures, brain-behavior findings have allowed quite distinct neuropsychological interpretations. Reitan (1955) further refined Halstead's original battery with the replacement of the Henmon–Nelson Intelligence Test with the Wechsler–Bellevue Scale and later the Wechsler Adult Intelligence Scale (WAIS). As was true for the inclusion of other tests in the battery, the addition of the Wechsler Scale was prompted by data showing the incremental validity of this measure in diagnosing neuropathology while expanding comprehensiveness of the assessment system (Reitan, 1955, 1966). Although data support the sensitivity of the Wechsler Scales to cerebral impairment in their own right (see McFie, 1975), Reitan (1959, 1966) offers rather convincing evidence that combined subtests from the Halstead–Reitan Battery (referred to as the Halstead Impairment Index) are more sensitive to brain damage in adults than subtests of the Wechsler Scale (Wechsler–Bellevue, Reitan, 1955; Wechsler Adult Intelligence Scale, Reitan, 1974). A good deal of evidence suggests that verbal and nonverbal factor scores are selectively affected by left and right hemispheric lesions respectively (Matarazzo, 1972). Thus, Reitan's adoption of the Wechsler Scale represented an improvement in psychometric sophistication and comprehensiveness of the battery (Reitan, 1955, 1966, 1974).

Other adjuncts to the battery have been the inclusion of the Minnesota Multiphasic Personality Inventory (MMPI) and the Wide Range Achievement Test (WRAT) (see Reitan & Davison, 1974). Two related factors led to the inclusion of the MMPI. First, because the referral question often asked relates to the discimination of neurological involvement from psychogenic syndromes. Second, the Halstead–Reitan Battery is oriented toward more intellectual, cognitive, and the sensorimotor end of the measurement spectrum often to the exclusion of the affective factors. The MMPI offered a more comprehensive picture of the respondent's level of emotional functioning and provided a number of specialized scales of some utility in differential diagnosis of brain damage from various forms of psychopathology (Horton & Timmons, 1982; Jortner, 1965). The Wide Range Achievement Test is often included in the battery with children and when one's premorbid occupation appears related to academic skills.

Halstead's (1974) and Reitan's (1955) methods for the choice and development of neuropsychological assessment devices typifies the quantitative–actuarial approach. Moreover, the emphasis has been the diagnosis and subdivision of brain damage using psychological measures known to have a relationship to specific brain functions (Dean, 1982a). In sum, the method could be characterized as the development and adaptation of experimental procedures which offer information relevant to psychological deficits shown to result from cortical lesions (Boll, 1981; Halstead, 1947; Klove, 1974; Reitan, 1974). It should be clear that such an approach that relies on a correlational data base may be given to over-interpretation (Klove, 1974).

It is not the intent here to duplicate a number of indepth examinations of the Halstead–Reitan Battery (e.g., Boll, 1981; Dean, 1982a; Reitan & Davison, 1974), but rather as an overview of the present status of neuropsychological assessment. To this end, Table 8.2 is offered as an overview of the battery in terms of the source of specific measures, constructs measured, and the technical adequacy. It is hoped that this summary will provide the reader unfamiliar with neuropsychological assessment a reference for the remainder of the presentation. This recapitulation also makes the distinction between tests included in the Adult Battery (> 15 years) and those of the two children's versions of the assessment system.

Generally speaking, the greater opportunity to validate brain-behavior relationships during surgery and autopsies and the large number of neurological referrals during the post World War II era has led to greater sophistication in neuropsychological assessment with adults than that with children and adolescents (Boll, 1974; Obrzut, 1981). Therefore, the empirical approach that has characterized adult neuropsychological assessment is a comparatively recent endeavor with children and adolescents who have suffered brain damage. For a number of methodological reasons, clinical and experimental inferences concerning brain-behavior relationships are lacking with children (Benton, 1974; Boll, 1974). Recognizing the possible clinical utility of neuropsychological assessment procedures appropriate with children and early adolescents, Reitan (1969) administered the Halstead–Reitan Battery to a sample of normal children under the age of 15 in an attempt to establish a downward revision of the battery. This modified battery was administered to successively younger children until the lower age limit of 9 years was established based on psychometric properties of tests and predictive statements which could be made for children. In what has become known as the Halstead Neuropsychological Test Battery for Children (9–14 years), most of the individual tests of the battery have remained as a downward revision of the adult battery. Other than the substitution of the Wechsler Intelligence Scale for Children for the WAIS and alteration in the number of items and their organization, few substantive changes were made in the original battery.

With the collection of developmental data with the Halstead Battery for Children, it became clear that major revisions would be necessary in the assessment of younger children (< 9 years) (Boll, 1974; Reitan, 1969). The Reitan Indiana Neurological Test Battery for Children (5—8 years) represents major revisions in the tests of the Halstead Battery for Children and the addition of a number of

TABLE 8.2
Tests Included in the Halstead-Reitan Neuropsychological Battery

Name	Origin	Format	Technical Information
Category Test	Halstead (1947)	Semiautomated visual presentation of slides with underlying concept. Feedback provided as to correctness after depression of one of four levers. Controlled learning experiment.	Reitan (1955) Klove (1974)
Tactual Performance Test	Seguin-Goddard	Shapes placed in form board without aid of vision (dominant, nondominant, both hands). Recall of shapes and location of board	Reitan (1974) Boll (1974)
Speech-Sounds Perception	Halstead (1974)	Select paralog presented auditorially from four (three) alternatives, total of 60 items	Reitan (1974) Klove (1974)
Rhythm Test	Seashore Tests of Musical Talent	Identification of 30 pairs of rhythmic beats as being "same" or "different"	Reitan (1955)
Trail Making Test (A & B)	Army Individual Test Adjutant Generals Office (1944)	A--Connect 25 (15) numbered circles in numeric order B--Connect 25 (15) numbered or alphabetic circles alternating between numeric and alphabetic order (i.e., 1-A-2-B...)	Reitan (1955, 1974) Boll (1974)
Finger Oscillation Test	Halstead (1947)	Measure of number of taps with index finger in 10 seconds for each hand. Dominant, nondominant scores = means of five trials for each hand	
Adjunct Measures[1]			
Reitan-Indiana Aphasia Screening Test	Halstead & Wepman (1952) Reitan (1969)	Thirty-two (22) items requiring naming, spelling, reading, write, math, calculations, enunciate, identify body parts, pantomime actions, perform acts, draw, shapes, and identify directions	Reitan (1969) Dean (1982) Klove (1974)
Reitan-Klove Sensory Perception Examination	Reitan (1969)	Accuracy of unilateral and bilateral simultaneous tactile, auditory, and visual imperception, finger localization from tactile stimulation without vision. Tactile number perception on fingertips. Tactile recognition of shapes without vision	Reitan (1969) Klove (1974)
Strength of Grip Test (Hand Dynamometer)	Reitan (1955)	Measure of strength of grip. Alternating measures for preferred and nonpreferred hands	Reitan (1955)

[1]In addition to the above, the age appropriate Wechsler Intelligence Scale, Minnesota Multiphasic Personality Test, Lateral Dominance, and Wechsler Memory Scale are most often incorporated.

newly developed measures (Reitan, 1969). New tests were included to measure gross skeletal function, abstraction, and a number of visual—spatial relationships.

Since Halstead's (1947) early efforts to operationalize his theory of biological intelligence and later with Reitan's (1955) refinements and extensions has come a plethora of research concerning the utility and technical adequacy of these measures. The adult battery has undergone repeated cross validation since Reitan's (1955) early report (e.g., Reitan, 1958, 1959, 1960, 1964; Vega & Par-

Explicit Constructs	Implicit Constructs	Forms	Scores
Concept Formation Abstraction Integration	Visual Acuity Attention Concentration	Adult--208 slides 9-15--168 slides 5-9--80 slides	Total Errors
Tactual Discrimination, Manual Dexterity, Kinesthesis Incidental Memory Spatial Memory	Kinesthesis Tactual-motor Integration	Adult--10 shapes 9-15--6 shapes 5-9--6 shapes	(1) Time (2) Memory (3) Location
Verbal Auditory Discrimination Auditory-Visual Integration Phonetic Skills	Auditory acuity Language Attention	Adult--4 alternatives 9-15--3 alternatives	Total Errors
Nonverbal Auditory Discrimination Auditory Perception	Attention Concentration	--	Total Errors
A--Motor speed, visual scanning, visual-motor integration B--Above and mental flexibility, integration of alphabetic and numeric systems	Visual Acuity Attention Concentration	Adult--25 circles 9-15--15 circles	Separate Time A & B
Motor speed Dexterity	Distractibility Concentration	Same	Mean Taps Each Hand
Wide Band Language and Nonverbal Functions	Education Occupation Concentration	Adults--32 items 9-15--32 items 5-9--22 items	By Item, Total Error Score
Lateralized Sensory Perception (Visual, Auditory, and Tactile)	Distractibility	--	Errors Left vs. Right side of body
Motor Strength	--	--	Left and Right hand Scores

sons, 1967, Wheeler, Burke, & Reitan, 1963, Wheeler & Reitan, 1962). In general, using various test combination procedures, the battery has been shown to discriminate brain damaged groups from normal controls with considerable accuracy (84–98%) (e.g., Boll, Heaton, & Reitan, 1974; Reitan, 1976; Wheeler, Burke, & Reitan, 1963). The localization accuracy in the discrimination of left from right hemispheric damage and localized from diffuse involvement has been consistently shown to be of clinical utility (> 70% accuracy) (Vega & Parsons, 1967; Wheeler, Burke, & Reitan, 1963; Wheeler & Reitan, 1963). For various reasons, the battery has fared less well in diagnosing brain damage when psychiatric groups were included as controls (e.g., Watson, Thomas, Anderson, &

Felling, 1968). This area of research is discussed at length later in this chapter, but it suffices to say that at this point the HRB suffers its sharpest drop in predictive efficiency when utilized in an attempt to distinguish brain damage from schizophrenia (Lacks, Harrow, Colbert, & Levine, 1970; Watson et al., 1968) and other psychotic disturbances (Dean, 1982b).

Similar to the findings for the adult HRB, research has shown sensitivity to brain damage for the children's and adolescent versions of the battery (e.g., Boll, 1974; Dean, 1982c; Klonoff & Low, 1974; Reed, Reitan, & Klove, 1965; Reitan, 1971). Although, as is pointed out further, while the research with adults is far from unequivocal in a number of areas, research with children on the HRB, and for that matter neuropsychology in general, is far more tentative (Boll, 1972b; Dean, 1982a; Hartlage, 1981; Reitan, 1974). Problems notwithstanding, the finding of impaired performance for brain-damaged children when compared with normal cohorts seems robust (Boll, 1974; Day & Ulatowska, 1979; Reed, Reitan, & Klove, 1965; Reitan, 1974). Although this is true, differential diagnoses of location with brain-damaged groups seem a far more tenuous procedure with children than with adults. Moreover, brain damage of "middle" childhood onset produces a more global deficit in cognitive ability than that found for adults (Ernhart, Graham, Eichman, Marshall, & Thurston, 1963; Reitan, 1974). As Dean (1982a) points out, "older brain-damaged children (> 9 years) display greater discrepancy from normals on high level language and motor functions than is usually found with younger brain-damaged children" [p. 190]. In this regard, Reed and Reitan (1969) offer data favoring fewer lateralized effects from frank neurological damage with children in general than that found in adults.

Qualitative Approach. Numerous authors have characterized the strict empirical emphasis of neuropsychological assessment in North America as being so quantitative that it has become atheoretical (Luria, 1966; Luria & Majovski, 1977; Rourke, 1976). Rourke (1976) argues that with efforts to reduce a plethora of behavioral information to a differential diagnosis, much of the comprehensiveness necessary in understanding the individual patient and planning rehabilitation is lost. Moreover, many neuropsychologists stress the comprehensive understanding of the patient's functioning in favor of the more quantitative information (e.g., Lezak, 1976; Luria, 1966; Rourke, 1976). Rourke (1976) suggests a dynamic interaction between neuropsychological assessment and rehabilitation. An implicit assumption here is that diagnosis or syndrome identification is heuristic only to the extent to which differential treatment or understandings are conveyed. For many neurological, nosological categories, little is gained in our appreciation of the individual patient's capacity or, in fact, the patient's needs in rehabilitation planning (Luria & Majovski, 1977). From this point of view, the actuarial approach which has characterized neuropsychological assessment in North America is seen as investing itself in the development of standardized batteries to the decrement of an understanding of the patient's functional capacities (e.g., Luria, 1966).

A more qualitative approach to neuropsychological assessment had been articulated by Luria (1963, 1966, 1973). Rather than the neo-structural point of view which has typified the quantitative approach as reflected in the HRB, Luria (1966) argued in favor of a theoretically consistent overview of *functional systems* of the brain. Specifically, Luria (1963, 1966, 1973) envisioned neuropsychological assessment as a dynamic procedure, the objective of which is the understanding of the integrity of functions as a prelude to the development of rehabilitation strategies. Thus, the selection of specific assessment techniques is based on pathognomonic signs, medical history, and an extensive interview with the patient. A clinical procedure, assessment involves definition of specific symptoms. In this way, specific behaviors are investigated which allow the formulation of hypotheses concerning the brain functions for individual patients (Luria, 1970). It will be recognized that not only does this orientation represent a departure from the psychometric concerns of a qualitative approach, but also represents a reconceptualization of inferences concerning the brain. As opposed to the neo-structural assumptions implicit in quantitative assessment, this approach has at its foundation an orientation to behavior as it may relate to functional systems of the brain. Luria (1966) long maintained that higher forms of human psychological activity are based on the participation of all levels of cortical activity. Therefore, behaviors are more heuristically related to functional systems of the brain than specific structural components (Luria, 1966; Luria & Majovski, 1977). During assessment, specific techniques are formulated to examine patterns of "functioning" (e.g., Luria, 1973), rather than relying on standardized methods and comparisons with normative samples. Thus, specific tests utilized vary from patient to patient depending on the early hypothesis of the patient's impairment. Such flexibility in assessment would most often be considered more consistent with a "clinical technique" than what we in the West have considered as assessment (Dean, 1983a). This qualitative emphasis is most often faulted as being far too subjective an approach to allow opportunities to establish other than clinical norms (Luria, 1973). consistent with a clinical orientation, interpretation relies upon the knowledge and experience of the clinician. Unlike the psychometric orientation adapted in the construction of the HRB, in which assessment methods were selected on their predictive efficiency, Luria's (1966, 1973) methodology relied heavily on his theory of brain functioning in the choice of tasks during assessment. "In sum then, the quantitative school is often portrayed as being atheoretical and ignoring descriptive data; whereas, the qualitative approach has often been faulted for its reliance on case study methods and a failure to systematically evaluate assessment methods" [Dean, 1982a, p. 18].

Luria's Neuropsychological Assessment Strategies. Within his functional theory of the brain, Luria (1973) characterized neuropsychological assessment as involving attempts to analyze "zones" of the brain which were "responsible for the performance of complex mental activities" [p. 33]. Rejecting the notions of direct cerebral localization and equipotentiality, Luria (1966, 1973) argued in favor of the evaluation of "zones" of the brain only for their relevant contribution to more complex, functional systems. Early in his theoretical formulation, Luria distinguished between different definitions of "function." Function in neuropsychological assessment was seen as "the complicated adaptive activity of the organism . . . [and was] determined by a specific operation [Luria, 1973, p. 16]. Individual functions were not seen to be localized to specific cortical locations; rather, such functions were seen as being "distributed in a complex system, or constellation of cooperating zones of the cerebral cortex and subcortical structures. . . each of the areas makes a highly specific contribution to ensue the operation of a functional system" [Luria, 1973, p. 17].

It soon becomes obvious that the knowledge of Luria's functional theory of the operation of the brain is the *sine qua non* of the assessment system. This theory is organized around three "blocks" of the brain which incorporates its basic functions. Consisting of the reticular formation and association areas of the brain stem, the arousal block is viewed as the regulator of the cortex energy level. Input and integration of sensory stimuli forms a second block and is seen as consisting of the temporal, parietal, and occipital lobes. The frontal lobe is attributed with primary responsibility for the behavioral planning and execution (Luria, 1966, 1973). Functional blocks are seen as containing specific zones responsible for individual abilities. Importantly, damage to any one block is thought to affect a number of functional systems (e.g., reading, speech, and writing) (Luria, 1966). Table 8.3 presents the areas of functioning most often assessed within this theoretical model.

Christensen (1975) has emphasized that neuropsychological tasks represent but one element in the evaluation of an individual patient in Luria's system. Patient evaluation involves three phases; each attempt to define intact and impaired functions. Brain lesions are seen as disturbing groups of psychological functions rather than a single ability to the exclusion of others within a given block. The initial phase of the neuropsychological assessment procedure involves a clinical interview with the patient. This portion of the evaluation is seen as important, for it allows the clinician an opportunity to generate hypotheses relevant to the nature and location of the lesion. In what is similar to the initial portion of a medical examination, information regarding the patient's history, personality, level of consciousness, and attitude toward illness is gathered.

The second stage of assessment involves the administration of a number of brief tasks meant to outline how multiple functions relate to general mental activity. The item selection, administration, and performance on these tasks are done in a quite subjective fashion (Luria, 1973). By analogy to a more quantitative approach, this phase may well be viewed as a screening procedure. Areas of deficit performance are investigated further in the final phase of the assessment procedure. The objective at this point is to explicate and expand on areas of performance not fully understood in the second phase of assessment. Here, the clinician is encouraged to modify and adapt tasks to maximize the functional

TABLE 8.3
Summary of Individual Scales of the Luria-Nebraska Neuropsychological Battery

Individual Scale	Description	Items	Technical Reference
Motor Functions	Assess a number of motor skills for left and right sides of body. Unilateral and simultaneous simple and complex motor movement	51	Golden et al. (1978)
Acoustico-Motor (Rhythm Scale)	Evaluate similarity of tones, reproduce tones orally and motorically. Rhythmic patterns generated from verbal description	12	McKay & Golden, (1979)
Higher Cutaneous Kinesthetic (Tactile Scale)	Without aid of vision must identify where touched, head and point of pin, direction of movement, geometric and alpha-numeric symbols traced on wrist, matching movements, and item identification	22	Golden et al. (1982)
Spatial (Visual Functions)	Visual recognition of common objects, pictures with obscurity and disorganization. Matrices tasks and complex block count	14	Golden (1981)
Receptive Speech Scale	Requires written, oral, and motoric response to spoken speech	33	Golden (1981)
Expressive Speech Scale	Items involve orally repetition words, increasingly complex sentences, name, count, recite, offer missing words, and organize mixed-up sentence	42	Golden (1981)
Writing Scale	Involves basic writing skills--spelling, copying words and letters from cards and memory. Writing words and letters from dictation and spontaneously	13	Golden et al. (1978)
Reading Scale	Range from reading letter sounds, syllables, words, sentences, and a short story	13	Golden et al. (1978)
Arithmetic Scale	Involves simple number identification, writing, and reading series of numbers, simple skills to more complex skills	22	Golden et al. (1981)
Memory Scale	Requires learning word list, picture memory, rhythmic pattern, hand positions, sentences, story, and paired associate task	13	Golden et al. (1982)
Intellectual Scale	Sequencing pictures, abstract theme of pictures, identify picture absurdities, proverbs, definitions, opposites, and analogies, and word problems	34	Golden & Berg (1983)

performance of the patient. Of interest to rehabilitation recommendations are the types of compensations which are possible within the respondent's spectrum of abilities. Luria (1966) encourages the examiner to at times change the tempo and adapt the tasks such that an appreciation of the patient's performance may be gained under alternative conditions. Although these procedures may allow clinical insights into intraindividual differences useful in rehabilitation, they do not allow the investigation of the validity of either the theoretical underpinnings or the assessment procedures in a public fashion. Therefore, a discrimination of the utility if Luria's procedures from his clinical expertise had been impossible.

Recently, attempts have been made to operationalize Luria's methods (Christensen, 1975, 1980; Golden, Hammeke, & Purisch, 1978; Majovski, Tanguay, Russell, Sigman, Crumley, & Goldenberg, 1979a, 1979b) and aspects of this theory (e.g., Das, 1973; Kaufman & Kaufman, 1983). The degree to which these attempts reflect Luria's (1966) original thinking is controversial, and the implications of such models are examined at length later in this chapter.

Critical Issues and Future Areas of Emphasis

The previous section hopefully has provided an overview of the current status of neuropsychological assessment with an examination of two distinctly different systems that exemplify the major approaches to the field. For the reader unfamiliar with the area of neuropsychology, these systems would best be considered as opposite poles of a qualitative–quantitative dimension. Both hold implications for the future complexion of neuropsychological assessment and will influence how critical questions which face neuropsychology will be approached. It should be noted that in the clinical setting evaluation of the individual patient often involves a combination of these approaches.

Among the most salient influences in the future of neuropsychological assessment, certainly research in the clinical and basic neurosciences would seem to rank high. Obviously, adaptation of advances in basic psychometrics to neuropsychological questions will continue. A less obvious and infrequently articulated influence is what could be termed as a "paradigm shift" in American Psychology. Moreover, the recent polemics in psychology could hardly be seen as fertile ground for the growth of neuropsychology. Clearly, the late 1970s heralded a more flexible, cognitive approach to our understanding of human behavior. Since this time, the number of systematic investigations which have begun to outline neurological correspondence to complex human behaviors have grown geometrically (Dean, 1983a). The remainder of this chapter attempts to define a number of critical issues in neuropsychological assessment, relate presently available research in the area, and offer tentative projections.

Interpreting Neuropsychological Assessment Results

Despite the sensitivity with which neuropsychological assessment procedures have been shown to have in distinguished brain damage from normal functioning (e.g., Reitan, 1955; Wheeler et al., 1963) and psychiatric disturbances (Golden, 1977; Klonoff & Low, 1974), the interpretation of individual results remains obscure. That is to say, it is often difficult to base the interpretation of a patient's protocol on the available research for groups. Moreover, the individual patient's premorbid history, age at onset of disturbance, time from onset to assessment, and behavioral reaction to impairment all interact to make findings for groups obscure for the individual case (see Dean, 1983a). This state of affairs seems reasonable, for the luxury of control of extraneous variables found in research is not possible in individual interpretations. Reitan (1974) has offered some direction in the use of inferential techniques. These have involved (a) evaluation of the patient's "level of performance" with the use of cutting scores and normative standards; (b) pathognomonic constellations of behaviors indicative of neuropathology; (c) comparisons of sensory and motor performance for left and right sides of the body, and (d) comparisons of relative strengths of neuropsychological functioning for the individual in light of group patterns of dysfunction. Each of these methods has been shown to have utility in predicting the individual's neurological status. However, two obvious questions arise in the interpretation of the results for the individual patient. First, how are these methods combined to make a specific conclusion concerning the individual's test findings? Second, what weight should be given to such moderating variables as education, age at onset, and the like? Such questions are not unique to neuropsychological assessment but, rather, are critical where clinical decisions must be made for the individual client based on vast numbers of variables. Often then, the psychologist is left with few alternatives to "clinical prediction" in the interpretation of an individual's performance (Boll, 1981).

The Impairment Index, originally offered by Halstead (1947) and later refined (Reitan, 1955; Klove, 1974), is the proportion of a series of HRB tests found to be in the impaired range. This rather simple actuarial measure has shown considerable accuracy in predicting the degree of cortical impairment (Halstead, 1947; Klove, 1974; Reitan, 1966). Kiernan and Matthews (1976) also provide some evidence favoring actuarial prediction using the simple average of standard scores of the subtests of the HRB. It should be recognized that while both methods provide the psychologist some rudimentary information concerning overall impairment, each falls short of aiding in the interpretation of an individual's performance.

Certainly, more extensive normative data which controls for age, sex, race, and socioeconomic status would aid in individual interpretation. Such a data base would provide the clinician with the prerequisites to estimate premorbid functions based on the patient's demographics. Wilson, Rosenbaum, Brown, Rourke, Whitman, and Grisell (1978) and Reynolds and Gutkin (1979) offer evidence of a considerable relationship for demographic variables on general cognitive ability scores. More important, this research provides a means with which to estimate "expected premorbid levels" of functioning with a given demographic profile. Although more extensive normative data which accounts for demographic variables is basic to the future of actuarial prediction, the problem of data integration for the individual goes beyond demographics.

The clinical procedures necessary for data integration have remained a rather *mystical* procedure for the neophyte. A number of authors have argued that attempts to quantify the interpretative process beyond basic level of performance statements would do more to obscure the understanding of the results for the individual patient than to edify. Embroyonic research in mechanical interpretation has usually taken one of two approaches in the development of objective means of data integration. One approach has been the development of algorithms for making interpretative decisions. Russell, Neuringer, and Goldstein (1970) have attempted to establish taxonomic keys with the HRB for the diagnosis of brain damage in general and the localization and identification of the organic process (i.e., acute, static, or congenital). Russell et al., (1970) in an initial validation provided encouraging data with diagnostic hit rates ranging from 62% to 94% when neurological group formation was based on clinical judgment. More recently, Finkelstein (1976) developed a computer based refinement of this algorithmic approach (BRAIN 1). This program allows direct input of raw data from the HRB and attempts to predict the presence of brain damage, location, acuteness, and provides a "most consistent neurological diagnosis." Finkelstein (1976) reports impressive accuracy in predicting the presence of brain damage (96%), the lateralization of the lesion (75%), acuteness (83%), and neurological diagnoses (64%). Anthony, Heaton, and Lehman (1980) have compared the utility of Russell et al.'s (1970) and Finkelstein's (1976) methods to the predictions made on the basis of the Halstead Impairment Index. Consistent with Klove's (1974) data, this investigation showed superior classification of brain damaged and normals with the use of the Impairment Index.

Another system which has been seen as having actuarial potential is a multivariate statistical approach (Wheeler & Reitan, 1963). Using a priori defined neurological groups, a number of investigators have demonstrated the utility of a cross validated discriminant function in the prediction and localization of brain damage from HRB data (Golden, 1977; Goldstein & Shelly, 1972; Struss & Trites, 1977; Wheeler, 1964). Wedding (1983) compared the classification accuracy for a number of actuarial methods for a sample of 263. His results indicated the superiority of discriminant functions in accuracy and localization when compared with both Russell et al.'s (1970) key approach and Finkelstein's (1976) BRAIN 1. Consistent with research in other areas of applied assessment, these data supported the superiority of a mechanical/statistical methodology to the interpretation of neuropsychological test data.

Although not without methodological problems (e.g., Willson & Reynolds, 1982), such reports are encouraging and indicate considerable potential in statistical/mechanical methods in the interpretation of neuropsychological test data and the development of extensive code books. Although few attempts have been made to predict the degree of return of function or interface predictions with rehabilitation efforts, certainly the technical procedures exist to provide such information. Efforts to statistically estimate premorbid levels of functioning and in turn compare these data with actual reports will begin to provide the methodology necessary to objectively outline the degree of loss of function for the individual patient.

Attempts to Objectify Luria's Theory

Recognizing the heuristic value of a theoretically consistent view of the functioning of the brain, a number of recent attempts have been made to operationalize Luria's neuropsychological theory. Other than some basic research supporting Luria's theory (see Luria, 1965, 1966, 1973), the approach has appeal of providing a theoretically consistent template to aid in the development of neuropsychological assessment procedures. It should be recognized that the radical quantitative approach taken in the choice of neuropsychological tests in the West while increasing prediction has often alluded interpretation (Dean, 1983a). For a familiar example, the choice of subtests comprising the Wechsler Scales was made on quantitative grounds and Wechsler's (1958) broad notion of aggregates of abilities. The Wechsler Scales have been shown to offer one of the best single predictors of extra-test behaviors (see Dean, 1982a). The interpretation of subtest scores for individuals and groups has been one of the most researched and yet controversial topics in applied assessment (Dean, 1982a; Kaufman, 1979). This is less of an indictment of the quantitative approach than an overview of the primitive status of our knowledge of brain-behavior relationships. Thus, while Luria's assessment procedures would not satisfy elementary psychometric criteria (Dean, 1982a) necessary to be termed tests, his theoretical model of the functional organization of the brain provided a distinct advantage in the interpretation of performance and the planning of rehabilitation strategies.

Christensen (1975, 1980) has recently attempted to operationalize Luria's neuropsychological assessment system by objectifying the tasks used by Luria. In this way, Christensen (1975) has stimulated the examination of Luria's assessment tasks in a fashion which is more amenable to psychometric investigation. This is important and should allow investigation of the heuristic value of Luria's assessment system free of the clinical expertise which had been developed by Luria and his associates. Although Christensen's (1975, 1980) contribution to neuropsychological assessment is clearly recognized, these operationalized versions of Luria's methods are now concrete, many of the techniques remain more subjective than would be desired from a quantitative point of view.

Majovski et al. (1979a, 1979b) have recently attempted an independent translation of Luria's assessment system. Designed as an adolescent screening measure of cortical dysfunction, the measure focuses on the same categories proposed by Christensen (1975) (see Table 7.3) with the exclusion of the kinesthetic and cutaneous areas. Majovski et al. (1979a, 1979b) see the inclusion of these functions to be redundant with information gleaned from the neurological examination which generally precedes the neuropsychological referral. Containing 72 items, the instrument requires approximately one hour to administer to children 10 years of age and older. Although this time element represents a distinct advantage over other neuropsychological batteries, to date only minimal validation has been attempted. In sum, with the portability and relatively short administration time, the measure may hold utility for child psychologists. Although this is true, the psychometric properties of the instrument remain to be investigated.

In an ''attempt 'to wed' Luria's techniques with American clinical neuropsychology'' [Golden, 1980, p. 517], Golden, Hammeke, and Purisch (1978) have constructed an objective neuropsychological battery adapted from Luria's functional theory of the brain. Of the 282 items comprising the battery, the majority of tasks are refinements of those operationalized by Christensen (1975). The objective here has been to retain many of the positive features of Luria's functional theory of the brain while providing a standardized administration and scoring procedure (Golden, 1980). These items, referred to as the Luria–Nebraska Neuropsychological Test Battery (LNNB) (Golden et al., 1978) offer both portability and considerable reduction in the time necessary (approximately 2.5 hours) for assessment with more traditional, neuropsychological batteries (e.g., HRB—> 4 hours). As suggested in Table 7.3, the authors offer 10 discrete subscales which represent a nominalization of each functional zone proposed by Luria (1966) and operationalized by Christensen (1975, 1980).

Taking a quantitative posture in examining the adequacy of the measure, Golden and his associates (Golden & Berg, 1983; Golden, Fross, & Graber, 1981; Golden, Hammeke, & Purisch, 1978; McKay & Golden, 1979; Moses & Golden, 1979; Moses, Johnson, & Lewis, in press; Purisch, Golden, & Hammeke, 1978) reported a number of studies relative to the psychometric adequacy of the LNNB. The majority of these investigations report adequate test–retest reliability estimates for summary, localization, and individual ''factor'' scales. The exception being rather unacceptable reliability for two receptive speech factors (Golden, 1981). Interrater scoring agreement was also seen as adequate (see Golden, 1981). From a qualitative viewpoint, the battery has been shown to be sensitive in discriminating adult brain damaged patients from both normals (Golden et al., 1978) and hospitalized psychiatric patients (Moses et al., in press). Using combinations of the 10 factors of the LNNB, Golden and his associates (Golden et al., 1981; McKay & Golden, 1979) report localization value for subscales. For example, in response to needs in the applied setting, McKay and Golden (1979) report the development of hemispheric localization scales shown to be of clinical significance in diagnosing lateralized brain damage. The majority of these findings appear to be robust with investigations from other laboratories reporting at least partial replication (e.g., Malloy & Webster, 1981; Shelly & Goldstein, 1982a, 1982b). Although impressive at first blush, extensive research will be necessary to completely define and assess the technical adequacy and utility of the LNNB.

Adams (1980) has argued that with the quantitative approach taken in the development and standardization of the Luria–Nebraska Neuropsychological Test Battery, much of the appeal of Luria's theory of the functioning of the brain is lost. Indeed, more work with the LNNB has been done to offer psychometric credibility than to offer insights how the measure can make use of its theoretical underpinnings in understanding the patient and structuring rehabilitation plans. Although recognizing the contributions of Luria (1966) and Christensen (1975), Golden's (1980, 1981) emphasis in the LNNB is not an attempt to develop a standardized version of Luria's neuropsychological assessment system per se, but rather a quantitative sound system with a theoretically consistent foundation. Spiers (1981, 1982) argues that nonverbal abilities are difficult to assess with the LNNB because of the role played by language in the presentation and response requirements to nonverbal items. The nominal scoring system adapted in the LNNB has been seen to reduce the sensitivity of individual items (Spiers, 1981, 1982). Indeed, it would seem difficult to glean information relative to a patient's performance in which his/her scores were found to be within normal limits. Golden, Ariel, Moses, Wilkening, McKay, and MacInnes (1982) argue that a patient's scaled scores, item patterns, qualitative data, and the medical history must all be involved in an analysis of an individual's neuropsychological functioning. In this regard, it should be noted that considerable sophistication with Luria's (1966, 1973) model of brain functioning and technical aspects of the LNNB are necessary for the competent administration, scoring, and interpretation of this battery. Controversy may well continue concerning the extent to which Golden and his associates have been successful in melding both the quantitative–qualitative and functional–structured points of view in neuropsychology. Clearly, the LNNB does provide a coherent group of tasks which are theoretically consistent.

The Kaufman Assessment Battery for Children (Kaufman & Kaufman, 1983) (K-ABC) is a wide-band cognitive battery based on a functional theory of the brain. Unlike the atheoretical approach taken in the development of cognitive measures such as the Wechsler Intelligence Scales, the Mental Processing Scales of the K-ABC were developed to conform to the sequential–simultaneous processing dichotomy articulated by Luria (1966) and operationalized by Das (e.g.,, 1973). Although interpretations may differ, Kaufman and Kaufman (1983) present a good deal of factor-analytic evidence consistent with a sequential-simultaneous distinction for subtests of the battery. The K-ABC also makes a very useful distinction between school related achievement and more basic elements of mental processing. Because less reliance is placed on past experience in scales of cognitive efficiency, the authors have argued in favor of a reduction in assessment bias and a wide range of information useful in remedial planning (Kaufman & Kaufman, 1983). Although it is not clear how the LNNB and K-ABC will stand the scrutiny of time, it seems apparent that Luria's functional theory of the brain will accent neuropsychological assessment as practiced in North America.

Neuropsychological Assessment of the Child

The early success in defining the behavior deficits following brain damage in adults stimulated such study with children and adolescents (Boll, 1974). By comparison, though, efforts to define brain-behavior relationships with children are even a more recent pursuit. Greater numbers of adult referrals with discretely localized lesions have provided a rather large research base on which detailed neuropsychological studies could be made. Moreover, adult neurological disorders, as opposed to those seen in children, are more likely to provide the opportunity to validate assessment techniques during surgery and autopsies (Dean, 1983a; Reitan, 1974). As mentioned previously, a number of moderator variables exist which may modify the interpretation of adults' performance on neuropsychological tests (history, demographics, time from onset to assessment, and the like). Clearly, such factors play an important role in psychological findings for children with cerebral dysfunction. In addition, the interpretation of children's performance must account for a number of potential confounding variables not usually considered with adults (Dean, in press). The child's premorbid developmental history (Benton, 1974), brain development at onset (Boll, 1974), the chronicity of the disorder (Hartlage & Hartlage, 1977), and environmental background (Dean, 1983a) combine interactively to make reliable inferences about the functioning of the child's brain far more tenuous an undertaking than that with adults. Similar lesions in mature and developing nervous systems have been shown to produce far different neuropsychological test patterns and expectations for recovery of function (Dean, in press; Hartlage, 1981; Reed, Reitan & Klove, 1965).

Klonoff and his associates (Klonoff, Low, & Clark, 1977; Klonoff & Robinson, 1967; Klonoff & Thompson, 1969) present evidence that the type of brain damage most frequently occurring in childhood is responsible in part for adult–child differences in test impairment shown in these groups. Klonoff's group reports epidemiological estimates showing brain damage resulting from *falls* was approximately three times more prevalent in children than adults. Therefore, it seems our knowledge of children's neuropsychological assessment data will involve a greater understanding of the effects of closed head injury in the developing child.

Although definitive conclusions concerning brain-behavior relationships remain to be made for adults, the neuropsychological interpretation of behavior with children is far more equivocal. Similar to findings for adults, a good deal of research indicates a sensitivity of neuropsychological batteries to brain dysfunction (Boll, 1972b; Dean, 1982c; Golden, 1981; Hartlage, 1981; Klonoff & low, 1974; Reed, Reitan, & Klove, 1965; Reitan, 1971; Reitan & Heineman, 1968). Reed, Reitan, and Klove (1965) and Boll (1974) using extensive neuropsychological batteries have provided evidence that patterns of deficits for adults and children who had sustained head injuries were much different. Brain damage in children seems to have more global effects than that found in adults with similar lesions. Children are more likely to exhibit a general depression of those cognitive abilities which have been shown to be rather resistant to impairment with adults (Ernhart, Graham, Eichman, Marshall, & Thurston, 1963; Reed, Reitan, & Klove, 1965). Using the HRB, Reitan (1974) reports scores of the Wechsler Intelligence Scale for Children to be more salient in predicting brain damage in children (6–8 years) than other measures of the battery. As pointed out earlier in the chapter, for adults, the Halstead Impairment Index is most often found to be a more sensitive indicator of brain damage than that offered by the Wechsler Scales (Reitan, 1955). Older, well practiced skills or, if you will, *overlearned* material is more resistant to cerebral insult in adults than that which has been found in children. In sum, the effects of brain damage in childhood are more profound than that found in adults; and, the earlier the insult occurs, the more generalized will be the resulting impairment on measures of neuropsychological functioning (Boll, 1972b; Fitzhugh & Fitzhugh, 1965; Reed & Fitzhugh, 1966, Reitan, 1974). Problems in translating adult neuropsychological research conclusions to children involves the ongoing autonomical changes in the child's brain (Boll, 1974). Indeed, some research indicates that age-related differences in the psychological sequelae of childhood brain damage may be due to developmental changes in functional lateralization in children (Dean, 1982a).

Related to the neuropsychological assessment of children's learning disorders is the relative lack of data regarding development trends for normally functioning children on tests of neuropsychological measures. Such data will begin to portray the development of cortically related behaviors and the amount of normal variability which can be expected. Crockett, Klonoff, & Bjerring (1969) emphasized the importance of this area with data showing that increasing neuropsychological complexity is necessary to describe the within-subject variability on some 32 neuropsychological variables with descending age. Whether these findings represent invariant developmental trends or fluctuations between development levels remains to be investigated. This area seems most promising in providing advances in both our understanding of the development of brain-behavior relationships and a base from which to evaluate the individual child's performance from an individual difference approach.

Factors other than cortical functioning seem to play a more salient role in test results for children than adults. Indeed, the naive application of aberrant behavioral patterns found in adult neurological disorders (encephalitis) to children exhibiting functional neurological disorders (e.g., Strauss & Lehtinen, 1948) may well be responsible for the rejection of neuropsychological studies with children. For, while consistent behavioral patterns may well be reliable predictors of adult cortical impairment, the assumption that these same behaviors reflect underlying cortical dysfunctions in children is tenable (Dean, 1978a, 1982a). This seems particularly true in our understanding of children's learning disorders. For over a century attempts have been made to explain children's classroom learning problems in terms of neurological aberrations. Although early overzealous attempts to explain all children's learning problems in terms of neurological dysfunction have been rejected by most serious reviewers of the available literature, problems remain in nosological classification of children's learning disorders.

Attempts have begun to isolate behavioral constellations from neuropsychological measures which may encourage further understanding of categories of learning problems (e.g., Dean, 1978a; 1983b, in press; Fisk & Rourke, 1979; Rourke, 1975, 1976, 1979). Clearly, some forms of children's learning disorders relate to neurological dysfunction (Dean, 1982a). This conclusion is as obvious as the fact that many childhood learning problems may more heuristically be related to an interaction of environmental and developmental factors. Although a number of authors have argued in favor of a neuropsychological perspective which goes beyond the diagnosis of impaired neurological processes to the structuring of educational programs which maximize the child's assessed strengths (Dean, 1982a; Hartlage & Reynolds, 1981), future attempts to statistically segregate behaviors for children with learning problems hold considerable promise in our understanding and treatment of children's learning disorders.

It has also become apparent that children with histories of classroom failure retain an underlying emotional reaction to school-related tasks even after obvious success (Dean, 1983b; Lang, 1964; Severson, 1970). Therefore, attempts to isolate subtypes of learning disorders and offer treatment of these problems require a focus on an interaction of emotional and neuropsychological factors. Dean (1983b) offered early data favoring the combination of neurological and emotional factors in what amounts to systematic desensitization procedures to treat emotional reactions to failure. This procedure was effective in increasing school-related achievement while it had little effect on the child's measures of neuropsychological functioning. Interestingly, Dean (1983b) also reported a reduction in behavioral problems concomitant with gains in classroom related achievement.

The issue of the relationship between emotional disturbance and brain damage with children is a complex one. Emotional disturbance has been reported with a significantly greater frequency in children who have suffered brain damage than normals (see Shaffer, 1974). Another line of research also indicates neuropsychological impairment occurs significantly more frequently in emotionally disturbed children without an obvious neurological abnormality than that found in samples of normal children (Reed & Reitan, 1963). Although brain damaged children may well react to a perceived loss of function in an emotional fashion (Dean, 1982a), as plausable is the onset of emotional symptoms concomitant with damage to specific areas of the brain (Shaffer, 1974). The truth, of course, will probably be most clearly understood on a multivariate continuum involving etiological and environmental factors. Such study would seem of value when examined in conjunction with developmental trends of lateralization of functions. Although information necessary for a basic understanding of these issues remains to be reported, numerous laboratories have begun intensive projects which will allow a greater understanding of the basic issues in neuropsychological assessment with children.

Neuropsychological Assessment in the Later Years

Dementing disorders are syndromes originating from a rather wide variety of causes. The etiology of these disorders has been attributed to intrinsic changes in the brain cells, vascular insufficiencies, cardiovascular accidents, metabolic disturbances, and hormonal imbalances. Although some forms of dementia have been shown to be treatable, the majority is not (Barbizet, 1970). The course of dementia involves the deterioration of higher cognitive functioning, emotion, motoric behavior, and eventually death (NIH, 1981).

It has been estimated that 15% of the population of the United States older than 65 years has some form of dementing disorder, amounting to some three million persons (NIH, 1981). Clearly, then, while the vast majority of elderly do not develop such disorders, it does represent a significant health problem. This problem becomes acute when projections for increased life span and the growth of the number of individuals who will exceed 65 years in the early 21st century are examined. Thus, our ability to diagnose dementia and the understanding of the normal aging process is fast becoming even more critically important than it has in the past.

With the essential symptom of dementing disorders being the "loss of intellectual abilities of sufficient severity to interfere with social or occupational functioning" [DSM-III, 1980, p. 58], the neuropsychologist is often called upon to make a judgment on the extent of impairment from neuropsychological assessment procedures. Although memory losses are the most prominent feature of dementia (Barbizet, 1970), the neuropsychological measures recommended are "multifaceted and involve memory, judgment, abstract thought, and a variety of other higher cortical functions" [DSM-III, 1980, p. 159].

Dementia is one of the most *overdiagnosed* disorders in the elderly (DSM-III, 1980). Part of the problem inherent with the diagnosis of this group of disorders relates to the fact that there are numerous other syndromes (e.g., affective disorders) with symptoms which resemble the early stages of true dementias. A critical difference between these disorders is that many of the nondementing disorders (e.g., depression) may be successfully treated (Dean, in press). Confusion presently exists in the diagnosis of dementia which is due, in part, to a lack of our understanding of the effects normal aging has on the majority of the available neuropsychological measures. Second, yet no less important, is the lack of knowledge of how neuropsychological tests correspond to adaptive behaviors and the emotional stability necessary for independent living (Dean, in press).

Investigations have shown a substantial, negative relationship between age and performance on a number of measures of cortical functioning with healthy older adults (Reed & Reitan, 1963; Reitan, 1955; Vega & Parsons, 1967). Using the presently available adult norms, the performance on neuropsychological measures of healthy individuals above the age of 50 frequently falls into the range seen as reflective of organic impairment (Davies, 1968; Prigatano, 1978; Reed & Reitan, 1963). Bak and Greene (1980), using standard scoring procedures and available adult norms, showed 85% of a sample of healthy, normal functioning adults between 50–86 years of age would be classified as impaired. Of interest, the neuropsychological measures used in this investigation (i.e., Wechsler Memory Scale, Wechsler Adult Intelligence Scale, and the Halstead–Reitan Neuropsychological Battery) were those most commonly employed in the differential diagnosis of dementia. Although some authors (e.g., Finlayson, Johnson, & Reitan, 1977) have aruged that many of the findings of impairment in normal older adults are compromised by educational level and omnibus intelligence, numerous investigators have reported concomitant decline for various memory and intellectual functions with increasing age (Matarazzo, 1972). Such findings call into question the utility of standardized neuropsychological measures in the diagnosis of impairment and as a reflection of adaptive functional behavior. Moreover, many normal older individuals classified as impaired on neuropsychological measures have been reported to be functioning normally in their environment in terms of daily living skills (Bak & Greene, 1980; Davies, 1968; Reitan, 1955). Obviously, the lack of information relative to the effects of normal aging on measures of neuropsychological functioning represents a major obstacle to their utility with this population.

Although it has been known for some time that age (> 45) interacts with the performance on measures of higher cortical functioning (Horn & Cattell, 1966, 1967; Matarazzo, 1972) and psychomotor speed of operation (McFarland, 1968), little normative data exist with which the clinician can compare obtained results on the majority of measures of neuropsychological functioning. Such investigations are rudimentary to our understanding of the aging process and vital in the diagnosis of dementia. Future normative data will aid in differential diagnosis of dementias and, more importantly, reduce misdiagnoses of dementia when other disorders exist which are more amenable to treatment. Clearly, the interaction of neuropsychological functions, emotional stability, and daily living skills remains to be investigated with adults in their later years.

Neuropsychological Assessment of Adult Psychiatric Patients

Neuropsychological assessment procedures have in the main been developed after extended observations of patients with neurological disorders and more often than not documented brain lesions (e.g., Halstead, 1947; Luria, 1966; Reitan, 1955). The choice of items and the validation of resulting measures have often emphasized the discrimination of neurological groups from normal controls (e.g., Halstead, 1947; Reitan, 1955). As previously mentioned, such a methodology has served the field well with the predictive accuracy most often exceeding 85% when intact groups of normals and brain damaged patients are used (Klove, 1974; Reed & Reitan, 1963; Reitan, 1955, 1971; Wheeler, Burke, & Reitan, 1963). So too, the prediction of diffuse, left and right hemispheric lesions is fairly well established (Boll, 1972; Filskov & Goldstein, 1974; Reitan, 1966).

Because the most often experienced referral question asked involves differential diagnosis, a number of problems arise when attempts are made to infer the source of a behavioral disturbance. That is to say, in the medical setting, a diagnostic question often asked involves whether the patient's behavioral disturbance has an "organic" base or if it may more heuristically be attributed to a "functional psychiatric" disorder. Here, "functional mental disorder" would imply a disturbance without a known physical abnormality of the brain; whereas, "organic mental disorder" has traditionally implied a behavioral aberration resulting from a biochemical or structured lesion. Of particular interest to the practicing neuropsychologist has been the accuracy of assessment procedures in the differential diagnosis of "functional" and "organic disorders." Moreover, the differentiation of functional psychosis with concomitant aberrations in thought processes from patients with neurological disorders remains a difficult diagnostic problem (Dean, 1983a).

The research objectives in this area have focused on the ability of neuropsychological assessment procedures to differentially diagnose functional psychiatric and organically related disorders (e.g., Golden, 1977; Matthews, Shaw, & Klove, 1966; Reitan, 1976) and the establishment of performance patterns which would aid in differential diagnoses (Klonoff, Fibiger, & Hutton, 1970; Parsons & Klein, 1970; Watson, Thomas, Anderson, & Felling, 1968). A more basic question that is beginning to receive attention involves a neuropsychological understanding of the etiology and effects of functional mental disorders (Dean, in press; Flor-Henry, Fromm-Auch, Tapper, & Schopflocher, 1981; Rockford, Detre, Tucker, & Harrow, 1970; Taylor, Greenspan, & Abrams,

1979). The previously suggested intervening variables of medical history, age at onset, educational level, chronological age, site of dysfunction, premorbid environment, and individual differences in anatomical structures are prominent factors in the evaluation of research in this area. In addition, a number of variables which are more unique to a psychiatric setting. often accent neuropsychological assessment data and obscure their interpretation. The most consistently mentioned factors involved:

(a) the chronicity of the disorder (see Watson, 1974);
(b) a failure to rule out an organic involvement for groups diagnosed with a functional mental disorder (see Dean, in press);
(c) apparent lack of appreciation for treatment effects associated with somatic interventions (e.g., medication, electro-convulsive therapy) in psychiatric disorders (see Baker, 1968);
(d) length of hospitalization (see Klonoff, Fibiger, & Hutton, 1970);
(e) failure to consider the severity of psychiatric symptoms (see Goldstein & Halperin, 1977);
(f) a failure to examine response style as an obviating variable of performance (see Hirsch & DeWolfe, 1977); and, finally,
(g) contamination of diagnosis with neuropsychological assessment data (see Golden, 1977).

Although the majority of studies in this area have utilized some combination of tests of the Halstead–Reitan Battery (Golden, 1977; Heaton, Vogt, Hoehn, Lewis, Crowley, & Stallings, 1979), a number have examined more specialized measures (e.g., Bender Visual Motor Gestalt Test; Benton Visual Retention Test; Birkett & Boltuch, 1977; Holland, Wadsworth, & Royer, 1975; Luria Nebraska Neuropsychological Battery; Memory-for-Designs Test; Minnesota Percepto-Diagnostic Test; Moses, Cardellino, & Thompson, 1983; Optimal Symbol Digit Modalities; Watson & Benton, 1976). It is not the objective here to attempt a duplication of a number of excellent reviews of this area of research (e.g., Heaton, Baade, & Johnson, 1978; Heaton & Crowley, 1981; Reitan, 1976) but, rather, to identify research developments as they may relate to new areas of emphasis in the field of neuropsychological assessment.

With this objective in mind, it seems clear that the diagnostic accuracy in differentiating functional, psychiatric, and organic disorders clearly suffers when chronic schizophrenics are included as a functional disorder (Goldstein & Halperin, 1977; Heaton et al., 1979; Watson et al., 1968). After a review of the available literature, Heaton et al., 1978, reported a median correct classification of functional and organic patients of 75% when chronic schizophrenics were eliminated from consideration. With the inclusion of chronic schizophrenics, the median "hit rate" dropped to a chance level in prediction (54%). These findings in combination with the research since 1975 indicate that the rates of correct diagnosis between normals and brain damaged are not significantly different than rates found between organics and patients with functional psychiatric disorders when chronic or process schizophrenics are eliminated from consideration (Heaton et al., 1978; Heaton & Crowley, 1981). On both the HRB (e.g., Chelune, Heaton, Lehman, & Robinson, 1979) and LNNB (e.g., Moses, Cardellino, & Thompson, 1983) chornic schizophrenia seems more clearly related to an overall level of impaired performance than any specific test pattern of deficits. As is true with brain damaged patients (Doehring & Reitan, 1960), there appears to be a relationship between impaired, neuropsychological assessment findings, and the degree of emotional disturbance with schizophrenics (Schwartzman, Douglas, & Muir, 1962).

At this point in the discussion, it should be noted that relatively few attempts have been made to compare neuropsychological assessment results for different functional psychiatric disorders (e.g., depression, schizophrenia, etc.). Moreover, the focus of the bulk of the studies in this area has been the differentiation from organically related behavioral disturbances and psychiatric patients with the aid of neuropsychological assessment data. The tacit assumption here has often been that "functional disturbances" were less related to abnormal brain functioning than to psychosocial influences. However, recent evidence indicates that biochemical (Fish, 1977; Glassman, Perel, Shostak, Kantor, & Fleiss, 1977; Young, Taylor, & Holmstrom, 1977) and structural abnormalities (Andreasen, Olsen, Dennert, & Smith, 1982; Johnstone, Crow, Frith, Husband, & Kreel, 1976) exist in the brains of patients diagnosed with disorders hitherto considered to be functional. Therefore, difficulties in the past in differentiating organics from some forms of functional, psychiatric patients may relate to an underlying neurological substrate for some functional, psychiatric disorders (Dean, in press). Clear differences in brain chemistry have been identified for patients with affective disorders (Glassman et al., 1977; Jarvik, 1977; Young et al., 1977) and forms of schizophrenia (Fish, 1977; Goodman & Gilman, 1975). Similarly, although some methodological problems exist, data have been reported showing abnormalities in the brain structure and function for patients diagnosed as schizophrenic (Andreasen et al., 1982; Haug, 1963; Mirsky, 1969). The force of these data indicates a greater probability for abnormal electroencephalograms (EEG) (Lester & Edwards, 1966) and enlargement of ventricular structures (Luchins, 1982) concomitant with more debilitating forms of schizophrenia. A fair amount of evidence exists across investigations to hypothesize that patients diagnosed with a primary affective–depression show an abnormal decrease in activity of the right hemisphere (d'Elia & Perris, 1974); and, specifically, abnormal EEG findings have been reported in the area of the right temporal lobe (Flor-Henry, 1976).

The neuropsychology of differing "functional" psychiatric disorders has only recently begun to be investigated in a systematic fashion. Studies have, in the main, attempted to define patterns of neuropsychological test scores which would differentiate schizophrenia and affective disorders (bipolar and unipolar).

Generally, patients of both nosological groups show significant impairment on measures of neuropsychological functioning (Flor-Henry, 1976; Golden, Moses, Zelazowski, Graber, Zatz, Horvath, & Berger, 1980; Miller, 1975). Flor-Henry (1976), using a stepwise discriminant analysis, was able to correctly classify some 90% of the schizophrenic and patients suffering from affective disorders (unipolar-manic) using neuropsychological test results. Interestingly, although both groups exhibited what was interpreted as frontal–temporal dysfunction, the abnormalities found in the group with affective disorders were significantly more right hemispheric. Specific impairment on neuropsychological tests of the right hemispheric functions seems robust (Taylor, Greenspan, & Abrams, 1979). Abrams and Taylor (1980) have since extended these data and noted deficit right hemispheric performance on measures of neuropsychological functioning for affective disorders in general (both unipolar and bipolar disorders). Although the neuropsychological test findings for schizophrenia seem to be more generalized frontal–temporal dysfunction, at least one study reported data favoring left hemispheric lateralization of neuropsychological deficit performance for schizophrenics (Rockford, Detre, Tucker, & Harrow, 1970).

As alluded to earlier, the level of chronicity and specific schizophrenia classification play a major role in the interpretation of data in this area. Some of the confusion here may well relate to the manner in which psychiatric groups were formed and the diagnostic criteria used. Although in some investigations the clinical impressions of a psychiatrist are used, most, more recent studies have begun to rely on criteria offered by the American Psychiatric Association (DSM–III, 1980), Research Diagnostic Criteria (Spitzer, Endicott, & Robins, 1977), or those known as Feighner Criteria (after Feighner, Robins, Guze, Woodruff, Winokur, & Munoz, 1972). This point of diagnostic criteria seems important to this area of research because of what amounts to a tacit assumption in the past that patients' nosological placement was without error variance. Indeed, recent research indicates that while sets of diagnostic criteria may have similar interrater reliability, patients diagnosed as schizophrenic by one system may receive an entirely different diagnosis when the criteria of another system are employed (e.g., Endicott, Nee, Fleiss, Cohen, Williams, & Simon, 1982).

Abnormal neuropsychological assessment findings have been shown with psychiatric disorders other than schizophrenia and manic-depression; but, these data seem less striking and have yet to be replicated across laboratories (see Flor-Henry, Fromm-Auch, Tapper, & Schopflocher, 1981; Flor-Henry, Yeudall, Koles, & Howorth, 1979). The extent to which such findings of specific neuropsychological dysfunction for "functional psychiatric disorders" are amenable to psychotherapeutic or psychopharmacologic manipulations is unclear. Future research which compare consistently diagnosed psychiatric groups while controlling for somatic treatment modalities (see Heaton & Crowley, 1981) will be necessary to define whether distinctive neuropsychological test profiles exist which are consistent with specific diagnoses. Continued findings in a positive direction would lead to the reconsideration of the locus of a number of psychiatric disorders which have been assumed to be "functional." While the amount of future research will need to be considerable, future and present neuropsychological assessment procedures would seem to hold promise in our understanding into the etiology of a number of psychiatric disorders and have the potential to aid in diagnosis and the planning of therapies for individual patients.

Neuropsychological Assessment in the Rehabilitation Process

Neuropsychological assessment has made its major contribution in providing diagnostic information for patients for which there was equivocal physical evidence of brain damage. The present diagnostic–etiological approach to assessment is more the result of pressures in the medical setting than a theoretical vision of the specialty (Dean, 1982a). However, the continuing sophistication of physical procedures may serve to de-emphasize the diagnostic role of neuropsychological assessment (Dean, 1983a).

The sophistication of noninvasive physical diagnostic techniques has grown geometrically in the past 20 years. The new generation of CT scanning equipment and more recent advances in positron emission tomography (Raichle, 1979) hold clear implications for the diagnosis of neurological disorders. Although problems presently exist in the reliability of these radiological techniques, the future extensions of these procedures are clear. In the past, the relative benign nature of neuropsychological assessment and a real mortality risk for some physical diagnostic methods made the utility of neuropsychological assessment as a diagnostic tool obvious. The future impact of more sophisticated radiological techniques seems clear. While presently providing a criteria for validating neuropsychological diagnostic procedures, continued refinement of radiological procedures may well reduce the dependence on neuropsychological assessment in diagnosis.

Although increasingly more accurate neurological knowledge will be available for the individual patient, rarely would it be possible for the neurosurgeon or neurologist to make specific predictions concerning the patient's behavior or functioning. Dean (1982a) argues that the direction of neuropsychological assessment will be influenced by the continuing need to understand the patient's behavioral deficits in the medical treatment of neurological disorders and in planning rehabilitation approaches. Neuropsychological assessment is seen by a number of researchers as offering a heuristic framework in which components of the patient's emotional, cognitive, and physical functioning can provide rehabilitation specialists an indepth view of the patient. However, few attempts have been made to interface our sophisticated psychometric techniques designed to outline subtle cognitive, perceptual, and motor consequences of neuropathology with rehabilitation strategies (Ben-Yishay, Gerstman, Diller, & Haas, 1970; Golden, 1978).

The issues posed by rehabilitation are quite different than those of diagnosis. Assessment with the objective of outlining rehabilitation goals goes beyond diagnosing impairment relative to normative group performance. The focus here is on definition of how impairments are expressed as disabilities in the patient's premorbid environment (Dean, 1983a). Unlike neurological diagnosis, which attempts to segregate a constellation of behaviors into a single disorder, rehabilitation works to define brain damage in terms of the medical, vocational, speech and hearing, and physical aspects in need of treatment. Moreover, brain damage seldom is expressed as a single impairment. Within the reality of the patient's environment, neuropsychological assessment seeks to: (1) define the cognitive, emotional, perceptual, and sensory aspects of the patient's functioning; (2) establish a baseline in which the rehabilitation can be followed; (3) predict the return of function in light of damage; (4) given multiple impairments and limited resources, what should be the priorities for a given patient; and (5) define in what ways will the patient's impairments be expressed as disabilities in their premorbid environment (see Diller & Gordon, 1981, for a review). It seems fair to say that neuropsychological assessment in the past has focused on the acute phase of brain damage and, as such, represented a psychometric extension of the clinical neurologic examination. The research and applied emphasis in neuropsychological assessment has been within the areas numbered one and two above. Continuing rehabilitative needs and more reliance upon physical, diagnostic procedures will be expressed in greater concern for rehabilitative outcomes.

Although psychologists have contributed to theories of impairment and rehabilitation following brain damage (Goldstein, 1979), relatively few attempts have been made to use neuropsychological test results to answer basic, remedial questions. The extent to which neuropsychological assessment may predict different remedial outcomes has not been considered in any continuing fashion (Dean, in press). Ben-Yishay, Gerstman, Diller, and Haas (1970) have reported some evidence that left hemiplegics (right hemisphere damage) who exhibit less impairment in a number of spatial skills have shorter rehabilitation stays and learn activities necessary for daily living more quickly than left hemiplegics with more grossly impaired spatial skills. This finding seems congruent with other data showing positive relationships between performance on cognitive and perceptual measures and success in training daily living skills (e.g., Twitchell, 1951; Weisbroth, Esibill, & Zuger, 1971; Williams, 1967). Certainly, such work is seminal in an area where the relationship between neuropsychological tests and success in rehabilitation is unknown.

Which rehabilitative approach is more heuristic given a pattern of neuropsychological impairment is an empirical question which has only begun to be addressed (Diller & Gordon, 1981). A need also seems acute in our ability to predict activities of daily living (after Brown, 1960). For, it seems that the relationship between neuropsychological test findings and the ability of the patient to perform in their premorbid environment is incompletely understood (Ben-Yishay et al., 1970). In short, the future of neuropsychological assessment will depend, in part, on our ability to go beyond diagnosis. Goldstein (1979) and Horton (1979) argue rather convincingly in favor of an interface between neuropsychological assessment and behavior therapy in the prediction and evaluation of rehabilitation procedures.

The neuropsychological assessment in the coming decade may well involve traditional tests as well as measures designed to assess the acquisition rate of skills which in turn would allow prediction of plateaus in return of functions. Such measures will also allow an examination of the interaction between emotional responses to loss (e.g., depression, denial, and the like) and cognitive/perceptual factors in the rehabilitative process. This information would seem to be a prerequisite to cuing and cognitive retraining procedures presently being examined in the rehabilitation settings (e.g., Cermak, 1975; Crovitz, 1979; Hartlage & Reynolds, 1981; Meier, 1963).

From a rehabilitative point of view, the neurological diagnosis serves as an intervening variable. Diagnosis does not address how the patient should be treated differently because of membership in a given nosological category. Future efforts which attempt to relate neuropsychological test results directly to the differential outcomes from multiple remedial approaches, without the undue reliance on diagnosis which would improve the utility of assessment. Attempts to employ an actuarial approach to treatment outcomes have the potential of reducing our reliance on a pure clinical level of inference.

Summary

Neuropsychological assessment was portrayed as an attempt to psychometrically describe behavior change in patients following brain damage. This chapter has provided an overview of the present approaches to neuropsychological assessment. Although necessarily brief, a review of the research concerning the rationale and utility of the quantitative and qualitative schools of thought in neuropsychology were presented. Congruent with the rapid and extensive increase in our understanding of brain behavior relationships has come the ability to make valid inferences regarding brain functions for individual patients. Neuropsychology, in general, and neuropsychological assessment, specifically, are in their infancy and face a number of problems which will influence development. The rapid evolving knowledge base in the neurosciences was seen as a challenge to the conceptualization of the functioning of the brain and the methodology to be used in assessment. Future trends in neuropsychological assessment were drawn from a review of the literature which defines a number of critical issues in the area. Framing points within a historical context, theoretical extensions of Luria's functional theory of the brain, and future assessment interpretive schemes were examined. Successes in the interpretation of neuropsychological measures in early and middle adulthood were contrasted with the need for a developmental, individual difference approach in research in childhood and the later years. A reduction in neuropsychological diagnosis concomitant with advances in physical diagnostic techniques was offered as the stimulus to further define adaptive behavior and predict outcomes in the rehabilitation setting. Developments in the neurology of some "functional" mental disorders were seen as offering a future potential for neuropsychological test results as diagnostic markers of some psychiatric disorders.

REFERENCES

Abrams, R., Taylor, M. A. (1980). A comparison of unipolar and bipolar depressive illness. *American Journal of Psychiatry, 137,* 1084-1087.

Adams, K. M. (1980). In search of Luria's battery: A false start. *Journal of Consulting and Clinical Psychology, 48,* 511-516.

American Psychiatric Association (1980). *Diagnostic and statistical manual of mental disorders III,* Washington, D.C., APA.

Andreasen, N. C., Olsen, S. A., Dennert, J. W., & Smith, M. R. (1982). Ventricular enlargement in schizophrenia: Relationship to positive and negative symptoms. *American Journal of Psychiatry, 139,* 297–302.

Anthony, W. Z., Heaton, R. K., & Lehman, R. A. W. (1980). An attempt to cross-validate two actuarial systems for neuropsychological test interpretation. *Journal of Consulting and Clinical Psychology, 48,* 317-326.

Bak, J. S., & Greene, R. L. (1980). Changes in neuropsychological functioning in an aging population. *Journal of Consulting and Clinical Psychology, 48,* 395-399.

Baker, R. R. (1968). The effects of psychotropic drugs on psychological testing. *Psychological Bulletin, 69,* 77-87.

Barbizet, J. (1970). *Human memory and its pathology.* San Francisco: Freedman and Company.

Baron, I. S. (1978). Neuropsychological assessment of neurological conditions. In P. R. Magrab (Ed.), *Psychological management of pediatric problems (Vol. 2).* Baltimore: University Park Press.

Ben-Yishay, Y., Gerstman, L., Diller, L., & Haas, A. (1970). Prediction of rehabilitation outcomes from pyschometric parameters in left hemiplegics. *Journal of Consulting and Clinical Psychology, 34,* 436-441.

Benton, A. L. (1974). Clinical neuropsychology of childhood: An overview. In R. M. Reitan & L. A. Davison (Eds.), *Clinical neuropsychology: Current status and applications.* New York: Wiley.

Birkett, D. P., & Boltuch, B. (1977). Measuring dementia. *Journal of the American Geriatrics Society, 25,* 153-156.

Boll, T. (1972a). Right and left cerebral hemisphere damage and tactile perception: Performance of the ipsilateral and contralateral sides of the body. *Neuropsychologia, 12,* 235-238.

Boll. T. J. (1972b). Conceputal vs. perceptual vs. motor deficits in brain-damaged children. *Journal of Clinical Psychology, 28,* 157-159.

Boll, T. J. (1974). Behavioral correlates of cerebral damage in children nine through fourteen. In R. M. Reitan & L. A. Davison (Eds.), *Clinical neuropsychology: Current status and applications.* New York: Wiley.

Boll, T. J. (1981). The Halstead—Reitan neuropsychology battery. In S. B. Filskov & T. J. Boll (Eds.), *Handbook of clinical neuropsychology.* New York: Wiley.

Boll, T. J., Heaton, R. K., & Reitan, R. M. (1974). Neuropsychological and emotional correlates of Huntington's chorea. *Journal of Nervous and Mental Disease, 158,* 61-69.

Boring, E. G. (1942). *Sensation and perception in the history of experimental psychology.* New York: Appleton-Century.

Broca, P. (1861/1960). Remarks on the seat of the faculty of articulate language, followed by an observation of apheasia. In G. von Bonin (trans.), *Some papers on the cerebral cortex.* Springfield, IL: Charles C. Thomas.

Brown, M. E. (1960). Patient's motionability: Evaluation methods, trends, and principles. *Rehabilitation Literature, 21,* 46-58, 78-96.

Cermak, L. S. (1975). Imagery as an aid to retrieval for Korsakoff patients. *Cortex, 11,* 163-167.

Chelune, G. J., Heaton, R. K., Lehman, R. A., & Robinson, A. (1979). Level versus pattern of neuropsychological performance among schizophrenic and diffusely brain damaged patients. *Journal of Consulting and Clinical Psychology, 47,* 155-163.

Christensen, A. L. (1975). *Luria's neuropsychological investigation.* New York: Spectrum.

Christensen, A. L. (1980). *Luria's neuropsychological investigation* (Rev., 2nd ed.). Copenhagen: Munksgaard.

Crockett, D., Clark, C., & Klonoff, H. (1981). An overview of neuropsychology. In S. B. Filskov & T. J. Boll (Eds.), *Handbook of clinical neuropsychology.* New York: Wiley.

Crockett, D., Klonoff, H., & Bjerring, J. (1969). Factor analysis of neuropsychological tests. *Perceptual and Motor Skills, 29,* 791-802.

Crovitz, H. F. (1979). Memory rehabilitation in brain damaged patients: The airplane list. *Cortex, 15,* 131-134.

d'Elia, G., & Perris, C. (1974). Cerebral functional dominance and depression: An analysis of EEG amplitude in depressed patients. *Acta Psychiatrica Scandinavia, 49,* 191-197.

Das, J. P. (1973). Structure of cognitive abilities: Evidence for simultaneous and successive processing. *Journal of Educational Psychology, 65,* 103-108.

Davies, A. D. M. (1968). Measures of mental deterioration in aging and brain damage. In S. S. Chown & K. F. Reigel (Eds.), *Interdisciplinary topics in geronotology, Vol. 1.* Basel, New York: S. Karger.

Day, P. S., & Ulatowska, H. K. (1979). Perceptual, cognitive and linguistic development after early hemispherectomy: Two case studies. *Brain and Language, 7,* 17-33.

Dean, R. S. (1978a). The use of the WISC-R in distinguishing learning disabled and emotionally disturbed children. *Journal of Consulting and Clinical Psychology, 46,* 381-382.

Dean, R. S. (1978b). *Laterality preference schedule.* Madison, WI: University of Wisconsin.

Dean, R. S. (1982a). Neuropsychological assessment. In T. Kratochwill (Ed.), *Advances in school psychology (Vol. 2).* Hillsdale, NJ: Lawrence Earlbaum Associates.

Dean, R. S. (1982b). Cognitive neuropsychological differences in schizophrenia and primary affective depression. *Proceedings of the Society of Behavioral Medicine, 2,* 72-72.

Dean, R. S. (1982c). Neuropsychological correlates of total seizures with major-motor epileptic children. *Clinical Neuropsychology, 5,* 1-4.

Dean, R. S. (1983a). Neuropsychological assessment. In Staff College (Ed.), *Handbook of diagnostic and epidemiological instruments.* Washington, D. C.: National Institute of Mental Health.

Dean, R. S. (1983b, August). *Integrating neuropsychological and emotional variables in the treatment of children's learning disorders.* Presented at the Annual Convention of The American Psychological Association, Anaheim, CA.

Dean, R. S. (in press). Psychiatric aspects of neuropsychological assessment. In J. D. Cavenar, R. Michaels, H. K. H. Brodie, A. M. Cooper, S. B. Guze, L. L. Judd, G. L. Klerman, & A. J. Solnit (Eds.), *Psychiatry.* Philadelphia, PA: J. B. Lippincott Company.

Dean, R. S., Schwartz, N. H., & Smith, L. S. (1981). Lateral preference patterns as a discriminator of learning difficulties. *Journal of Consulting and Clinical Psychology, 49,* 227-235.

Diller, L., & Gordon, W. A. (1981). Rehabilitation and clinical neuropsychology. In S. B. Filskov & T. J. Boll (Eds.), *Handbook of clinical neuropsychology.* New York: Wiley.

Doehring, D. G., & Reitan, R. M. (1960). MMPI performance of aphasic and nonaphasic brain-damaged patients. *Journal of Clinical Psychology, 16,* 307-309.

Endicott, J., Nee, J., Fleiss, J., Cohen, J., Williams, J. B. W., & Simon, R. (1982). Diagnostic criteria for schizophrenia: Reliabilities and agreement between systems. *Archives of General Psychiatry, 39,* 884-889.

Ernhart, C. G., Graham, F. K., Eichman, P. L., Marshall, J. M., & Thurston, D. (1963). Brain injury in the preschool child: Some developmental considerations. *Psychological Monographs, General and Applied, 77*(11), 17-33.

Feighner, J. P., Robins, E., Guze, S., Woodruff, R. A., Winokur, G., & Munoz, R. (1972). Diagnostic criteria for use in psychiatric research. *Archives of General Psychiatry, 26,* 57-63.

Filskov, S., & Goldstein, S. (1974). Diagnostic validity of the Halstead—Reitan neuropsychological battery. *Journal of Consulting and Clinical Psychology, 42*(3), 382-388.

Finkelstein, J. H. (1976). BRAIN: A computer program for interpretation of the Halstead—Reitan neuropsychological test battery (Doctoral dissertation, Columbia University). *Dissertation Abstracts International,* 5349-B. (University Microfilms, No. 77-8864).

Finlayson, M. J. A., Johnson, K. A., & Reitan, R. M. (1977). Relationship of level of education to neuropsychological measures in brain-damaged and non-brain damaged adults. *Journal of Consulting and Clinical Psychology, 45,* 536-542.

Fish, B. (1977). Neurobiologic antecedents of schizophrenia in children; evidence for an inherited, congenital neurointegrative defect. *Archives of General Psychiatry, 34,* 1297-1313.

Fisk, J. L., & Rourke, B. P. (1979). Identification of subtypes of learning-disabled children at three age levels: A neuropsychological, multivariate approach. *Journal of Clinical Neuropsychology, 1,* 289-310.

Fitzhugh, K., & Fitzhugh, L. (1965). Effects of early and late onsets of cerebral dysfunctioning upon psychological test performance. *Perceptual and Motor Skills, 20,* 1099-1100.

Flor-Henry, P. (1976). Lateralized temporo-limbic dysfunction and psychopathology. *Annals of the New York Academy of Science, 280,* 777-797.

Flor-Henry, P., Fromm-Auch, D., Tapper, M., & Schopflocher, D. (1981). A neuropsychological study of the stable syndrome of hysteria. *Biological Psychiatry, 16,* 601-626.

Flor-Henry, P., Yeudall, L, T., Koles, Z. J., & Howorth, B. G. (1979). Neuropsychological and power spectral EEG investigations of the obsessive-compulsive syndrome. *Biological Psychiatry, 14,* 119-130.

Gibson, W. C. (1962). Pioneers of localization of function in the brain. *Journal of the American Medical Association, 180,* 944-951.

Glassman, A. H., Perel, J. M., Shostak, M., Kantor, S. J., & Fleiss, J. L. (1977). Clinical implications of imipramine plasma levels for depressive illness. *Archives of General Psychiatry, 34,* 197-204.

Golden, C. (1977). Validity of the Halstead—Reitan neuropsychological battery in a mixed psychiatric and brain injured population. *Journal of Consulting and Clinical Psychology, 45*(6), 1043-1051.

Golden, C. J. (1978). *Diagnosis and rehabilitation in clinical neuropsychology.* Springfield, IL: Charles C. Thomas.

Golden, C. J. (1980). In reply to Adams' "In search of Luria's battery: A false start." *Journal of Consulting and Clinical Psychology, 48,* 517-521.

Golden, C. J. (1981). A standardized version of Luria's neuropsychological tests: A quantitative and qualitative approach to neuropsychological evaluation. In S. B. Filskov & T. J. Boll (Eds.), *Handbook of clinical neuropsychology.* New York: Wiley.

Golden, C. J., Ariel, R. N., Moses, J. A., Wilkening, G. N., McKay, S. E., & MacInnes, W. D. (1982). Analytic techniques in the interpretation of the Luria—Nebraska neuropsychological battery. *Journal of Consulting and Clinical Psychology, 50,* 40-48.

Golden, C. J., & Berg, R. A. (1983). Interpretation of the Luria—Nebraska neuropsychological battery by item intercorrelation: Intellectual processes. *Clinical Neuropsychology, 5*(1), 23-28.

Golden, C. J., Fross, K. H., & Graber, B. (1981). Split-half reliability of the Luria—Nebraska neuropsychological battery. *Journal of Clinical and Consulting Psychology, 49,* 304-305.

Golden, C. J., Hammeke, T. A., & Purisch, A. D. (1978). Diagnostic validity of a standardized neuropsychological battery derived from Luria's neruopsychological tests. *Journal of Consulting and Clinical Psychology, 46,* 1258-1265.

Golden, C. J., Moses, J. A., Zelazowski, R., Graber, B., Zatz, L. M., Horvath, T. B., & Berger, P. A. (1980). Cerebral ventricular size and neuropsychological impairment in young chronic schizophrenics. *Archives of General Psychiatry, 37,* 619-623.

Goldstein, G. (1979). Methodological and theoretical issues in neuropsychological assessment. *Journal of Behavioral Assessment, 1,* 23-41.

Goldstein, G., & Halperin, K. M. (1977). Neuropsychological differences among subtypes of schizophrenia. *Journal of Abnormal Psychology. 86,* 34-40.

Goldstein, G., & Shelly, C. H. (1972). Statistical and normative studies of the Halstead neuropsychological test battery relevant to a neuropsychiatric hospital setting. *Perceptual and Motor Skills, 34,* 603-620.

Goldstein, K. H. (1942). *After effects of brain injuries in war.* New York: Grune and Stratton.

Goodman, L. S., & Gilman, A. (1975). *The pharmacological basis of therapeutics.* New York: MacMillan.

Halstead, W. C. (1947). *Brain and intelligence.* Chicago: University of Chicago Press.

Halstead, W. C., & Settlage, P. H. (1943). Grouping behavior of normal persons and of persons with lesions of the brain. *Archives of Neurology and Psychiatry, 49,* 489-506.

Hartlage, L. C. (1981). Neuropsychological assessment techniques. In C. R. Reynolds & T. Gutkin (Eds.), *Handbook of school psychology.* New York: Wiley.

Hartlage, L. C. & Hartlage, P. L. (1977). Psychological testing in neurological diagnosis. In J. Youman (Ed.), *Neurological surgery.* Philadelphia, PA: W. B. Saunders.

Hartlage, L. C., & Reynolds, C. R. (1981). Neuropsychological assessment and individualization of instruction. In G. W. Hynd & J. E. Obrzut (Eds.), *Neuropsychological assessment and the school-age child.* New York: Grune and Stratton.

Haug, J. O. (1963). *Pneumoencephalographic studies in mental disease.* Oslo, Norway: Universitetsforlaget.

Heaton, R., Baade, L., & Johnson, K. (1978). Neuropsychological test results associated with psychiatric disorders in adults. *Psychological Bulletin, 85*(1), 141-163.

Heaton, R. K., & Crowley, T. J. (1981). Effects of psychiatric disorders and their somatic treatments on neuropsychological test results. In S. B. Filskov & T. J. Boll (Eds.), *Handbook of clinical neuropsychology.* New York: Wiley.

Heaton, R. K., Vogt, A. T., Hoehn, M. M., Lewis, J. A., Crowley, T. J., & Stallings, M. A. (1979). Neuropsychological impairment with schizophrenia versus acute and chronic cerebral lesions. *Journal of Clinical Psychology, 35,* 46-53.

Hebb, D. O. (1949). *The organization of behavior.* New York: Wiley.

Hirsch, C., & DeWolfe, A. (1977). The associative interference and premorbid adjustment in schizophrenia. *Journal of Abnormal Psychology, 86*(6), 589-596.

Holland, T. R., Wadsworth, H. M., & Royer, F. L. (1975). The performance of brain-damaged and schizophrenic patients on the Minnesota percepto-diagnostic test under standard and BIP conditions of administration. *Journal of Clinical Psychology, 31,* 21-25.

Horn, J. L., & Cattell, R. B. (1966). Age differences in primary mental ability factors. *Journal of Gerontology, 21,* 210-220.

Horn, J. L., & Cattell, R. B. (1967). Age differences in fluid and crystallized intelligence. *Acta Psychologica, 26,* 107-129.

Horton, A. M. (1979). Behavioral neuropsychology: Rationale and research. *Clinical Neuropsychology, 1,* 20-24.

Horton, A. M., & Timmons, M. (1982). Separation of brain damaged from psychiatric scales. *Perceptual and Motor Skills, 55,* 755-758.

Hunt, H. F. (1943). A practical clinical test for organic brain damage. *Journal of Applied Psychology, 27,* 375-386.

Jackson, J. H. (1874, 1932). On the duality of the brain. Medical Press, 1874, 1:19. Reprinted in J. Taylor (Ed.), *Selected writings of John Hughlings Jackson (Vol. II).* London: Hodder & Stoughton.

Jarvik, M. E. (1977). *Psychopharmacology in the practice of medicine.* New York: Appleton-Century-Crofts.

Johnstone, E. C., Crow, T. J., Frith, C. D., Husband, J., & Kreel, L. (1976). Cerebral ventricular size and cognitive impairment in chronic schizophrenia. *Lancet, 2,* 924-926.

Jortner, S. (1965). A test of Hovey's MMPI scale for CNS disorders. *Journal of Clinical Psychology, 21,* 285.

Kaufman, A. S. (1979). *Intelligent testing with the WISC-R.* New York: Wiley.

Kaufman, A. S., & Kaufman, N. (1983).*Kaufman Assessment battery for children.* Circle Pines, MN: American Guidence, Inc.

Kiernan, R., & Matthews, C. (1976). Impairment index versus T-score averaging in neuropsychological assessment. *Journal of Consulting and Clinical Psychology, 44*(6), 951-957.

Klonoff, H., Fibiger, C., & Hutton, G. (1970). Neuropsychological patterns in chronic schizophrenia. *Journal of Nervous and Mental Disease, 150*(4), 291-300.

Klonoff, H., & Low, M. (1974). Disordered brain function in young children and early adolescents: Neuropsychological and electroencephalographic correlates. In R. M. Reitan & L. A. Davison (Eds.), *Clinical neuropsychology: Current status and applications.* New York: Wiley.

Klonoff, H., Low, M., & Clark, C. (1977). Head injuries in children, a prospective 5 year follow-up. *Journal of Neurosurgery and Psychiatry, 12,* 1211-1219.

Klonoff, H., & Robinson, G. (1967). Epidemiology of head injuries in children. A pilot study. *Canadian Medical Association Journal, 96,* 1308-1311.

Klonoff, H., & Thompson, G. (1969). Epidemiology of head injuries in adults: A pilot study. *Canadian Medical Association Journal, 100,* 235-241.

Klove, H. (1974). Validation studies in adult clinical neuropsychology. In R. M. Reitan & L. A. Davison (Eds.), *Clinical neuropsychology: Current status and applications.* New York: Wiley.

Knights, R. M. (1973). Problems of criteria in diagnosis. A profile similarity approach. *Annals of the New York Academy of Sciences, 205,* 124-131.

Lacks, P. B., Harrow, M., Colbert, J., & Levine, J. (1970). Further evidence concerning the diagnostic accuracy of the Halstead organic test battery. *Journal of Clinical Psychology, 26,* 480-481.

Lang, P. (1964). Experimental studies of desensitization psychotherapy. In J. Wolpe, A. Salter, & L. J. Peyna, *The conditioning therapies.* New York: Holt, Rinehart, and Winston.

Lester, B. K., & Edwards, J. J. (1966). EEG fast activity in schizophrenics and control subjects. *International Journal of Neuropsychiatry, 2,* 143-156.

Lezak, M. D. (1976). *Neuropsychological assessment.* New York: Oxford University Press.

Luchins, D. J. (1982). Computed tomography in schizophrenia: Disparities in the prevalence of abnormalities. *Archives of General Psychiatry, 39,* 859-860.

Luria, A. R. (1963). *Restoration of function after brain injury.* New York: MacMillan.

Luria, A. R. (1965). Neuropsychology in the local diagnosis of brain damage. *Cortex, 1,* 2-18.

Luria, A. R. (1966). *Higher cortical functions in man.* New York: Basic Books.

Luria, A. R. (1970). The functional organization of the brain. *Scientific American, 3,* 66-78.

Luria, A. R. (1973). *The working brain.* London: Penguin Press.

Luria, A. R., & Majovski, L. V. (1977). Basic approaches used in American and Soviet clinical neuropsychology. *American Psychologist, 32,* 959-968.

Majovski, L. V., Tanguay, P., Russell, A., Sigman, M., Crumley, K., & Goldenberg, I. (1979a). Clinical neuropsychological screening instrument for assessment of higher cortical deficits in adolescents. *Clinical Neuropsychology, 1,* 3-8.

Majovski, L. V., Tanguay, P., Russell, A., Sigman, M., Crumley, K., & Goldenberg, I. (1979b). Clinical neuropsychological evaluation instrument: A clinical research tool for assessment of higher cortical deficits in adolescents. *Clinical Neuropsychology, 1,* 9-19.

Malloy, P. F., & Webster, J. S. (1981). Detecting mild brain impairment using the Luria—Nebraska neuropsychological battery. *Journal of Consulting and Clinical Psychology, 49,* 768-770.

Matarazzo, J. D. (1972). *Wechsler's measurement and appraisal of adult intelligence.* Fifth Edition. Baltimore: Williams and Wilkins.

Matthews, C. G., Shaw, D. J., & Klove, H. (1966). Psychological test performances in neurologic and "pseudoneurologic" subjects. *Cortex, 2,* 244-253.

McFarland, K. A. (1968). The sensory and perceptual processes in aging. In K. W. Schaie (Ed.), *Theory and methods of research on aging* (pp. 9-52). Morgantown, WV: West Virginia University.

McFie, J. (1975). *Assessment of organic intellectual impairment.* New York: Academic Press.

McKay, S., & Golden, C. J. (1979). Empirical derivation of neuropsychological scales for the lateralization of brain damage using the Luria—Nebraska neuropsychological test battery. *Clinical Neuropsychology, 1*(2), 1-5.

Meier, N. R. F. (1963). Selector-integrator mechanisms in behavior. In N. R. F. Meier & T. C. Schneitlar (Eds.), *Principles of animal psychology* (pp. 621-649). New York: Dover Books.

Miller, W. R. (1975). Psychological deficit in depression. *Psychological Bulletin, 82,* 238-260.

Mirsky, A. F. (1969). Neuropsychological bases of schizophrenia. In P. H. Mussen & M. R. Rosenzweig (Eds.), *Annual review of psychology, Vol. 20.* Palo Alto, CA: Annual Reviews.

Moses, J. A., Cardellino, J. P., & Thompson, L. L. (1983). Discrimination of brain damage from chronic psychosis by the Luria—Nebraska neuropsychological battery: A closer look. *Journal of Consulting and Clinical Psychology, 51,* 441-449.

Moses, J. A., & Golden, C. J. (1979). Cross validation of the discriminative effectiveness of the standardized Luria neuropsychological battery. *International Journal of Neuroscience, 9,* 149-155.

Moses, J. A., Johnson, G. L., & Lewis, G. P. (in press). Reliability analyses of the Luria—Nebraska neuropsychological battery summary, localization, and factor scales. *International Journal of Health.*

NIH Publication No. (81-2252). (1981). The dementias: Hope through research. Bethesda, MD: National Institute of Health.

Nielsen, J. M. (1953). Spontaneous recovery from aphasia. Autopsy (report of a case). *Bulletin of the Los Angeles Neurological Society, 28,* 147-148.

Obrzut, J. E. (1981). Neuropsychological procedures with school-age children. In G. W. Hynd & J. E. Obrzut (Eds.), *Neuropsychological assessment and the school-age child.* New York: Grune and Stratton.

Parsons, O. A. (1970). Clinical neuropsychology. In C. D. Spielberger (Ed.), *Current topics in clinical and community psychology (Vol. 2).* New York: Academic Press.

Parsons, O., & Klein, H. (1970). Concept of identification and practice in brain damaged and process reactive schizophrenic groups. *Journal of Consluting and Clinical Psychology, 35*(3), 317-323.

Parsons, O., & Prigatano, G. (1978). Methodological considerations in clinical neuropsychological research. *Journal of Consulting and Clinical Psychology, 46*(4), 608-619.

Prigatano, G. P. (1978). Wechsler memory scale: A selective review of the literature. *Journal of Clinical Psychology, 34,* 816-832.

Purisch, A. D., Golden, C. J., & Hammeke, T. A. (1978). Discrimination of schizophrenic and brain-injured patients by a standardized version of Luria's neuropsychological tests. *Journal of Consulting and Clinical Psychology, 46,* 1266-1273.

Raichle, M. E. (1979). Quantitative in-vivo autoradiography with positron emission tomography. *Brain Research Reviews, 1,* 47-68.

Reed, H., & Fitzhugh, K. (1966). Patterns of deficits in relation to severity of cerebral dysfunctioning in children and adults. *Journal of Consulting Psychology, 30,* 98-102.

Reed, H. B. C., & Reitan, R. M. (1963). Intelligence test performance of brain damaged subjects with lateralized motor deficits. *Journal of Consulting Psychology, 27,* 102-106.

Reed, H. B. C., & Reitan, R. M. (1969). Verbal and performance differences among brain injured children with lateralized motor deficits. *Perceptual and Motor Skills, 29,* 747-752.

Reed, H. B. C., Reitan, R. M., & Klove, H. (1965). The influence of cerebral lesions on psychological test performance of older children. *Journal of Consulting Psychology, 29,* 247-251.

Reitan, R. M. (1955). An investigation of the validity of Halstead's measures of biological intelligence. *Archives of Neurology and Psychiatry, 73,* 28-35.

Reitan, R. M. (1958). The validity of the trail making test as an indicator of organic brain damage. *Perceptual and Motor Skills, 8,* 271-276.

Reitan, R. M. (1959). The comparative effects of brain damage on the Halstead impairment index and the Wechsler—Bellevue scale. *Journal of Clinical Psychology, 15,* 281-285.

Reitan, R. M. (1960). The significance of dysphasia for intelligence and adaptive abilities. *Journal of Psychology, 50,* 355-376.

Reitan, R. M. (1964). *Manual for administering and scoring the Reitan—Indiana neuropsychological battery for children (aged five through eight).* Indianapolis: University of Indiana Medical Center.

Reitan, R. M. (1966). Problems and prospects in studying the psychological correlates of brain lesions. *Cortex, 2,* 127-154.

Reitan, R. M. (1969). *Manual for administration of neuropsychological test batteries for adults and children.* Indianapolis, published by author.

Reitan, R. M. (1971). Sensorimotor functions in brain damaged and normal children of early school age. *Perceptual and Motor Skills, 33,* 655-664.

Reitan, R. M. (1974). Methodological problems in clinical neuropsychology. In R. M. Reitan & L. A. Davison (Eds.), *Clinical neuropsychology: Current status and applications.* New York: Wiley.

Reitan, R. M. (1976). Neurological and physiological bases of psychopathology. *Annual Review of Psychology, 27,* 189-216.

Reitan, R. M., & Davison, L. A. (1974). *Clinical neuropsychology: Current status and applications.* New York: Wiley.

Reitan, R. M., & Heineman, C. E. (1968). Interactions of neurological deficits and emotional disturbances in children with learning disorders: Methods for their differential assessment. In J. Hellmuth (Ed.), *Learning disorders, Vol. 3.* Seattle: Special Child Publications.

Reynolds, C. R., & Gutkin, T. B. (1979). Predicting the premorbid intellectual status of children using demographic data. *Clinical Neuropsychology, 1,* 36-38.

Rockford, J. M., Detre, T., Tucker, G. T., & Harrow, O. K. (1970). Neuropsychological impairments in functional psychiatric diseases. *Archives of General Psychiatry, 22,* 114-119.

Rourke, B. P. (1975). Brain-behavior relationships in children with learning disabilities: A research program. *American Psychologist, 30,* 911-920.

Rourke, B. P. (1976). Issues in the neuropsychological assessment of children with learning disabilities. *Canadian Psychological Review, 17*(2), 89-102.

Rourke, B. P. (1979). Conference overview. In R. M. Knights & D. J. Bakker (Eds.), *Treatment of hyperactive and learning disordered children.* Baltimore, MD: University Park Press.

Russell, E., Neuringer, C., & Goldstein, G. (1979). *Assessment of brain damage: A neuropsychological key approach.* New York: Wiley.

Satz, P., Fennell, E., & Reilly, C. (1970). Predictive validity of six neurodiagnostic tests: A decision theory analysis. *Journal of Consulting and Clinical Psychology, 34,* 375-381.

Schwartzman, A. E., Douglas, V. I., & Muir, W. R. (1962). Intellectual loss in schizophrenia: Part II. *Canadian Journal of Psychology, 16,* 161-186.

Selz, M., & Reitan, R. M. (1979). Rules for neuropsychological diagnosis: Classification of brain function in older children. *Journal of Consulting and Clinical Psychology, 47,* 258-264.

Severson, R. A. (1970). *Behavior therapy with severe learning disabilities.* Unpublished manuscript, University of Wisconsin.

Shaffer, D. (1974). Psychiatric aspects of brain injury in childhood: A review. In S. Chess & A. Thomas (Eds.), *Annual progress in child psychiatry and child development.* New York: Brunner Mazel.

Shelly, C., & Goldstein, G. (1982a). Psychometric relations between the Luria—Nebraska and Halstead—Reitan neuropsychological test batteries in a neuropsychiatric setting. *Clinical Neuropsychology, 4*(3), 128-133.

Shelly, C., & Goldstein, G. (1982b). Intelligence, achievement, and the Luria—Nebraska battery in a neuropsychiatric population: A factor analytic study. *Clinical Neuropsychology, 4*(4), 164-169.

Smith, A. (1975). Neuropsychological testing in neurological disorders. In W. J. Friedlander (Ed.), *Advances in neurology (Vol. 7).* New York: Raven.

Sperry, R. W., Gazzaniga, M. S., & Bogen, J. H. (1969). Interhemispheric relationships: The neocortical commissures; syndromes of hemisphere disconnection. In P. Vinken & G. W. Bruyn (Eds.), *Handbook of clinical neurology (Vol. 4).* New York: Wiley.

Spiers, P. A. (1981). Have they come to praise Luria or to bury him? The Luria—Nebraska battery controversy. *Journal of Consulting and Clinical Pyschology, 49,* 331-341.

Spiers, P. A. (1982). The Luria—Nebraska neuropsychological battery revisited: A theory in practice or just practicing? *Journal of Consulting and Clinical Psychology, 50,* 301-306.

Spitzer, R. L., & Endicott, J. (1974). Can the computer assist clinicians in psychiatric diagnosis? *American Journal of Psychiatry, 131,* 523-530.

Spitzer, R. L., Endicott, J., & Robins, E. (1977). *Research diagnostic criteria (RDC) for a selected group of functional disorders (Third Edition).* New York: New York State Psychiatric Institute, Biometries Research.

Strauss, A. A., & Lehtinen, L. E. (1948). *Psychopathology and education of the brain-injured child (Vol. 1).* New York: Grune and Stratton.

Struss, D. T., & Trites, R. L. (1977). Classification of neurological status using multiple discriminant function analysis of neuropsychological test scores. *Journal of Consulting and Clinical Psychylogy, 45,* 145.

Taylor, M. A., Greenspan, B., & Abrams, R. (1979). Lateralized neuropsychological dysfunction in affective disorder and schizophrenia. *American Journal of Psychiatry, 136,* 1031-1034.

Twitchell, T. E. (1951). The restoration of motor function following hemiplegia in man. *Brain, 74,* 443-480.

Tymchuk, A. (1974). Comparison of Bender error and time scores for groups of epileptic, retarded, and behavior-problem children. *Perceptual and Motor Skills, 38,* 71-74.

Vega, A., & Parsons, O. A. (1967). Cross-validation of the Halstead—Reitan tests for brain damage. *Journal of Consulting Psychology, 31,* 619-625.

Watson, C. (1974). Chronicity and the Halstead battery in psychiatric hospitals: In reply to Levine and Feirstein. *Journal of Consulting and Clinical Psychology, 42*(1), 136-138.

Watson, C. G., & Benton, A. L. (1976). Copying and reproduction from memory tasks in differentiation of brain-damaged and control psychiatric patients. *Perceptual and Motor Skills, 42,* 1107-1112.

Watson, C. G., Thomas, R. W., Anderson, D., & Felling, J. (1968). Differentiation of organics from schizophrenics at two chronicity levels by use of the Reitan—Halstead organic test battery. *Journal of Consulting and Clinical Psychology, 32,* 679-684.

Wechsler, D. W. (1958). *The measurement and appraisal of adult intelligence. (Fourth Edition).* Baltimore: Williams and Wilkins.

Wedding, D. (1983). Comparison of statistical and actuarial models for predicting lateralization of brain damage. *Clinical Neuropsychology, 5,* 15-22.

Weisbroth, S., Esibill, N., & Zuger, R. (1971). Factors in the vocational success of hemiplegic patients. *Archives of Physical Medicine and Rehabilitation, 52,* 441-447.

Wernicke, C. (1874). *Der aphasische symptomencomplex.* Breslau, Poland: Cohn and Weigert.

Wheeler, L. (1964). Complex behavioral indices weighted by linear discriminant functions for the prediction of cerebral damage. *Perceptual and Motor Skills, 19,* 907-923.

Wheeler, L., Burke, C. J., & Reitan, R. M. (1963). An application of discriminant functions to the problem of predicting brain damage using behavioral variables. *Perceptual and Motor Skills, 16,* 417-440 (Monograph Supplement).

Wheeler, L., & Reitan, R. M. (1962). The presence and laterality of brain damage predicted from responses to a short aphasia screening test. *Perceptual and Motor Skills, 15,* 783-799.

Wheeler, L., & Reitan, R. M. (1963). Discriminant functions applied to the problem of predicting cerebral damage from behavior tests: A cross-validation study. *Perceptual and Motor Skills, 16,* 681-701.

Williams, N. (1967). Correlation between copying ability and dressing activities in hemiplegia. *American Journal of Physical Medicine, 46,* 4.

Willson, V. L., & Reynolds, C. R. (1982). Methodological and statistical problems in determining membership in clinical populations. *Clinical Neuropsychology, 4*(3), 134-138.

Wilson, R. S., Rosenbaum, G., Brown, G., Rourke, D., Whitman, D., & Grisell, J. (1978). An index of premorbid intelligence. *Journal of Consulting and Clinical Psychology, 46,* 1554-1555.

Young, L. D., Taylor, I., & Holstrom, V. (1977). Lithium treatment of patients with affective illness associated with organic brain symptoms. *American Journal of Psychiatry, 134,* 1405-1407.

9 New Directions for Interest Testing

John L. Holland
Johns Hopkins University

This chapter is an attempt to outline where interest testing may be or should be in the near future: What changes will be seen in the development or revision of inventories, what new areas of application will occur, and what technical, social, and professional problems need resolution to get to a more desirable future.

This sounds like a rational task. I have been asked to describe a desirable future by canoeing through the rapids of psychometric fashions, disgruntled test takers, passive publishers, worried professionals and their righteous associations, and future islands of unpredictable theory. To make this task easier, the sponsor cautioned me to rely on empirical data, not daydreams.

Fortunately, I can recognize an impossible task without the aid of consultants. For several reasons, it appears helpful to redefine the task. Earlier opinions by distinguished pioneers in interest measurement have occasionally been off the mark. For example, Kuder (1954) suggested that occupational titles made poor items and that activity items would be the wave of the future. News item: Occupational items continue to be useful and popular in most inventories. And inventories that use only activity items usually include occupational titles disguised as "Be an accountant," or "Be a counselor." Developers apparently get tired of looking for good items by following a restrictive rule. At an earlier time, Strong (1943) and others dismissed a person's vocational aspiration as a weak index of the occupation a person would actually enter because this index did not have a substantial correlation with a person's measured interests. However, Dolliver (1969) started a cottage industry of research by demonstrating that aspirations and measured interests have about equal predictive validity. Later, we learned that the use of interest inventories and aspirations in tandem produced very substantial predictions.

These events and the work of futurists imply that it is helpful to see future developments not only as the continuation of current trends but also as developments that will be shaped by economic, social, technological, and theoretical forces that we cannot always anticipate or control. Consequently, I attempt to relate current developments to future developments, but my forecasts will surely be deflected by unanticipated events. I also try to distinguish long-term trends that I believe are desirable and helpful and those that may be undesirable and not helpful.

My reservations about this task have considerable empirical support. I have multiple conflicts of interest. I am the author of two interest inventories that have been the object of close scrutiny for 10 years. I am familiar with the evidence and issues about inventory biases, development, effects and usage, but my beliefs about these matters have received only mixed reviews (Gottfredson & Holland, 1978; Tittle & Zytowski, 1978). The most constructive outcome of this experience for me has been to perceive interest inventories in the context of usefulness, validity and reliability—and about in that order.

USEFULNESS, VALIDITY, AND RELIABILITY

Before I get to the main topic, it appears helpful to define usefulness and to indicate its relation to validity and reliability, for usefulness plays a key role in the evaluation of inventories and in their future development.

More explicitly, I view inventories as both interventions and assessment devices. They are interventions, because they are used to create one or more beneficial effects. Interest inventories can support a person's vocational aspiration, stimulate a comprehensive exploration of occupational possibilities, or provide a structure for understanding vocational interests and the occupational world.

They are also assessment devices, because they provide the test taker and counselor with a systematic account of a person's interests. Consequently, the validity and reliability of that information contributes to an inventory's usefulness.

Despite some interdependence of these constructs, it seems desirable to restrict validity and reliability to the data about an inventory's psychometric characteristics and to the literal interpretation of these data rather than to assume that these constructs can also account for all of an inventory's effects on the test-taker as well as its usefulness from the test-taker's point-of-view. Usefulness would then apply only to the effects of an inventory on the test-taker and to those aspects of an inventory that are assumed to enhance effects—clarity of language, influential interpretive materials, linkage to other interventions, and other sources of occupational information.

Some obscure research and development illustrate how loosely coupled these constructs—validity, reliability, and usefulness—can be. For instance, dissimilar interest inventories, which have been assumed to have divergent predictive validities, appear to have very similar beneficial effects on the test-taker (Pallas, Dahmann, Gucer, & Holland, 1983). Still other evaluations have indicated that many special experiences with obvious unreliable and invalid qualities such as interviewing an occupational representative or filling out a simple unresearched form about a particular vocational aspiration are frequently rated by test-takers as more helpful than taking the Kuder Occupational Interest Survey, the Strong-Campbell Interest Inventory, or the Self Directed Search (Evans & Rector, 1978; Rayman, Bernard, Holland, & Barnett, 1983).

These evaluations imply a need to perceive inventories in two contexts:

1. Does the information provided for decision making have at least a moderate degree of construct validity and retest reliability?
2. Are the effects of taking the inventory consistent with one or more test-taker goals?

This two-fold perspective appears helpful for understanding where we have been in interest measurement and where we appear to be going. It also provides a structure for resolving old and potential controversies about the biases or effects of inventories.

This perspective clarifies some issues by suggesting a more analytical and pragmatic approach to evaluation and inventory development. In the same way that we have learned that validity and reliability are ambiguous words that require elaboration into several elements, it is also helpful to divide usefulness into elements. In this instance, the most common uses are: (a) exploration or increasing the range of occupations a person will consider, (b) reassurance or providing support for a person's vocational aspiration or potential job, a goal that depends on an inventory's predictive validity, and (c) self-understanding or providing structured information for comprehending the character of one's interests. The uses or purposes of interest testing overlap those that Sundberg (1977) has proposed for personality assessment. He proposes "image forming," analogous to self-understanding; and "decision-making," analogous to reassurance and exploration. Sundberg also proposes "theory building" or the use of assessment data to develop theoretical concepts. Most people would consider "theory building" as a secondary goal of interest testing, although the data generated from interest inventories have stimulated some theoretical work.

With this perspective and my reservations, I outline the current status of interest assessment, some critical issues and developments, and finally, what the future may be or should be like.

CURRENT STATUS OF INTEREST TESTING

The current status of interest measurement is characterized by great diversity. More than 200 interest inventories (Buros, 1978, pp. 1549–1649) have been developed by following a wide range of strategies and methods: homogeneous scaling, defined criterion groups, factor analysis, and other psychometric methods or theoretical models of vocational interests.

The diversity in inventory construction also is matched by an enormous increase in the diversity of test takers. Ten years ago, inventories were most popular among high school and college students along with scattered industrial applications—largely in research or selection. Now inventories are being used by adults of all ages in career, educational, retirement, and recreational planning. Some inventories have been adapted for use with special groups—notably the visually-impaired; many are in one or more foreign languages, and several interest inventories have been developed for particular groups. The most striking example is the development of inventories for people with skilled trades interests.

The current status of interest assessment is also characterized by a shift from the traditional preoccupation with an inventory's psychometric characteristics to a preoccupation with its therapeutic or not so therapeutic influences (Zytowski & Borgen, 1983). This shift has been stimulated by multiple social events, evaluative research, incidental psychometric findings, and hard times.

Until 1972, there were no experimental studies on the effects of taking an interest inventory. Developers and counselors were concerned primarily with construct validity but especially an inventory's predictive validity. At that time, inventories were imbedded in the information for decision-making model. The counselor's role was to help a person obtain an accurate interpretation of the information in the interest profile and to integrate that information with other personal and occupational data. Counselors and psychologists were typically more concerned with therapeutic techniques than with the independent influence of interest inventories.

The women's movement, the attacks on the Strong Vocational Interest Blank (SVIB) and the Self Directed Search (SDS), the Association for Measurement and Evaluation in Guidance Commission report (AMEG, 1973), and a National Institute of Education Conference (Diamond, 1975) stimulated a strong interest in the influence of interest inventories on women. This controversy eventually generalized to the effects of inventories on blacks and whites, poor and rich, and other groups and stimulated about 100 pieces of research performed to determine the effects of test taking on a person's vocational aspirations, self-understanding, and other criteria. In response, developers and publishers have made multiple revisions to cope with criticism and new data, although no consensus was ever reached on a definition of sex bias or equity in interest measurement.

Hard times have also had an impact on test construction by stimulating the creation of inventories with low development costs. If inventories with high development and maintenance costs such as the SCII or KOIS are no more helpful than inventories with low development and maintenance costs such as the SDS or its clones, then inventories with brief homogeneous scales become attractive. This trend may have been reinforced by scattered and obscure investigations which show that short scales can be developed whose construct validity and retest reliability approximate those obtained by complex and lengthy scales (Meehl, 1972).

Several other influences appear to have accelerated the shift to simpler inventories and a concern with their effects. The need for career services has expanded at a rate that professionals could not or would not cope with so that people with little or no appropriate training have rushed in. Consequently, they found simpler inventories easier to use. In addition, graduate students in counseling psychology, the largest group of professional users of interest inventories, have over the last several decades become more interested in therapy than in vocational tests. Consequently, a test's apparent influence is seen as more important than its psychometric properties.

This increase in test users, whose training ranges from no special training to the Ph.D. in psychology or education, has spawned a large unsophisticated market for vocational tests. And among this growing army of self-taught career counselors is a burgeoning group of entrepreneurs who use bits and pieces of popular inventories to create brief inventories of interests, skills, or workbooks. These people are filling the need for career services, because standard inventories and tests are usually not available to them or because they wanted a special work kit of their own.

The shift to simpler inventories and a concern with effects have also been stimulated by the positive effect of immediate feedback when a person takes a self-scored inventory. Immediacy appears to generate satisfaction and acceptance of the results and to relieve practitioners of some scheduling and coordinating problems such as collating client appointments and test results. Simple inventories also cost less to develop and purchase than inventories with occupational keying and complex scoring.

Finally, the movement toward simpler devices has been accelerated by career theories and classifications (Holland, 1973; Roe, 1956) that use a small number of scales and categories to assess interests and to organize occupational data. The Holland typology led to the SDS and the reorganization of the SVIB as the SCII (Campbell & Hansen, 1981). The Roe theory led to Lunneborg's (1975) Vocational Interest Inventory. Before this theoretical and classificatory work, developers were faced with ad hoc classifications or infinite scale building. Now we have several models for establishing validity generalizations for a limited number of occupational categories and the kinds of interests associated with each. Gottfredson, Holland and Ogawa (1982) have applied the Holland typology to the *Dictionary of Occupational Titles* (U.S. Department of Labor, 1977) so that the SDS and similar inventories can be coordinated with 12,099 occupations. McCormick (1979) has also developed an interest inventory and a parallel occupational classification. These classification systems provide practical tools for interpreting inventories, for applying them to nearly all occupations, and for the study of person-job interactions.

The current status of interest testing is also characterized by an increasing use of inventories in computerized career exploration and assessment systems and in micro-computers employed by individuals and institutions. This usage is expected to continue to increase rapidly and to become more decentralized. At present, most usage entails scoring and the production of interpretative reports. One of the most helpful aspects of these developments is that some systems can cumulate test records and can revise scales at a rapid rate. Another positive outcome is the high ratings that computers receive relative to professional counselors (Wagman, 1980; Wagman & Kerber, 1980). The outcomes of these developments appear unusually unpredictable. At this time, computerized test administration and interpretation competes with as well as supplements the work of counselors, and these new services also compete with and supplement printed materials produced by publishers.

Last, the current collection of interest inventories still follows one of two measurement strategies: The first strategy entails an empirical comparison of a person's responses to a reference group. The SCII and KOIS accomplish this task with different methods, but both involve a comparison of a person's responses with a number of occupational samples. These well-established models have also been used in Clark's Minnesota Vocational Interest Inventory (1961) and Johansson's (1982) Career Assessment Inventory. The second strategy entails the development of homogeneous scales defined by factor analyses, typologies, or classifications of interests and occupations. The Holland and Roe typologies have been the most common models for homogeneous scaling. The Self Directed Search (Holland, 1979) and the Vocational Interest Inventory (Lunneborg, 1975) exemplify these applications.

There is a massive amount of data about the merits of individual inventories contained in manuals and in the literature, but there are only a few unequivocal studies of the relative merits of building an inventory following a particular strategy. In addition, a few inventories follow both strategies. For example, the SCII and the Jackson Vocational Interest Survey (JVIS) (Jackson, 1977) use both strategies.

Even a casual review of the evidence implies that different construction strategies result in inventories with similar concurrent and predictive validity. Earlier work (Clark, 1961; Reilly & Echternacht, 1979) suggests that empirical and homogeneous scales have similar predictive and concurrent validity, although each is occasionally superior to the other for special purposes. Differences are typically very small.

When the evidence for the relative concurrent and predictive validity of inventories constructed by different strategies is added to the evidence for the effects of different inventories on the test-taker, it is difficult to see that one construction strategy is generally superior to another. It is possible that there are some consistent differences for selected purposes, but the kind of research needed to identify such differences has rarely been performed. More important, it is very clear that much higher predictive validity can be obtained by using inventories *and* vocational aspirations in combination. Likewise, the popular strategy of making interest inventories more accessible and interpretable appears equally desirable, and it is a task that has been pursued and studied for only a few years. So I expect the shift in research effort from predictive validity to inventory usefulness to be a trend that will persist for at least a few more years.

CRITICAL ISSUES AND OPPORTUNITIES

At this time the development and use of interest inventories is faced with four major issues. They include how to make inventories available to more clients versus how to maintain professional incomes and standards, how to create inventories with more valid, influential and satisfying effects, how to insure equity in testing and how to integrate interest testing with other interventions. I found it difficult to label these topics as problems, issues, or golden opportunities. How you perceive these issues depends to some degree on your role (practitioner, publisher, developer, test taker), on your values, on your interpretation of the interest literature, and on your views of career development theory. Consequently, more data will not always create a consensus.

Improving Psychometric Characteristics

The development of new inventories or the revision of old ones appears to have reached two plateaus. No matter what the method of construction, the reliabilities of individual scales hover around .80 to .90. And the concurrent and predictive validities of diverse inventories appear very similar. This interpretation is not obvious, because the definition of what constitutes a predictive hit or error shifts from one inventory to another, because comparisons of different inventories using the same sample and the same criteria are rare, and because we must piece together a collage of concurrent and predictive studies.

The SCII and KOIS (Kuder & Diamond, 1979) present special problems in evaluation. The SCII uses two kinds of scales—homogeneous and occupational. The six homogeneous scales (GOT) are analogous to the six scales of the SDS and the UNIACT (Lamb & Prediger, 1981). A review of the manuals for these inventories reveals very similar hit rates for similar and dissimilar criteria. The occupational scales of the SCII have predictive hit rates (Hansen & Swanson, 1983) that closely resemble those for inventories that use homogeneous scales.

Studies of the concurrent and predictive validity of the KOIS also yield hit rates that resemble those of other inventories (Zytowski, 1976). Occasionally, an inventory will stand above the pack, but a careful review will usually disclose a propitious sampling: predictions were performed for very divergent occupations, a rectangular distribution of subjects was employed, small samples in selected categories were omitted, or only a single kind of interest was studied (O'Neil, Magoon, & Tracey, 1978).

It is conceivable that there are some significant differences in inventory validities, but they are difficult to cull out of the literature. If it is correct to assume that we have reached plateaus of retest reliability and predictive validity, we could instead concentrate on making inventories more accessible and more useful to the test-taker. Our psychometrically-oriented colleagues will prefer to spend more time searching for better scaling techniques. However, they will have difficulty reaching the levels of predictive validity attained by combining inventories and vocational aspirations. For example, interest inventories that use a six-variable structure have concurrent or predictive validity hit rates of 39 to 55%. In contrast, when a person's vocational aspiration and interest profile share the same category (person aspires to teacher, and his or her social or educational score is the high point of the profile) predictions of correct identification of future aspiration or occupation range from 60 to 85% for intervals of 1 to 11 years for the SDS, SVIB, or VPI (Bartling & Hood, 1981; Borgen & Seling, 1978; Holland & Gottfredson, 1975; Touchton & Magoon, 1977; Holland & Lutz, 1968). Other work (Holland, Gottfredson, & Nafziger, 1975) implies that the congruency of aspiration and interests is an alternative measure of identity (Holland, Gottfredson, & Power, 1980) and may be a useful predictor of career stability.

Increasing Inventory Influence

An alternative to the search for better scaling techniques is to continue to make inventories more influential and useful. Nearly all inventories in recent years have been revised to make them easier to interpret and apply. Most publishers have prepared more extensive interpretive materials and have tried to link inventory results to a wide range of occupational information or interventions. They have also attempted to increase the exploration effect by using balanced scales, male and female norms for the same occupation, more occupational scales, norms and raw scores for the same homogeneous interest scale, or by providing comprehensive occupational classification systems that are related to interest profiles. These revisions and the evaluative studies of the effects of interest inventories and other interventions point to a technology of career instruction or development that may be more productive than the polishing of inventory scales. In short, evaluative studies should lead to inventories that are more influential and practical, although their predictive validity and retest reliability may remain unchanged.

Giving Psychology Away

The creation of inventories that are more useful to the client also creates some professional conflict. "Giving psychology away" is consistent with our professional ideals. Everyone should have access to an interest inventory if they so desire. Unfortunately, the advent of hard times and self-scored inventories threatens the livelihood of selected professionals. Brief inventories that have obtained positive evaluations threaten some professionals by destroying the mystic of testing and imply the usefulness of inexpensive psychological services. But the old test standards frown on testing without the services of a professional. This situation for psychology is analogous in several ways to the use of cancer, diabetes, and pregnancy diagnostic tests for home use that bypass the family physician, but do not suggest a special treatment.

Research studies that compare the diagnostic services of physicians and do-it-yourself diagnostic kits support the value of many of these simple tests. Likewise, there are at least three experimental evaluations (Avallone, 1974; Krivatsky & Magoon, 1976; Nolan, 1974) that suggest that the usefulness of an interest inventory equals the value of seeing a professional career counselor (i.e., one with a Ph.D. in counseling psychology). In two other experiments (Wagman, 1980; Wagman & Kerber, 1980), an interactive computer system for personal counseling produced positive effects that lasted one month. And, some clients reported that they felt more at ease (42%) and more independent (45%) on the computer than if they saw a counselor. These evaluations and other evaluations imply that interest inventories could be made available to more people with no more expectation of harm or disservice than they would receive from a professional.

Insuring Equity

Perhaps the most complex, difficult, social-emotional issue is how to insure equity in interest inventory assessments within the constraints of current beliefs, knowledge, test standards, and financial resources. The 10-year controversy about the sex biases in interest testing illustrates the difficulties in arriving at a consensus.

The controversy began in the period 1971 to 1973, when individuals, groups, committees, and commissions charged that interest inventories served to keep women and men in traditional occupations (Diamond, 1975). The belief was that interest inventories by virtue of their items, instructions, interpretative materials, scoring and normative procedures helped to maintain the sex-segregated character of the work force. This belief was reinforced by numerous plausible hypotheses and some plausible data. For example, when males and females of any age take interest inventories, the distributions of suggested occupations for males and females are usually divergent (Lamb & Prediger, 1981, p. 29). According to interest inventories a small proportion of females have skilled trades and factory-oriented interests, whereas males have those interests in large proportions.

Now, after more than 10 years of discussion and research, there is still no obvious consensus (Tittle & Zytowski, 1978). Perhaps the only area of agreement is that we should be concerned with the influence of inventories on females and all other groups. There is also much disagreement about the interpretation of the more than 100 empirical studies that have evaluated the outcomes of different normative procedures, scaling techniques, special directions, interpretive materials, or the pre-post experiments on the vocational aspirations of females and males.

For example, Zener and Schnuelle (1976) stimulated more than 25 experimental evaluations of the effects of different inventories and other vocational interventions. These investigations suggested that all inventories had beneficial effects on one or more criteria and that the effects appeared similar for inventories constructed and interpreted in different ways.

However, test-taker goals were related to the outcomes of testing in only one investigation (Power, Holland, Daiger, & Takai 1979). That study was important, for it made clear that high school students' expectations for test-taking varied by sex and were not always in accord with the typical counselor goal of exploration. Both boys and girls wanted most of all "reassurance" about a vocational aspiration they already had. "Wanting more alternatives" ranked well below this desire. High school girls wanted more or fewer options to the same degree. Boys wanted fewer options rather than more. Although this study was only a beginning, it supports the experience of practitioners that their clients come to test-taking with different goals. Future evaluations should link test-taker goals to the outcomes of testing. In the Power et al. experiment the necessary analyses were performed, but the N was too small for dependable results.

Another program of studies by Prediger and his colleagues (Lamb & Prediger, 1981) provides another kind of evidence that should lead to a more comprehensive picture of the test-taking experience. Stimulated by a dissertation (Rayman, 1976), this group has demonstrated in numerous studies that it is possible to design an inventory that will suggest similar distributions of vocational alternatives to females and males. They accomplish this by writing items that females and males respond to at about equal rates. The assumption is that this kind of construction will encourage females and males to explore a greater range of options. This is a provocative idea that should be tested by a demonstration that inventories constructed by the balanced scale method have more exploratory effect than inventories developed by more conventional methods. So far, no one has demonstrated that any inventory has more exploratory effect than another. To the contrary, the exploratory effects of the SCII (Cooper, 1976), SDS (Holland, 1979), Kuder (cited in Zytowski, 1977) and ACT IV (Prediger, McLure, & Noeth, 1976) appear to be very similar; negligible differences are obtained if any.

There are only a few studies of the effects of editorial or psychometric revisions. Experimental revisions of items to determine the effects of gender-neutral words indicate that such revisions have little or no effect on item responses (Boyd, 1976; Gottfredson, 1976; Holland & Gottfredson, 1976). Extensive liberalized directions have failed to increase a person's satisfaction with the outcomes of taking the SDS and have produced only minor profile differences (Siebel & Walsh, 1977). In addition, extensive liberalized directions have failed to increase the number of options or the number of nontraditional options a college woman considers (Lawler, 1977).

In short, it is difficult to demonstrate that any inventory characteristic has a clear impact on the test-taker. One experiment (Holland, Takai, Gottfredson, & Hanau, 1978), of borderline statistical significance, does imply that if an inventory is arranged so that the test-taker can comprehend the scoring, and is provided an instructional booklet as well as a booklet of many options, then it is rated higher on several criteria by test-takers than an inventory with disguised scoring, few options, and no instructional booklet. Hardly a surprise.

I assume that current research activity has dwindled to an occasional lonely article because researchers have slowly realized that only a few read or care about the controversy. We cannot agree because we do not share a common set of beliefs about the purposes of career counseling. Some believe that the only purpose is exploration—especially the exploration or consideration of non-traditional occupations. Others, including myself, believe that career counseling has at least three purposes—exploration, reassurance, and self-understanding—and that these goals vary from person to person. And, because we do not agree on the goals of career counseling and testing, we do not evaluate the research literature in the same way. There are a host of other disagreements: psychometric or internal definitions of bias versus experimental evaluations of actual effects on the test-taker; different views of sex-role socialization (it is an unmitigated evil to some and is celebrated by others); and finally, different goals for interest inventories—social action interventions versus assessment devices.

This controversy and its ambiguous outcomes could be repeated for any new group-retirees, reentry women, Hispanics, blacks, and so on. Somehow we need a strategy of test evaluation and revision that would be generally fair, practical and scientifically sound.

DESIRABLE DEVELOPMENTS

Now I outline some desirable or needed developments for improving the quality and usage of interest inventories. Many of these developments are underway; and others are potential strategies for which there are sometimes data. Despite my doubts about finding new methods for increasing the validity or reliability of current inventories, only a few inventories appear to take full advantage of current knowledge. There is much room for improvement with standard scaling and validation research. Likewise, if conventional inventories are to compete with vocational workbooks, card sorts, and assorted homemade interest assessments, developers must continue the attempt to make interest inventories more amenable to a wider range of professional and nonprofessional users.

Improving Psychometric Characteristics

Although a few inventories appear to have reached plateaus of retest reliability and predictive validity, the majority could probably reach the same plateaus with a more active research effort. Many deficiencies could be easily remedied in the case of inventories with computerized scoring services in which substantial data collection is possible. Representative or not, such data would usually be superior to the very small homogeneous samples reported in many test manuals. In addition, Gottfredson, Holland, and Holland (1978) and others have demonstrated how unrepresentative data can be weighted to approximate general population data.

The usefulness of homogeneous and occupational scales needs more examination with more comprehensive occupational samples: occupations of high and low status, and more heterogeneous samples. So far, most comparisons of these scaling methods have been performed on very homogeneous samples—primarily skilled trades or military technical specialities. One or more comprehensive studies might settle the relative merits of these scaling methods. At this time, the evidence implies small differences, but different criteria favor different methods. In short, it looks like a tie.

A variation of the homogeneous versus occupational keying question is how to distinguish among occupations or occupational specialities. Strong (1943) employed a men-in-general group with considerable success, but had difficulty in distinguishing among skilled trades and lower level occupations. Kuder (1966) used the lambda technique, which allowed him to omit a men-in-general group, and obtained efficient discriminations. And inventories that use homogeneous scales simply compare different occupations on a particular set of scales. Each method has produced impressive results, but it is still not clear that one method is generally superior, because there has been no unequivocal study in which all three methods were applied to the same occupational samples. Here another evaluation would be helpful, although a review of the literature suggests that the differences are again small.

The next stop in inventory refinement may be to explore Norman's (1972) strategy for better psychiatric diagnosis; that is, see interest assessment as a "multi-stage, sequential, and branching . . . enterprise in which the discriminations that one must make at one stage depend upon what . . . has already been learned." He uses some old interest inventory research to demonstrate that the selection of an appropriate reference group makes it possible to differentiate among psychologists in different specialities (Kriedt, 1949) and among nonprofessional occupations (Clark, 1961). This kind of work was abandoned a long time ago—probably because the sampling problems are difficult, expensive, and long-term, but Norman's lucid discussion outlines a persuasive strategy for new work.

Most inventories could be improved by a more systematic representation of the labor force. For example, some inventories sample largely professional level occupations; whereas others sample only the skilled trades. Even if a developer can justify the use of a single subsample, the sampling of the most populous occupations according to census data has merit. Occupational keys for every occupation are not feasible even with unlimited funds. In short, an inventory should represent the occupational domain covered in a more explicit and rational way. The same prescription applies to inventories with homogeneous scales. In this instance, the item content should be clearly related to the particular occupational classification scheme employed.

More explicit research might be performed to establish the "validity generalizations" implied by the General Occupational Theme (G.O.T.) scales of the SCII and the scales of the SDS. For example, if a scale is validated for a few occupations in an occupational category, the scale is probably valid for other occupations in the same category. The required research would entail validity studies for unexamined occupations in the same category to validate both the scales and the occupational classification. As data accumulate, scales and classifications can be revised so that the need for new validity studies will decrease. Some of the needed research has already been performed (Gottfredson et al., 1982), but a more explicit effort would accelerate progress for all inventories that use classifications to interpret inventory scales and profiles.

Developers might also consider the incorporation of aspirational and work history data within the inventory itself or in the interpretive materials. The inclusion of these materials will increase predictive validity substantially and should increase self-understanding. It appears more advantageous to exploit the virtues of self-expressed intentions and work history in combination with inventory scores than it would be to ignore the value of their joint use.

Finally, everyone might profit from reading David Campbell's (1972) book chapter, "The Practical Problems in Revising an Established Psychological

Test.'' His discussion of user acceptance (users may slow revisions), technical ignorance (we never have enough knowledge), and administrative arrangements (royalities, funds, and responsibilities of authors and publishers) is a helpful and amusing account of the problems that all developers face, but which are rarely discussed in the measurement literature.

Increasing Inventory Influence

Perhaps the most promising development would be to accelerate the attempts to make all inventories more useful to the test-taker. The revision of inventories and interpretive aids that began in the 1970s and the investigations of the effects of these revisions are moving us toward a psychology of instruction in career assistance. The simplification of interpretive profiles and the widespread use of homogeneous scales has increased communication between counselor, test-taker, and inventory, and it has probably increased the reliability of test interpretation.

Despite some work on the before-and-after effects of whole inventories (Holland, 1979), there has been relatively little investigation of the actual impact of different interpretive materials or formats. Likewise, most psychometric revisions—separate sex and combined sex norms, scales formed from balanced items (preferred about equally by females and males)—have been examined for concurrent or predictive validity but rarely for their effect on the client. Generally, we do not know if the typical client comprehends and uses the interpretive materials, notices the difference between various scaling methods, or even cares.

If we are to create more useful inventories, more programmatic and comprehensive research is required. The typical research about a test's psychometric characteristics is necessary but not sufficient. That work needs to be supplemented by more evidence about the effects of different interpretive materials, different scaling procedures, the effects of test-taking on a person's vocational aspirations, and so on. The assumption that different scaling techniques, formats, and interpretive materials lead automatically to different test-taker outcomes is not supported by the data.

For example, when the Kuder, SCII, and the SDS were administered in different orders to every student in three sections of a career course taught by three different instructors and rated by students on seven criteria (having more vocational alternatives, being reassured, having more self-understanding, and so on), there were no significant differences among the student ratings of these dissimilar inventories (Rayman, 1983). These and other experimental results (O'Neil, Price, & Tracey, 1978) make clear that psychometric and rational analyses of an inventory's impact are grossly inaccurate signs of actual impact.

It is unlikely that any journal would publish work of this kind or that publishers would share unpublished research. On the other hand, developers and publishers could be encouraged to substantiate claims of usefulness or test-taker effects in manuals or technical reports. It is conceivable that this kind of research might be published in one or more educational journals because instruction about one's interests is clearly a facet of education as well as career assistance.

The development of self-scored inventories has made inventories available to more clients and more diverse populations. This trend could be accelerated by the reduction of test-taker scoring errors and more assessments of the consequences of self-testing. So far, self-scored inventories compare favorably with professional counselors in three investigations (Avallone, 1974; Krivatsy & Magoon, 1976; Nolan, 1974). In addition, self-scored inventories are relatively neutral or unobtrusive assessment devices. Pallas et al. (1983) found that the SDS was rated as less distressing than 5 of 6 common experiences: "filling out income tax forms, visiting my doctor, taking the written drivers test, or worrying about making ends meet." The SDS was rated as equal to "trying to locate something in the yellow pages." It is reasonable to expect similar ratings for most interest inventories, because they have similar effects on the test-taker.

Self-scored forms of most inventories are possible and may be necessary for populations that require more rapid or inexpensive service. Because self-scored inventories and the research associated with their influence has removed much of the mystification in interest assessment, the old test standards (American Psychological Association, 1974) are out of phase with this recent development, and the recent drafts of the proposed test standards (American Psychological Association, 1984) take a similar obstructive stance by requiring professional supervision.

Insuring Equity

Insuring equity in interest assessment is a goal that most people endorse, but we lack a consensus on how to achieve that goal for the reasons I outlined earlier—differences in beliefs, values, purposes of interest testing. Accordingly, I have regarded the controversy about the alleged sex biases of interest inventories as not resolvable—at least by the collection of more data. For the same reasons, the potential controversies about the biases of inventories for blacks, Hispanics, aged and other groups have also appeared irresolvable.

Now I can imagine several strategies for coping with these controversies that might lead to more equitable inventory design and usage and that might attract a consensus. The main strategy would be to continue the work in which we are acquiring a more comprehensive and explicit account of what happens when a person takes an inventory. In evaluative studies, inventory purposes, test-taker and counselor goals could be linked to a full range of an inventory's psychometric characteristics and interpretive materials as well as counselor behaviors. Such investigations would help us sort out real from imagined effects; document the effects of revisions planned to increase occupational exploration or self-understanding; document what test-takers want, what they learn, and in follow-up studies, what they remember or how their lives may have been affected.

Although the earlier psychometric and influence studies are a good beginning, a more comprehensive and programmatic set of studies should have more positive outcomes:

1. We would have a short list of inventory characteristics that lead to a particular test-taker outcome.
2. We would have a long list of inventory characteristics that have no effect on tthe average person.
3. Counselors and test-takers could select inventories that coincide with their goals and values.
4. This information would contribute not only to the development of more helpful interest inventories but also to other interventions such as career courses, career counseling, vocational card sorts, and workbooks.

I see this venture as a win-win-win situation for test-takers, developers, and counselors. If it can be demonstrated that specific inventory characteristics create specific outcomes, then the debates about inventories will be clarified, and counselors can deliver treatments that are more consonant with client goals. If current inventories do not generate differential effects, we can turn to more promising areas of research and development. Finally, no matter what outcomes are obtained—clear or mushy—we will have generated a more rational and comprehensive knowledge of inventory influence that can be readily applied to new questions of equity and old questions of career assistance.

I have emphasized influence studies, because I believe tested effects (external criteria) are more defensible than the growing list of psychometric signs of bias. These special internal analyses form a heterogeneous collection of intermediate criteria that have a special set of ambiguities—especially if a developer performs two or more tests for bias. I am reminded of a news item in which two MIT aeronautical engineering students demonstrated in a computer analysis that the common bee should not be capable of flight.

Having said this, I would still advocate a review of test items for extreme endorsement rates according to age, sex, SES, race and other attempts (judgmental and statistical) to identify items that did not function in the same way across different groups. My experience is that tests of this kind reidentify bad items identified earlier by traditional item analysis.

The problems inherent in the use of internal analyses as a means for detecting bias are apparent in a recent *Handbook of Methods for Detecting Test Bias* (Berk, 1982). Sixteen authors outline more than sixteen methods (sometimes contradictory) of detecting bias. This book is a valuable source of defensive maneuvers for developers and publishers. At the same time, Burrill's (1982) chapter, "Comparative Studies of Item Bias Methods" summarizes the ambiguities and practical difficulties that need to be overcome before some useful consensus is obtained.

PROBABLE DEVELOPMENTS

My sketch of desirable developments is only one of several. Different people would have a different list of desirable developments or different emphases for the same topics. On the other hand, a list of probable developments based on the strength and popularity of particular current developments should result in more agreement and is probably more predictive of what will happen.

Two current developments appear likely to survive and flower: (a) the creation of inventories that are easier to develop, interpret, and use with a wider range of people, and (b) the pursuit of evaluative and experimental research that may lead to a clearer understanding of how inventories achieve their effects. These trends have little opposition and are fueled by the demands of special groups, the needs of publishers, developers, practitioners and researchers.

In contrast, future work on equity in testing looks like a no-win situation for developers, publishers, and reformers who have already been through 10 years of controversy in which both defenders and reformers have come away bruised and confused. Similar controversies about the interests of reentry women, retirees, blacks, and Hispanics do not appear more rewarding, for they would contain all the ambiguity and complexity that the sex bias controversy exhibited. Instead I believe the question of equity should shift from vague or debatable beliefs about equity to the study of particular effects so that people can then select inventories according to their demonstrated influences. Likewise, the development of inventories with more construct validity and reliability appears unlikely, because inventories developed by long-term expensive methods are not clearly superior to inventories developed from vocational theory or homogeneous scaling techniques.

This interpretation is strengthened by Meehl's (1972) long-term attempts to build a better MMPI. His work suggests that ideal scales are brief (15–20 items), homogeneous, and easily interpreted by both clients and theorists. In addition, ideal scales should function effectively "in subpopulations homogeneous with regard to age, sex, education, IQ, race, social class. . . ." Much of the research about old and new interest inventories leads to the same interpretations.

More Practical Inventories

The development of more practical inventories has been underway for more than a decade and appears to be accelerating. By "practical" I mean inventories that are inexpensive to develop, interpret and use, but are also scientifically sound. Such inventories are characterized by homogeneous scales, self-scoring, many occupational options, and readable, comprehensive interpretive materials.

I would also include several other trends in the movement toward more practical inventories. They include the growth of computer-interpreted inventories and the inclusion of standard inventories within computer-assisted career systems. The multiple attempts to make inventories more helpful to females, people with skilled trades interests, and to older Americans are also attempts to make inventories more useful to old and new clients. The provision of more vocational information and treatment ideas in interpretive materials as well as the publication of inventories in comprehensive workbooks or along with a compatible set of related ability, values or other tests is another facet of this trend. We need research evaluations that compare the effects of workbooks with more labor-intensive treatments such as one-to-one career counseling, seminars, and courses.

Inventories as Interventions

The study of inventories as interventions should continue to increase because it is consistent with the attempts to create more useful inventories as well as with the diffuse research and theoretical effort devoted to increasing the quality of career assistance. This kind of work will not proceed at a rapid rate, for it has the support or interest of only a few publishers, developers, and researchers. It is easier to perform psychometric analyses and revise interpretive materials without testing their impact, for the research required to assess inventory effects or special revisions is characterized already by some negative results, difficulties of execution, and a lack of publication outlets. On the positive side, this research requires little funding, but like many evaluations, it requires considerable cooperation, coordination, and negotiation. Consequently, one research team is lucky if they can do one evaluation a year. In the meantime, critics have developed a new list of outcome criteria.

INNOVATIONS AND CONSTRAINTS

Finally, it is important to acknowledge that inventories will be affected by future technological and theoretical events as well as by multiple environmental constraints that will alter any predictions of desirable or probable developments. Some recent events illustrate these unpredictable forces.

A recent Ph.D. thesis (Monahan, 1983) has demonstrated that a simple 60-item inventory of a student's environmental preference was more predictive of a student's current vocational aspiration (81% agreement) than the Vocational Preference Inventory (Gottfredson, Holland, & Holland, 1978), (55%) composed of 84 items. This work implies a new assessment strategy that is probably more efficient, and perhaps more conducive to learning. It could shift all controversy about bias to questions of objective environmental descriptions because test-takers express their preference for a variety of occupational environments rather than their preferences for different activities, occupational titles, and other traditional test items. In another innovation, Prediger (1982) has reduced the measurement of interests to two dimensions and has shown how all occupations can be displayed in a single plane divided into 12 regions.

Both innovations represent important developments, but both face strong environmental constraints—largely the traditional beliefs that professionals have about interest measurement, and the aversion that many people have to change. For example, the major revisions of the SCII have left some old timers unhappy. The publication of the SDS in 1971 with self-scoring and transparent scales also offended some professionals.

The proliferation of computers and microcomputers has the potential for making everyone a test publisher. This task can be accomplished by simply inserting an inventory into a computer and by adding a scoring and interpretive program, so that the computer becomes the publisher. These new publishers cannot be easily monitored by professional associations, traditional test publishers, or the public. When these technological developments are coupled with the proliferation of self-help work books, card sorts, and plausible but home-made interest inventories, they create a threat to traditional test publishers who are coerced to some degree by the old test standards.

The new test standards are still in revision, but the first four drafts would have required test publishers to conduct more research prior to publication than many, perhaps most, publishers would risk. The proposed standards imply not only more financial but also more legal risk than in the past. Old inventories should find compliance easier, but the legal risk may be similar for old and new inventories. We shall have to wait and see, but in no way can the proposed standards be regarded as a stimulus to innovation. Some compromises are needed to thread new and old inventories between idealistic standards and no standards. These might include selected exemptions for new inventories for a limited time and similar exemptions for inventories with small sales but again for a limited time. Likewise, exemptions should be possible for inventories developed for special populations, or it should be possible to advise professionals not to use inventories with selected groups to avoid the need to turn inventories into omnibus devices.

CONCLUSION

I have attempted to summarize the current status of interest testing, to outline some critical issues and to project what may or should happen next. Because some of my interpretations of this literature deal with very controversial topics, it is desirable to emphasize some interpretations, hopes, or trends for which there is less controversy.

First, I conclude that the need to understand how inventories create their effects is still with us and is an ideal research enterprise for developing more helpful inventories, for coping with equity questions, and for understanding inventories as interventions.

Second, I conclude that a greater effort needs to be made to stimulate authors and publishers to reach higher levels of validity where there has been little or insufficient research. Although I am pessimistic about attempts to increase predictive validity by new psychometric methods, many developers and publishers have not taken advantage of old methods. I have doubts about more psychometric research to develop scales with more validity, but I still hope that a limited effort would be devoted to this strategy. My rationale for these interpretations is two-fold: (a) More than 40 years of scaling research has not resulted in higher levels of predictive validity, but (b) only 10 years of research have resulted in inventories that are more interpretable, easier to score, and have lead to a clearer understanding of interest inventories as career treatments. There comes a time in research when you must cut your losses and invest in more promising strategies or at least redistribute your resources. That time seems now.

ACKNOWLEDGMENT

I am indebted to Fred H. Borgen, Gary D. Gottfredson, Linda S. Gottfredson, Jo-Ida Hansen, Thomas M. Magoon, and Jack R. Rayman for their counsel about earlier drafts of this chapter.

REFERENCES

American Psychological Association. (1974). Standards for educational and psychological tests and manuals. Washington, D.C.: Author.

American Psychological Association. (1984). Joint technical standards for educational and psychological testing. (Fifth draft). Washington, D.C.: Author.

Association for Measurement and Evaluation in Guidance. (1973). AMEG commission report on sex bias in interest measurement. *Measurement and Evaluation in Guidance, 6,* 171–177.

Avallone, V. L. (1974). A comparative study of the effects of two vocational guidance systems: The Self-Directed Search and a traditional vocational guidance model. *Dissertation Abstracts International,* 35, 2670A–2671A. (University Microfilms No. 74–24, 472).

Bartling, H. C., & Hood, A. B. (1981). An 11-year follow-up of measured interest and vocational choice. *Journal of Counseling Psychology, 28,* 27–35.

Berk, R. A. (Ed.). (1982). *Handbook of methods for detecting test bias.* Baltimore: Johns Hopkins Press.

Borgen, F. H., & Seling, M. J. (1978). Expressed and inventoried interests revisited: Perspicacity in the person. *Journal of Counseling Psychology, 25,* 536–543.

Boyd, V. S. (1976). Neutralizing sexist titles in Holland's Self-Directed Search: What difference does it make? *Journal of Vocational Behavior, 9,* 191–199.

Buros, O. K. (Ed.). (1978). *The Eighth Mental Measurements Yearbook* (pp. 1535–1649). Highland Park, NJ: Gryphon Press.

Burrill, L. E. (1982). Comparative studies of item bias methods. In R. A. Berk (Ed.), *Handbook of methods for detecting test bias* (pp. 161–179). Baltimore: Johns Hopkins Press.

Campbell, D. P. (1972). The practical problems of revising an established psychological test. In J. N. Butcher (Ed.), *Objective personaltiy assessments* (pp. 117–130). New York: Academic Press.

Campbell, D. P. & Hansen, J. C. (1981). Manual for the SVIB-SCII, 3rd Edition. Stanford: Stanford University Press.

Clark, K. E. (1961). *The vocational interests of nonprofessional men.* Minneapolis: University of Minnesota Press.

Cooper, J. F. (1976). Comparative impact of the SCII and the vocational card sort on career salience and career exploration of women. *Journal of Counseling Psychology, 23,* 348–352.

Diamond, E. E. (Ed.). (1975). *Issues of sex bias and sex fairness in career interest measurement.* Washington, D.C.: National Institute of Education.

Dolliver, R. H. (1969). Strong Vocational Interest Blank versus expressed occupational interests: A review. *Psychological Bulletin, 72,* 94–107.

Evans, J. R., & Rector, A. P. (1978). Evaluation of a college course in career decision-making. *Journal of College Student Personnel, 19,* 163–168.

Gottfredson, G. D. (1976). A note on sexist wording in interest measurement. *Measurement and Evaluation in Guidance, 8,* 221–223.

Gottfredson, G. D., & Holland, J. L. (1978). Toward beneficial resolution of the interest inventory controversy. In C. K. Tittle & D. G. Zytowski (Eds.), *Sex-fair interest measurement* (pp. 43–51). Washington, D.C.: National Institute of Education.

Gottfredson, G. D., Holland, J. L., & Holland, J. E. (1978). The Seventh revision of the Vocational Preference Inventory., *Psychological Documents, 8,* 98 (Ms. 1783)

Gottfredson, G. D., Holland, J. L., & Ogawa, D. K. (1982). *Dictionary of Holland Occupational Codes.* Palo Alto, CA: Counsulting Psychologists Press.

Hansen, J. C. & Swanson, J. L. (1983). Stability of interests and the predictive and concurrent validity of the 1981 Strong-Campbell Interest Inventory for college majors. *Journal of Counseling Psychology, 30,* 194–201.

Holland, J. L. (1973). *Making vocational choices.* Englewood Cliffs, NJ: Prentice-Hall.

Holland, J. L. (1979). *Professional manual for the Self-Directed Search.* Palo Alto: CA: Consulting Psychologists Press.

Holland, J. L., & Gottfredson, G. D. (1975). Predictive value and psychological meaning of vocational aspirations. *Journal of Vocational Behavior, 6,* 349–363.

Holland, J. L., & Gottfredson, G. D. (1976). Sex differences, item revisions, and the Self-Directed Search. *Measurement and Evaluation in Guidance, 8,* 224–228.

Holland, J. L., Gottfredson, G. D., & Nafziger, D. H. (1975). Testing the validity of some theoretical signs of vocational decision-making ability. *Journal of Counseling Psychology, 22,* 411–422.

Holland, J. L., Gottfredson, D. C., & Power, P. G. (1980). Some diagnostic scales for research in decision-making and personality: Identity, information, and barriers. *Journal of Personality and Social Psychology, 39,* 1191–1200.

Holland, J. L., & Lutz, S. W. (1968). The predictive value of a student's choice of vocation. *Personnel and Guidance Journal, 46,* 428–436.

Holland, J. L., Takai, R., Gottfredson, G. D., & Hanau, C. A. (1978). Multivariate analysis of the effects of the Self-Directed Search on high school girls. *Journal of Counseling Psychology, 25,* 384–389.

Jackson, D. N. (1977). Manual for Jackson Vocational Interest Survey. Port Huron, MI: Research Psychologists Press.

Johansson, C. B. (1982). Manual for The Career Assessment Inventory. Minneapolis: National Computer Systems.

Kriedt, P. H. (1949). Vocational interests of psychologists. *Journal of Applied Psychology, 33,* 482–488.

Krivatsy, S. E., & Magoon, T. M. (1976). Differential effects of three vocational counseling treatments. *Journal of Counseling Psychology, 23,* 112–118.

Kuder, G. F. (1954). Expected developments in interest and personality inventories. *Educational and Psychological Measurement, 14,* 265–271.

Kuder, G. F. (1966). The Occupational Interest Survey. *Personnel and Guidance Journal, 45,* 72–77.

Kuder, F., & Diamond, E. E. (1979). General Manual for the Occupational Interest Survey. Chicago, Il: Science Research Associates.

Lamb, R. R., & Prediger, D. J. (1981). *Technical report for the unisex edition of the ACT Interest Inventory.* Iowa City, IA: American College Testing Program.

Lawler, A. C. (1977). *Career exploration with women using the Non-sexist Vocational Card Sort and the Self-Directed Search.* Unpublished doctoral dissertation, College of Education, University of North Carolina, Chapel Hill.

Lunneborg, P. W. (1975). *Manual for the Vocational Interest Inventory.* Seattle: University of Washington, Educational Assessment Center.

McCormick, E. J. (1979). *Job analysis: Methods and applications.* New York: AMACOM, American Management Association.

Meehl, P. E. (1972). Reactions, reflections, projections. In J. N. Butcher (Ed.), *Objective personality assessment* (pp. 131–189). New York: Academic Press.

Monahan, C. J. (1983). *A decision theoretic approach to the predictions of vocational preference.* Unpublished doctoral dissertation, Iowa State University, Ames.

Nolan, J. J. (1974). The effectiveness of the Self-Directed Search compared with group counseling in promoting information-seeking behavior and realism of vocational choice. *Dissertation Abstracts International, 35,* 195A. (University Microfilms No. 74–16, 569)

Norman, W. T. (1972). Psychometric considerations for a revision of the MMPI. In J. N. Butcher (Ed.), *Objective personality assessment* (pp. 59–83). New York: Academic Press.

O'Neil, J. M., Magoon, T. M., & Tracey, T. J. (1978). Status of Holland's investigative personality types and their consistency levels seven years later. *Journal of Counseling Psychology, 25,* 530–535.

O'Neil, J. M., Price, G. E., & Tracey, T. (1978). The stimulus value treatment effects, and sex differences when completing two interest inventories. Counseling Center Research Report No. 78–2, *Resources in Education.* ERIC Document Reproduction Service, ED 158 208.

Pallas, A. M., Dahmann, J. S., Gucer, P. W., & Holland, J. L. (1983). Test-taker evaluations of the Self-Directed Search and other psychological tests. *Psychological Documents, 13,* 11. (Ms. No. 2550)

Power, P. G., Holland, J. L., Daiger, D. C., & Takai, R. T. (1979). The relation of student characteristics to the influence of the Self-Directed Search. *Measurement and Evaluation in Guidance, 12,* 98–107.

Prediger, D. J. (1982). Dimensions underlying Holland's hexagon: Missing link between interests and occupations? *Journal of Vocational Behavior, 21,* 259–287.

Prediger, D. J., McLure, G. T., & Noeth, R. J. (1976). Promoting the exploration of personally relevant career options in science and technology (NSF Grant No. SM175–18149 A01). Iowa City, IA: American College Testing Program.

Rayman, J. (1976). Sex and the single interest inventory: The empirical validation of sex-balanced interest inventory items. *Journal of Counseling Psychology, 23,* 239–246.

Rayman, J. R. (1983). *The effects of three career courses and selected career treatments.* Unpublished manuscript, Iowa State University, Ames.

Rayman, J. R., Bernard, C. B., Holland, J. L., & Barnett, D. C. (1983). The effects of a career course and some other popular treatments. *Journal of Vocational Behavior, 23,* 346–355.

Reilly, R. R. & Echternacht, G. J. (1979). Validation and comparison of homogeneous and occupational interest scales. *Applied Psychological Measurement, 3,* 177–185.

Roe, A. (1956). *The psychology of occupations.* New York: Wiley.

Siebel, C. E., & Walsh, W. B. (1977). A modification of the instructions to Holland's Self-Directed Search. *Journal of Vocational Behavior, 11,* 282–290.

Strong, E. K., Jr. (1943). *Vocational interests of men and women.* Stanford: Stanford University Press.

Sundberg, N. D. (1977). *Assessment of persons*. Englewood Cliffs, N.J.: Prentice-Hall.

Tittle, C. K., & Zytowski, D. G. (Eds.). (1978). *Sex-fair interest measurement: Research and implications*. Washington, D.C.: National Institute of Education.

Touchton, J. B., & Magoon, T. M. (1977). Occupational daydreams as predictors of vocational plans for college women. *Journal of Vocational Behavior, 10,* 156–166.

U.S. Department of Labor. (1977). *Dictionary of occupational titles* (Fourth edition). Washington, D.C.: U.S. Government Printing Office.

Wagman, M. (1980). PLATO DCS: An interactive computer system for personal planning. *Journal of Counseling Psychology, 27,* 16–30.

Wagman, M., & Kerber, K. W. (1980). PLATO DCS, an interactive computer system for personal counseling: Further development and evaluation. *Journal of Counseling Psychology, 27,* 31–39.

Zener, T. B., & Schnuelle, L. (1976). Effects of the Self-Directed Search on high school students. *Journal of Counseling Psychology, 23,* 353–359.

Zytowski, D. G. (1976). Predictive validity of the Kuder Occupational Interest Survey: A 12- to 19-year follow-up. *Journal of Counseling Psychology, 3,* 221–233.

Zytowski, D. G. (1977). The effects of being interest-inventoried. *Journal of Vocational Behavior, 11,* 153–157.

Zytowski, D. G., & Borgen, F. H. (1983). Assessment (pp. 5–40). In S. H. Osipow & W. B. Walsh (Eds.), *Handbook of Vocational Psychology* (Vol. 2). Hillsdale, NJ: Lawrence Erlbaum Associates.

10 Needed Directions for Measurement in Work Settings

Mary L. Tenopyr
The American Telephone and Telegraph Company

One of the greatest needs in psychology today is the establishment of more rigorous psychological measurement practices in the millions of work settings throughout the country. Today, any semblance of precise measurement appears to be limited to the largest of employers. Only the biggest corporations and the major governmental units, such as those in the federal government, have the scientific staffs to conduct the research and development work necessary to provide the type of measurement that is so needed.

Psychological measurement in work settings has a profound effect upon American society. Indeed, it affects almost all citizens' lives. Employees, job seekers, and their families all, to some extent, have their lives shaped by the psychological measurement practices of employers. How a breadwinner is appraised in a job application or a performance evaluation situation may have an impact on many lives. What job one works in, and even whether one works at all, are all decided mainly on the basis of some psychological measurement, however imperfect. The indirect effects of measurement also must be considered; many of those who have power over us, e.g., supervisors or government officials, were measured in some way when they were selected for their jobs, and also, they remain in their jobs as a result of some application of measurement.

The implications of psychological measurement in the workplace for the educational system cannot be lightly dismissed. Obviously, one major function of education is to prepare students for work and careers. Only through measurement in employment settings can the critical abilities and skills necessary to develop educational curricula be designated and defined. Only then can students be adequately prepared.

The relationship between psychological measurement and the economic health of the country is more nebulous, but probably should be considered to be more than nominal. The well-documented productivity declines of the 1970s were not entirely explainable by typical economic measures, such as amount invested in research and development (Dennis, 1979). The productivity of the individual worker may well have been partially responsible for this decline, and hence, by implication, the methods by which he or she was selected for and retained in the job may well have played a part.

One may well ask why, if measurement in the workplace has so many potential effects in our society, has it not been a subject of great concern in employing organizations. The answers do not emerge readily. There is probably no single explanation for the general lack of precise measurement in the employing community. Certainly, the legal climate for measurement is considered inhospitable by many employers. Results of a recent survey (Bureau of National Affairs, 1983) indicate that the little employee selection testing which has been going on is on the decline. It appears that about 5% to 9% of employers are doing any testing at all. Employers who are dropping testing have indicated that they are doing so because of fear of litigation. However, fear of legal difficulties is only a part of the story. The abuses of testing in business several decades ago became part of American folklore, mainly as a result of the activities of popular writers (Hoffman, 1962; Whyte, 1956). Despite the fact that the lay criticism was mainly of personality inventories, a dark cloud fell over all testing by employers. Many business people began to speak of testing in terms usually reserved for activities such as examining the entrails of birds. Unfortunately, those employers who did continue testing often did so without benefit of validation research. This type of testing culminated in a U.S. Supreme Court decision (*Griggs v. Duke Power Company,* 1971), which mandated a demonstration of job relatedness for any test having a disparate impact upon a minority group. The response to this decision and the many court decisions and administrative actions that have followed was two-fold. Most employers, troubled by the bad reputation of testing, coupled with the possibility of legal difficulties, fled from testing. At the other extreme, a few major employers began utilizing testing research staffs and tried to meet the provisions of the law. Thus, the situation we have today with less than 10% of employers testing (Bureau of National Affairs, 1983) has come to prevail. Most employees are selected by interviews and reference checks, both of which are usually of uncertain validity.

MAJOR CONSIDERATIONS

One of the most fruitful new directions that can be taken, relative to measurement in work settings, is to undertake a massive educational program, not only for those responsible for employment procedures, but also for those who make government policy and law. However, we must concurrently take some actions to ensure that our scientific house is in order. In fact, what is needed is a synergistic combination of educational and scientific considerations. For example, the research and development funds necessary for the scientific achievements we need are most logically supplied by employers, but this money will not be furnished unless employers recognize the value of sound psychological measurement. In particular, the relative merits of various alternate forms of measurement must become common knowledge in the employing and governmental communities. Reilly and Chao (1982) have pointed out that no alternatives to traditional tests are more valid, and most of them are less valid. A related important need is to help frame government policies so that the standards for use of tests are not so rigorous, that even more tests are abandoned in favor of techniques like unvalidated, unstructured interviews.

Also, those responsible for funding need to be aware that the development of reliable and valid measuring methods is not inexpensive. Concomitantly, these policy makers must become aware of the potential utility of sound measurement for increasing performance and productivity. In other words, these persons must come to know that the return on investment in sound measurement is usually substantial.

A third educational objective is to teach employers to recognize the difference between responsible experts in measurement and those with lesser skill or those who recommend measurement programs not based on sound research. It is the author's opinion that many of the difficulties employment testing faces today could have been averted if, in the past, employers had been trained to evaluate recommendations for testing programs on their merits instead of being unduly influenced by the salesmanship of those who proposed such programs.

Coupled with education, there are a number of scientific considerations that deserve attention. Although science should never be frozen in time, one cannot conduct an educational program relative to a scientific endeavor unless there are coherent principles underlying the science. There are a number of needs for research and development that would make the principle base for measurement more supportive. First, there needs to be a conceptualization of validity which is applicable in employment settings. Second, appropriate systems of constructs are required. Paralleling this need, is a need to reduce work requirements into meaningful and manageable dimensions; we need taxonomies of both abilities and work. Third, is a need for clarification regarding job analysis which is one of the major ancillaries to measurement. A fourth need is for performance measurement techniques which are reflective of performance and, at the same time, feasible to apply. Fifth, is a need for guidance in the development of alternatives to traditional paper-pencil tests, such as interviews and work samples. Finally, there is a need for clarification of the differential prediction area. In particular, there needs to be a meshing of theory with data.

Thus, we need combined educational and scientific efforts. Both must be multifaceted and coordinated. Measurement in employment settings cannot be improved without communications and education, on one hand, and scientific progress on the other.

CONCEPTUALIZATION OF VALIDITY

Validity, like Gaul, has been conceptually divided in three parts, since the publication in 1966 of *Standards for Educational and Psychological Tests and Manuals* (American Psychological Association, American Educational Research Association, and National Council on Measurement in Education, 1966). The division of validity into criterion-related, content, and construct parts has become standard in psychology. This conceptualization, however promising it may have appeared in the days before many of the practical issues of current concern had emerged, does not serve as well today. Possibly, the tripartite division of validity has had more relevance for educational and clinical settings than it has for employment situations.

Also, many persons and organizations (Equal Employment Opportunity Commission, Civil Service Commission, Department of Labor, Department of Justice) apparently have considered this division of validity more concrete than its framers intended. For example, *Standards for Educational & Psychological Tests* (American Psychological Association, 1974) spoke of criterion-related, content, and construct as "aspects" of validity and stresses their logical and operational interrelatedness. Certainly nothing in this document appears to warrant the stance that the government agencies (Equal Employment Opportunity Commission, Civil Service Commission, Department of Labor, Department of Justice, 1978) have taken, which so categorically applies different rules of evidence for criterion-related, content, and construct validity.

Various authors have taken issue with the rigid categorization of validity (Cronbach, 1980a, 1980b; Dunnette & Borman, 1979; Guion, 1977, 1978, 1980; Messick, 1975, 1980; Tenopyr, 1977; Tenopyr & Oeltjen, 1982). Moreover, in its statement of standards for selection procedures, the American Psychological Association, Division of Industrial and Organizational Psychology (1980) spoke in terms of *strategies* of validation and pointed out that the three traditional aspects of validity are really inseparable and do not necessarily represent differences in concept.

It appears that some conceptualization at a finer level than one major overall idea of validity is necessary to provide guidance for practitioners; the tripartite division does not appear to work well. Yet, at the same time, one must recognize that much of what has been said under the rubric of the three-category system has value and should not be lost.

Few would disagree that all validity is essential construct validity (Anastasi, 1976; Cronbach, 1980a; Guion, 1980; Loevinger, 1965; Messick, 1975, 1980; Tenopyr, 1977; Tenopyr & Oeltjen, 1982). However, what is needed is a conceptual framework to guide one to achieving construct validity. In developing such a framework, the following considerations are expounded upon as they relate to employment testing:

(a) validation strategies are largely situationally determined with the investigator's specific purpose being paramount,

(b) validity can be conceptualized along a continuum from *specific to general* without the imposition of rigid categories of validity,

(c) *content validity* is essentially a meaningless term,

(d) criterion-related and content-oriented strategies are closely interrelated, are only strategies and means to an end of contrust validity, and, depending upon the exact circumstances of test development and research, can fit at various points on the continuum from specific to general validity.

Situational Determination

For employee selection, in particular, it appears that the validation strategy, which will be optimal, is to a large extent situationally determined. It has long been held that it is the validity of inferences from test scores about which we should be concerned (Cronbach, 1971). One of the major problems in employment settings, is that such inferences usually must be made in a dynamic situation, whereas the typical modes of test validation to a great extent assume a static situation. For example, when one embarks upon a criterion-related study, one gets a criterion at a particular point in time. Tradition holds that the criterion must be maximally relevant for conditions that exist at that point in time. For example, if a criterion is a measure of job performance, the job duties involved in criterion measurement must be those which are actually done at that point in time. If, however, the job changes, as most jobs do, the criterion may no longer be relevant, and the validation study results and inferences based thereon will be, at the best, considered ambiguous. Either criteria must be broadened so that they become more general, such as substituting supervisor's ratings for work samples, or new validation studies must be done to accommodate ever changing criteria.

In a typical employment situation, jobs do not remain constant; the notion of a fixed job simply must be dismissed. One of the things personnel selection psychologists have to cope with is the ever-changing job. Sometimes it is found that in a long predictive study, the job involved changes so that the early subjects are not doing the same job as later subjects. Furthermore, job context factors are often changing. Although it is not likely, some of these may serve to alter validation study applicability. Applicant populations also change; although many applicant characteristics do not affect validation results (National Research Council, 1982), there may be some that do. Finally, in any employing organization, jobs must be grouped in some way for admistrative purposes. For example, most employers would not change a secretary's pay or cause him or her to be retested when moving from one supervisor to another, regardless of the differences in styles of supervisors and the ways they utilize their secretaries. In any validation effort, jobs also must be grouped. It is seldom one in practice encounters a situation in which everyone in a validation sample does exactly the same job. If one strictly followed typical validation tenets, one might be able to muster at the most *N's* of two or three. The situation with job grouping has many of the same effects as that with job changes. Narrow, job-specific criteria will not usually result in validation results that support the inferences one needs to make. What has been said of criteria also applies to predictors developed on the basis of content or psychological theory.

Specific and General Validity

It appears that, at least for employment settings, there must be some reconceptualization of validation. Other authors (Cronbach, 1971; Loevinger, 1957) have pointed out the *ad hoc* nature of most validation efforts and the need to extrapolate in all validation whether in an employment setting or not. The limits of permitted extrapolation depend on how one developed one's validity evidence in the first place. No precise rules for extrapolation can probably ever be developed, but some new ways of thinking about validity, which may aid in making judgments about inferences from tests or other measuring devices, appear to be in order.

It is proposed that there is a continuum on which, at one end, is *specific validity* and, at the other end, *general validity*. Neither of these two terms signifies a type or component of validity. They just represent extremes differentiated only by a shift in emphasis. Most validation results will fall somewhere between the two extremes. In many ways, the two terms denote many of the conditions Cook and Campbell (1979) described when they spoke of *internal* and *external* validity. The term *specific* roughly corresponds to the term *internal,* and *general* is close in meaning to *external*. The new terms have been chosen because the meanings do not exactly coincide with those of the older terms, and confusion with the teachings in experimental psychology might result were different terms not used.

Specific validity occurs when one designs a study so that the results will have a high fidelity in a given situation, in a given location, for a specific population, at a specific point in time. If one does his or her work well, inferences within the confines of the given situation will be relatively accurate. Yet, if the situation is at all dynamic and/or generalization to a similar situation is required, one has little evidence upon which to proceed. An example would be a job knowledge test for machinists, which would not be so applicable to stock clerks.

General validity occurs when one designs a study so that results will have generality for a number of situations. Usually, it can be expected that the inferences relative to any one situation in the set of situations covered will not be so accurate as they would be had the study been done using procedures more appropriate to the *specific* end of the continuum. An example would be a verbal aptitude test which could be expected to have some validity for both machinists and stock clerks.

A general hypothesis can be stated regarding *specific* and *general* validity. That is, both cannot be maximized at the same time. In general, to increase one is to decrease the other. As one moves away from the specific end of the continuum, one automatically moves toward the general end and vice versa. Ultimately, the continuum of test development depends highly upon one's purpose and the exact situation.

It is difficult to test this hypothesis, as most organizations will not support the type of research involved. For example, the typical development of highly specific work sample tests, e.g., data-entry tests, involves a situation in which tests and any appropriate cirteria are so similar that a criterion-related validation study would result in a validity coefficient which would approximate a test-retest reliability coefficient, e.g., (Tenopyr & Caire, 1966). Supporting content-oriented test construction is usually the only investment an organization will make in a situation of this sort. Also, an organization would not normally support efforts to show that a data-entry test is more valid for predicting data-entry performance than sheet-metal work performance. On the other hand, organizations will support the typical research that is reported in the literature, i.e., studies involving the same more general tests (aptitude tests) for a variety of jobs. It is also significant to note that even after adjustment for restriction in range and unreliability of the criteria (Schmidt & Hunter, 1977), predictive validities of these more general tests fall far short of their reliabilities.

A logical parallel may be drawn in the field of education. Despite the fact that it is known that a general scholastic aptitude test is a fair predictor of grades in many courses, it is a rare educator who would consider this general test to be more valid for assessing classroom performance than a specific test requiring mastery of what was taught in the class. Nor would an educator conduct research to determine whether a classroom algebra mastery test was as valid as a classroom English composition test in predicting performance in composition.

Because a research base will probably never be developed to determine the tenability of the hypothesis outlined, the notion of the incompatibility of specific and general validity will probably never achieve more than the status of a working hypothesis.

Also, it should be noted that the notion of specific v. general validity applies most logically in the context of predicting performance; whether it would apply in situations where criteria like tenure are predicted is a research question.

Specific Validity

An example near the specific end of the continuum would be a work simulation which had a high fidelity to the duties of a specific position. If a screw is to be turned to the left on the job, it is turned to the left in the simulation. However, the notion that one can exactly duplicate the work in a testing situation is a fiction. Every test is an abstraction. Some tests are just less abstract than others. The least abstract are probably the flight simulators, whose technology is far too expensive to duplicate in normal employment situations. Even the supposedly simple typing test is an abstraction. In fact, the typing test presents a good vehicle to demonstrate the necessity of abstraction in employment testing. First, there is the question of the material to be typed. In an organization of any size, one will find wide inter- and intraindividual differences in many characteristics of the material typed. For example, one person types only one- or two-paragraph memoranda. Another types a combination of memoranda and statistical reports. Some production typists may encounter all types of work. The work for an individual typist may vary from day to day. In developing material for a typing test, one is faced with a number of dilemmas. However carefully one samples the material typed in an organization, the resultant material selected for the actual test or tests will be a compromise of some sort and probably not reflect what any given typist in the organization actually types on a given day. Considerations relative to the job applicant population must also be taken into account. for example, in an engineering firm, does one include in the test technical words that a person in a high school typing course probably has never encountered? There are other considerations. If it is found that most typists type from handwritten copy or edited drafts, whose style of penmanship does one use for the test material? How clear and consistent should the editing be—like a professional editor's work or like the chicken scratches of a harried manager? Is spelling to be corrected? Are the length of the test and time limits to reflect the duties of a busy secretary who cannot type for more than 5 minutes without being interrupted, or the activities of a word processing production typist who is expected to type over long periods? How should speed and accuracy be weighted? In view of the employing organization's policies on job classification, pay, and employee mobility, can more than one test or a test with different critical scores be used?

Equipment and job applicant-equipment interactions must be considered. With all of the varieties of typewriters and word processors available today and often coexisting in a given employing organization, equipment choice is very difficult. Furthermore, one must consider that many applicants may not have had training on any of the equipment used in the organization, and one may wish to measure basic skills as opposed to equipment-specific skills.

Also, equipment considerations interact with content choice. For example, if hyphens are at different places on various keyboards, one may wish to eliminate typing of hyphens from the test content. Consequently, equipment considerations may serve to add to the abstract nature of the test, making it far from an actual job sample.

Perhaps the highest specific validity, at least in concept, is achieved by well designed probationary periods or documented experience in the work involved.

Also, in concept, it would be expected that lower internal validity would be associated with aptitude tests or general education requirements.

Criterion-related strategies may fall anywhere between the extreme of the two ends of the specific-general continuum. The exact placement depends on the nature of both the predictors and the criteria. If one uses as a predictor a very specific test, designed for the particular job, and employs a criterion which accurately reflects specific job requirements, one's validity will probably be nearer the specific than the general end of the continuum. Various combinations of specific and general predictors and criteria can exist; consequently, one has to examine the exact situation to estimate how general or specific one's criterion-related validity is.

Experienced researchers recognize that specific validity is not necessarily optimal, despite its intuitive appeal. For example, the more faithful a replica of a job a work sample is, the more likely it is to have to be changed constantly to accommodate changes in detailed job procedures. If, perchance, performance on one's detailed work sample involves constructs that have broad generality, one should have additional evidence to defend generalization. Furthermore, in employment settings, face validity takes on importance with the psychologist's clients. For example, a test battery for telephone operators once contained a test involving completing mark-sense cards. The job of telephone operator was changed to eliminate the use of such cards. Thereafter, the supervisors of telephone operators assumed that the whole test battery was not useful in selecting operators, despite the fact that the test was still valid. A more practical strategy might have been not to strive for less specificity in predictors.

Content Validity and Specific Validity

As every test, even the supposedly simple typing test is an abstraction; the very notion of content validity is called into question. Content sampling for the purpose of selection-device construction always results in something other than a job replica. The specific end of the continuum may be more easily approached in educational achievement testing, where sampling from what is taught is a somewhat simpler task than sampling in a dynamic job situation. However, even in educational testing, it is probable that true specificity is seldom achieved.

Content validity as a concept has been criticized for a variety of reasons (Guion, 1977, 1978; Messick, 1975; Tenopyr, 1977). Messick (1975), in particular, has proposed that what is typically called content validity is concerned with inferences about test construction, not individuals. Tenopyr (1977) has proposed that content only be considered one form of evidence for construct validity.

Nevertheless, in employment settings, content cannot be ignored in trying to achieve specific validity. If one wants a high fidelity selection procedure, even though it may have little generality, content-oriented strategies in test or criterion development must be used. However, one must always remember that the inferences one makes are on the basis of constructs, however narrow they might be. A limited concept like the "ability to type numerals" is indeed a construct. If one wants to infer constructs, not made obvious by the content of the test, other evidence such as results of a criterion-related study must be brought to bear.

It should be noted, however, that some tests developed for specific, narrow purposes may have more generality than is apparent. For example, performance in a drafting test may be related to performance in a drill press operator's job. This generality may be artifactual, e.g., both draftspersons and drill press operators are trained in the specifics of blueprint reading. However, there may be some commonality of more basic constructs between a draftsperson's and a drill press operator's job requirements. Space visualization is a likely candidate. Again, evidence other than content that generalization is possible should be developed.

This is not to say that content alone cannot be the only evidence of validity. There are many situations in which content-oriented evidence of validity is sufficient, despite the difficulties in moving from inferences about test construction to inferences about individuals. Most of these, however, will be toward the specific end of the continuum. Certainly the more general interpretations should be supported by more than content. No precise rules can or should be formulated to fit all situations. Whether one chooses to use content considerations alone requires the exercise of professional judgment, taking all situational factors into account.

General Validity

General validity refers to the end of the continuum where the inferences to be made are less situation-specific. At the specific end of the continuum, one might make inferences about the ability to enter numeric information in a computer terminal. At the general end, one's inferences would reflect abilities more like that to do general clerical work.

These inferences differ mainly in their specificity. They do not differ in kind. Both reflect constructs; the more specific inference reflects a narrow construct, presumably largely supported by a wide variety of evidence, which may include results of a criterion-related study and does not necessarily exclude content. However, when one is attempting to support a general inference, it appears that there would be few situations in which content alone would be sufficient evidence.

As mentioned previously, in most employment situations, it is the more general validity in which one is interested. One normally needs to make inferences about behavior in more than a narrow band of situations. How much validity can be extended to a variety of situations is a matter which has been discussed in the courts (*Douglas v. Hampton* et al., 1975). Pearlman (1980) has indicated how job grouping can be done on the basis of test validity. If employers did indeed usually group jobs on the basis of ability-related job requirements, the psychologists' situation relative to marshaling evidence of general validity would be much simpler. However, most employers do not have professional psychologists doing job analysis, job evaluation or job grouping for progression and pay purposes. These matters also are often bargained for, making them even farther from the psychologists' control. Also, there may be wide intercompany differences and, even within the same company, interdepartmental differences. What is considered a job in one company or department may be considered a group of several jobs in another department or company.

In addition, where systems of job grouping and progression are developed, in some companies it has not been uncommon for such systems to reflect biases of various sorts. For example, jobs normally populated by persons of one sex are grouped together, regardless of differences in skill and ability requirements.

Personnel psychologists find it easier to work within existing systems than to try to change them. The ethical dilemmas involved are not discussed here, but needless to say, they are many.

Working within these existing job systems, psychologists probably still can do much to effect valid selection procedures. The question of whether a given predictor has generality enough to be used for groups of jobs is largely, although not entirely, an empirical question. It is not feasible to attack the problem wholly by strict empiricism. For example, how much of a job change or difference in jobs dictates a new validation study is a judgment question. How much lowering of validity in the specific situations one is willing to tolerate when using a general predictor is also a matter of judgment.

Although empiricism can take many forms (Cronbach, 1971), it can consist of criterion-related studies relative to a sampling of jobs in the job group in question. Validity generalization then can be helpful in extrapolating to jobs not in the sample.

Validity generalization to date has been discussed in terms of which tests are valid for which jobs (Callender & Osbourn, 1980; Schmidt, Gast-Rosenberg, Hunter, 1980; Schmidt & Hunter, 1977; Schmidt, Hunter, Pearlman, & Shane, 1979). Another perspective on validity generalization is taken here. It is suggested that validity generalization research tells us as much about criteria as it does about tests. In discussing any relationship such as those indicated by coefficients of correlation, both variables underlying the relationship must be considered.

In particular, both predictors and criteria must be considered relative to their generality. Most of the validity generalization research has been done on aptitude tests which fall near the general end of the continuum. The criteria employed in these studies have been, for the most part, supervisors' ratings. These also are highly general. There has been little research involving either more specific tests or criteria.

The generally positive validity generalization results obtained to date, in the author's opinion, represent essentially the operation of only those abilities supervisors discern most readily in a work situation. The halo effect attendant upon supervisors' ratings is well known, and it appears that typical supervisors' ratings reflect only the grossest of behavior. Were one's criteria able to capture the nuances of behavior in given jobs, one might have a better basis for inferences about generality. This is not an easy task. Anyone who has attempted to develop tests to measure specific abilities knows well the large number of false starts associated with this effort. Test tasks which, on the surface, appear to measure the same ability many times, indeed do not. Other tests designed to measure different abilities instead measure the same ability. Except at the most rudimentary level, the endeavor to glean ability requirements from job duties is even more difficult. Experienced investigators doing validation research know well that, despite the results of job analyses, some ''shotgunning'' of predictors is still a viable research strategy. The problem is compounded with jobs in which the manner in which one performs the job is to any extent discretionary, and different abilities may be used by different persons to achieve the same performance levels. Also, improvement of prediction of behavior in employment settings is much needed. As Ghiselli (1966) pointed out, prediction of job performance has not been highly impressive. If we are to improve prediction, we must design any validity generalization research carefully. We should pay as much attention to the criterion-side as we pay to the predictor-side. It is the author's opinion that the most meaningful validity generalization research would be that in which criteria are relatively specific so that abilities that might be obscured by a more general criterion can be captured.

If we design our research this way, we may achieve a more optimal point on the specificity-generality scale than has been achieved by validity generalization research to date. By dealing with both general predictors and general criteria, this research has indicated that we can achieve moderate prediction of performance in a wide range of jobs with a few general types of tests. If we are to improve prediction, we are probably going to have to move toward the specific end of the scale in terms of both criteria and predictors and be satisfied with less generality. Again, how far one moves on the continuum is a matter of judgment and, to a large extent, influenced by situational factors.

Research of this sort would also enable one to do a better job of developing taxonomies. In this respect, it should be pointed out that the basis for job taxonomies also form a continuum from specific to general. Job taxonomies can be formed on anything from a narrow to a broad basis. For example, a job which involves turning a screw to the left instead of to the right as in another job may be put in a different family from the other job. At the other extreme, approaching jobs from a worker attribute rather than a task approach and assuming that all jobs involve some overall ability, all jobs could be grouped in one family. The level of generality one chooses as a basis for taxonomies and where one establishes taxonomic boundaries should, as in testing, have some empirical support, but are in the end judgment calls.

Content, Criterion-Related and Construct Strategies

Two of the three traditional strategies of validation, content and criterion-related, have largely been presented here as means to an end. That end is construct validity. The necessity of the use of professional judgment in determining which strategy or which combination of strategies one employs has been emphasized. The role of situational factors has been indicated to be important and a major basis for judgment.

The type of strategy one uses and, consequently, the evidence of validity one amasses cannot be dictated by precise rules. As has been indicated, content can be a form of evidence anywhere along the continuum from specific to general. However, it becomes the major form of evidence near the specific end of the scale. Criterion-related strategies cannot be divorced from content strategies. Content is usually a major consideration in criteria. Criterion-related strategies can form evidence anywhere along the continuum depending upon the generality of the predictors and the criterion.

Construct validity, which should be the basis of all inferences from psychological measurement, of course, cannot be separated from the strategies used to achieve it. Construct validation can draw from a number of lines of research and is not a simple matter (Cronbach, 1971). Defining precise measurement steps for achieving construct validity as some have done (Equal Employment Opportunity Commission, Civil Service Commission, Department of Labor, Department of Justice, 1978) tends to belie the complexity of all measurement situations and freeze the state of science at a rudimentary level.

Construct Interpretation

One of the greatest problems in industrial and organization psychology today is the inability of practitioners to make construct interpretations from their measurements. This is not a problem solely for this group of psychologists, but is a difficulty throughout psychology. Witness the number of tests in print (Buros, 1978). Were there available meaningful systems of constructs or, for that matter, any systematic efforts to establish construct validity except for a few major tests, there would be less test development. If test reviewers had meaningful construct relevant information with which to evaluate new tests, perhaps test authors would be less enthusiastic in making claims of having measured something new and different. Certainly, the "made-up-on-the-spot" construct has led to considerable confusion.

General validity could more easily be achieved were there the possibility for more well supported construct interpretations. If systems of constructs were available, the need for *ad hoc* evidence of general validity in each validation effort would be considerably lessened.

Science, of course, cannot progress without constructs; in fact, well developed systems of constructs are the mark of a fully developed science. If psychology and employment psychology, in particular, are to mature, they must do more to achieve the bases for construct interpretation of data. Meaning must be added to measurement. Numerous *ad hoc* studies, as has been the tradition of employment psychology, can result in the evolution of some principles, but seldom do they result in the explanations that are so much needed.

In the ability area, psychologists have the work of Ekstrom, French, and Harman (1979) to which to turn. This monograph covered well the status of aptitude constructs at its date of completion and can be used to support the construct interpretations of various types of aptitude test. The work can also be used judiciously in establishing general validity.

Unfortunately, in areas such as personality or character measurement, there appears to be no counterpart work to which to turn. In attempting personality measurement, one is faced with numerous unsupported and often contradictory claims of construct validity. Because of the various problems attendant upon personality measurement, e.g., invasion of privacy, low validity, inventories in this area are not used much by employers (Bureau of National Affairs, 1983). However, if some of the problems of construct confusion were eliminated, there might be some possibility that predictors in the noncognitive areas could become more useful in employment testing than they now appear to be.

Another area in which construct systems are needed is organizational psychology. This field is now characterized by a myriad of questionnaires and rating scales of various sorts, which are purported to measure things like job satisfaction and job commitment. These are part of the whole employment process and may be conceived of as potential criteria in certain situations. Such questionnaires and scales have most of the problems associated with personality measurement.

If we are to have general validity, not just in the limited sense of aptitude test validation, we must get better bases for construct interpretations through the whole range of measurement techniques. It has been said that no two psychologists could agree on systems of constructs, but independent investigators, free from the constraints of any one laboratory, or other organizational setting should attempt to bring more meaning to our measurements.

A related problem, as discussed by Pearlman (1980), is the need for meaningful taxonomies of work. Unless the bases for criteria can be organized in some meaningful fashion, there is little hope for achievement of the more general validity we usually need. Attempts to obtain an all-encompassing taxonomy of work probably would be disappointing. However, it is suggested that progress toward such a taxonomy can be expedited by first attempting to develop better taxonomies of human characteristics through traditional measurement. Test tasks, although they are necessarily limited in scope, can be the basis for systems of constructs which can be used to classify job tasks.

JOB ANALYSIS

There are many ways of analyzing jobs, depending upon one's purpose. Many of the better developed of these techniques have been reviewed by Pearlman (1980). However, it appears that practically every investigator uses somewhat different methods in analyzing jobs.

The fact that different job analysis techniques are needed for different purposes had led to some extent to the proliferation of these job analysis procedures. Also, considering that every job analytic situation is different, involving different jobs and different populations, the comparison of job analytical techniques from different investigations is made difficult. Few investigations involving use of different job analyses in the same situation seem to have been made; there are, however, some exceptions (Ghiselli, 1966).

Although a universal job analysis system is not advocated, it appears that there is a need for developing some principles for analyzing jobs. Despite the large number of job analyses which are being done today, there does not appear to be available the research base from which the needed principles can be drawn. The lack of principles to guide methodology, of course, hinders the development of the taxonomies relative to worker attributes and job requirements.

The major question of the validity of the masses of data which have been generated is of utmost importance. Pearlman (1980) has suggested that test validities and the results of validity generalization studies be used to form job groups based upon abilities required of the incumbents. Unfortunately, there are some problems associated with this approach. In particular, the job groupings afforded may be too broad to use for a particular purpose and may not reflect the more specific ability requirements in different jobs.

Approaches involving having supervisors or job incumbents rate jobs on construct-oriented scales were advocated by this author (Tenopyr, 1977). These methods do not seem so appealing upon reconsideration. The main problem is that there is a dearth of evidence that job experts can rate jobs validly in terms of their ability requirements. The often demonstrated finding that job experts can agree on ability constructs needed for job performance, of course, supports reliability for such ratings, but there seems to be no evidence that these ratings are related to validities of corresponding tests.

What is needed is a series of studies which attempt to determine validity of construct estimates by job experts. Various types of rater should be examined, e.g., supervisors, incumbents, psychologists. Different specificity levels of construct should be employed. Studies to determine the degree of response style associated with such ratings should be undertaken.

JOB PROFICIENCY MEASUREMENT

Methods for measuring job proficiency or performance have been a subject of study for many years. The studies in this area have been reviewed by Tenopyr and Oeltjen (1982). They represent special measurement problems of their own but become additionally problematic when used as criteria in validation.

There are essentially two categories of measurement involved in evaluating proficiency, the supervisors' rating and the objective record of performance. Some of the problems with supervisors' ratings have been discussed previously. Supervisors' ratings developed for the organization's administrative purposes pose special problems. The most important of these is the coupling problem. Rating results are often coupled with administrative actions such as salary increases, promotions, or personal development counseling. Rumors abound in every organization of supervisors who "back into" a rating, e.g., they decide on the amount of the raise first and then give a rating to justify it. When ratings are tightly coupled with one administrative purpose, they are often found useless for other purposes to which they are less tightly coupled. Unless operational ratings are tightly coupled to all the administrative purposes for which they will be used, including feedback to the employee, or are not tightly coupled with any administrative system, they will not be maximally useful as criteria. Most practicing personnel psychologists, therefore, appear to prefer not to use in-place rating systems as a basis for criteria. They instead rely on specially developed criteria for the study involved.

Objective records, despite their intuitive appeal, have many drawbacks. For example, for welders, error rate per inches of weld made might be considered for performance measurement. However, the most proficient welders might get the most difficult welding jobs, such as welding corners of boxes or joining materials that are difficult to weld. In a factory situation, the most senior, but not necessarily the most proficient, operator may get the newest and most efficient machine.

A major problem is that in place performance measurement systems are often gamed. A plant manager may turn out high, short-term profits by skimping on maintenance of the factory. His or her successor may then have to do the maintenance and, thereby, turn in poorer profit picture. The phenomenon of employees' paying attention to those phases of the job upon which pay and promotion are based and neglecting other job aspects is common. Unfortunately, those who have tried to develop operational performance measurement systems have generally found it impossible to cover all aspects of any job and thus reduce the possibility of "gaming."

Another problem with objective performance measurement is that of getting a large enough number of observations to get reliable measurement. This is particularly true when error rate is small. Also the task of obtaining and summarizing the data is often so administratively burdensome that employers avoid sophisticated performance measurement systems.

A final problem is that operational performance ratings are often not available in unionized operations where personnel decisions are made largely on the basis of seniority.

Thus, when personnel researchers want objective criteria, they are often forced to develop *ad hoc* measures for any study involved. Even then, there are often administrative difficulties in getting supervisors to make enough systematic observations to obtain reliable measurement.

Despite the serious problems in this area, research on performance measurement should continue. As more and more jobs are involving automated equipment, the probability of sufficient, accurate data in simpler jobs is increased. Also, larger computer-based measurement systems are being made possible. The many problems with supervisors' ratings will not be solved easily. Probably researchers, if forced to use supervisors' ratings as criteria, will continue to develop them on an *ad hoc* basis.

ALTERNATIVES TO TESTS

One of the needs for new direction for measurement in employment settings is to provide better development guidance for the alternatives to paper-pencil tests. In particular, the employment interview, which is in wide use (Bureau of National Affairs, 1983), needs further development (Reilly & Chao, 1982; Tenopyr & Oeltjen, 1982).

Much has been published about what goes on in the interview. Tenopyr and Oeltjen (1982) found that over a recent 3-year period, there were sixteen studies involving the effects of race and sex upon interview results. Most of these were of the "paper people"-type which involved identical descriptions of people, except for race or sex. Unfortunately, in this same review, only one validation study for an interview was found.

A dynamic situation like an interview is not an easy subject for study. Often when research is done, it is necessary to reduce this fluid situation to written form. Paper people is one vehicle; casting interview questions so that they are nothing more than an orally administered biodata blank is another.

It appears that much is known about the pitfalls of interviewing; now is the time to work on the development of valid interviews. It is a much easier task to examine minutely existing practices than to develop new practices which actually work.

Other widely used procedures for which there is little guidance for developmental practices are experience evaluation methods and work samples. The former are of as much importance as the interview because they are so widely used. Certainly their validity to date has not been impressive (Caplan & Schmidt, 1977).

DIFFERENTIAL PREDICTION

Possibly no discussion of new directions in measurement can ignore the question of group differences in regression systems. The research in this areas, as reviewed by Tenopyr and Oeltjen (1982) has been abundant, if controversial. This whole line of research has taken a course which is perplexing to a scientist. The long search for differential validity and differential prediction systems has taken place without any reasonable scientific hypotheses as to why such phenomena should be found. Unless science can be incorporated into this research and meaningful hypotheses generated, it is suggested that such research receive less emphasis. Any further research, such as that into sex differences in regression systems, should be more carefully designed so that artifacts, such as differences in exposure to certain kinds of training, do not lead to erroneous interpretations of results. Investigators should also be certain that criteria have the same meaning for all groups concerned. Too often, factors such as affirmative action programs or attitudes of supervisors and trainers may render ratings or, even training grades, unsuitable criteria. Certainly, in the absence of meaningful scientific hypotheses, researchers bear a heavy burden to prove any group differences found are not artifactual.

SUMMARY

Measurement in employment settings is fraught with many difficulties, some of which are unique to personnel selection. Unless the organizational support for sound measurement is obtained, these difficulties will be with us for many years to come. A synergistic combination of education for organizational personnel and application of science to measurement in organizations is needed.

A more flexible reconceptualization of validity coupled with a renewed emphasis on interpretation of data in terms of constructs is required. Methodology needs to be improved to achieve these ends. There needs to be a revitalized effort to improve interviews and other techniques which are far more widely used than paper-pencil tests. Finally, science should be incorporated into research on group differences and, unless rational scientific hypotheses can be generated and tested relative to differential prediction, such research should receive less emphasis.

REFERENCES

American Psychological Association, American Educational Research Association, & National Council on Measurement in Education. (1966). *Standards for educational and psychological tests and manuals*. Washington, D.C.: American Psychological Association.

American Psychological Association, American Educational Research Association, National Council on Measurement in Education. (1974). *Standards for educational and psychological tests*. Washington, D.C.: American Psychological Association.

American Psychological Association, Division of Industrial-Organizational Psychology. (1980). *Principles for the validation and use of personnel selection procedures*. Berkeley, CA: Author.

Anastasi, A. (1976). *Psychological testing* (4th ed.). New York: Macmillan.

Bureau of National Affairs. (1983). *ASPA-BNA survey no. 45. Employee selection*. Washington, D.C.: Author.

Buros, O. K. (Ed.). (1978). *The eighth mental measurements yearbook*. Highland Park, NJ: Gryphon.

Callender, J. C., & Osburn, H. G. (1980). Development and test of a new model for validity generalization. *Journal of Applied Psychology, 65*, 543-558.

Caplan, J. R., & Schmidt, F. L. (1977). *The validity of education and experience ratings*. Paper presented at Annual Meeting of the International Personnel Management Association. Assessment Council. Kansas City, Missouri.

Cook, T. D., & Campbell, D. T. (1979). *Quasi experimentation: Design and analysis issues for field settings*. Chicago: Rand McNally.

Cronbach, L. J. (1971). Test validation. In R. L. Thorndike (Ed.), *Educational measurement* (pp. 443-507). Washington, D.C.: American Council on Education.

Cronbach, L. J. (1980a). Selection theory for a political world. *Public Personnel Management, 9*, 37-50.

Cronbach, L. J. (1980b). Validity on parole: How can we go straight? *New Directions in Testing and Measurement, 5*, 99-108.

Dennis, E. F. (1979). *Accounting for slower economic growth—The United States in the 1970's*. Washington, D.C.: The Brookings Institution.

Douglas v. Hampton et al. (1975). 512 F2d. 976.

Dunnette, M. D., & Borman, W. C. (1979). Personnel selection and classification. In M. R., Rosenzweig, & L. W. Porter, (Eds.), *Annual review of psychology, 30*, 477-525.

Ekstrom, R. B., French, J. W., & Harman, H. H. (1979). Cognitive factors: Their identification and replication. *Multivariate Behavioral Research Monograph*, No. 79-2.

Equal Employment Opportunity Commission, Civil Service Commission, Department of Labor, Department of Justice. (1978). Adoption by four agencies of uniform guidelines on employee selection procedures. *Federal register, 43*, 38290-38310.

Ghiselli, E. E. (1966). *The validity of occupational aptitude tests*. New York: Wiley.

Griggs v. Duke Power Co., 401 U.S. 424 (1971). 3EPD P8137, 3FEP175.

Guion, R. M. (1977). Content validity, the source of my discontent. *Applied Psychological Measurement, 1*, 1-10.

Guion, R. M. (1978). Content validity in moderation. *Personnel Psychology, 31*, 205-214.

Guion, R. M. (1980). On trinitarian doctrines of validity. *Professional Psychology, 11*, 385-398.

Hoffman, B. (1962). *The tyranny of testing*. New York: Crowell-Collier.

Loevinger, J. (1957). Objective tests as instruments of psychological theory. *Psychological Reports, 3*, 635-694.

Loevinger, J. (1965). Person and population as psychometric concepts. *Psychological Review, 72*, 143-155.

Messick, S. A. (1975). The standard problem: Meaning and values in measurement and evaluation. *American Psychologist, 30,* 955-966.

Messick, S. A. (1980). Test validity and the ethics of assessment. *American Psychologist, 35,* 1012-1027.

National Research Council. (1982). *Ability testing: Uses, consequences, and controversies.* A. K. Wigdor & W. R. Garner (Eds.), Washington, D.C.: National Academy Press.

Pearlman, K. (1980). Job families: A review and discussion of their implications for personnel selection. *Psychological Bulletin, 87,* 1-28.

Reilly, R. R., & Chao, G. T. (1982). Validity and fairness of some alternative employee selection procedures. *Personnel Psychology, 35,* 1-62.

Schmidt, F. L., Gast-Rosenberg, I., & Hunter, J. E. (1980). Validity generalization results for computer programmers. *Journal of Applied Psychology, 65,* 643-661.

Schmidt, F. L., & Hunter, J. E. (1977). Development of a general solution to the problem of validity generalization. *Journal of Applied Psychology, 62,* 529-540.

Schmidt, F. L., Hunter, J. E., Pearlman, K., & Shane, G. S. (1979). Further tests of the Schmidt-Hunter Bayesian validity generalization procedure. *Personnel Psychology, 32,* 257-281.

Tenopyr, M. L. (1977). Content-construct confusion. *Personnel Psychology, 30* 47-54.

Tenopyr, M. L., & Caire, J. (1966). *The validity of a key punch test.* Unpublished research report. El Segundo, California: Rockwell International.

Tenopyr, M. L., & Oeltjen, P. D. (1982). Personnel selection and classification. In M. R., Rosenzweig, & L. W. Porter, (Eds.), *Annual review of Psychology, 33,* 581-618.

Whyte, W. H. Jr. (1956). *The organization man.* New York: Simon and Schuster.

Author Index

Italics denote pages with bibliographic information.

C

Y,Z

Subject Index

F

G

H

I

J

L